ELMER SPERRY

Thomas Parke Hughes

ELMER SPERRY
Inventor and Engineer

The Johns Hopkins University Press   Baltimore and London

Softshell Books edition, 1993

The Johns Hopkins University Press
2715 North Charles Street
Baltimore, Maryland 21218-4319
The Johns Hopkins Press Ltd., London

Library of Congress Catalog Card Number 71-110373

ISBN 0-8018-1133-3
ISBN 0-8018-4756-7 (pbk.)

A catalog record of this book is available from the British Library.

This biography is based on a research project completed
in the Center for the Study of Recent American History.

For Tom, Jr.,
who wanted to be in this book

Contents

PREFACE

This study was made possible by the generous support and encouragement of Professor Alfred Dupont Chandler's Center for the Study of Recent American History at The Johns Hopkins University and by the Sperry family. Inventors, engineers, and their families have not often taken care to preserve the sources that make possible the writing of history; nor have established historians and centers for historical research energetically cultivated detailed studies of the professional careers and works of inventors and engineers. Interested in technology as well as biography and history, I have been fortunate that the Sperry family and Professor Chandler do not conform to the general pattern.

Mr. and Mrs. Robert Lea made a generous grant to the Center for the Study of Recent American History, and the center supported my research and writing for a period of three years. Mrs. Helen Lea is the daughter of Elmer Sperry, and Robert Lea was one of the first group of engineers employed by the Sperry Gyroscope Company. One of Elmer Sperry's greatest admirers, Robert Lea kept fresh the memory of Sperry by tirelessly searching for information over many years and by enthusiastically encouraging the biographer. Elmer Sperry, Jr., one of the inventor's three sons, made an invaluable contribution to me—and perhaps it is not too much to say, to history—by preserving his father's papers. These include thousands of letters, both personal and professional, technical reports, notebooks, patents, photographs, and publications. He made these available to me without restriction. I am also grateful for the encouragement given by Mrs. Elmer Sperry, Jr., and the late Mr. William Goodman, Elmer Sperry's nephew, who also preserved Sperry papers. The Sperry family, however, exercised no editorial control over my book and is not responsible for the opinions expressed in it.

Professor Chandler, chairman of The Johns Hopkins University history department and founder of the Center for the Study of Recent American History, provided unwavering support and encouragement from the inauguration of my research and writing throughout the three years, when I gave my full time to the project as a fellow in his center. His strong and lucid concept of the process of historical change, his painstaking, encompassing, and fruitful methodology, and his dedication to the highest standards of the historian's craft greatly influenced my approach to, and interpretation of, Sperry, his work, and his place in history.

I am also indebted to those who read and made helpful comments about the manuscript as it evolved. Besides Professor Chandler, Professor Cyril Stanley Smith of the Massachusetts Institute of Technology, Preston R. Bassett, Jack Lea, Sperry Lea, Professor Francis Litz, Mary and Hunter R. Hughes read all or part of the manuscript. Professor Smith's detailed comments were especially helpful because of his years of rich experience in engineering and science and his insights into history. Preston Bassett's comments were of great value because he was a protégé of Sperry's, a chief engineer and then president of the Sperry Gyroscope Company. Jack Lea, a nephew, and Sperry Lea, a grandson, helped because each had cultivated the memory of Sperry and insisted that the biography "tell it as it really was." Jack Lea generously shared with me his analysis of Sperry's early years, one based upon his own first-hand studies.

There were others who kept fresh the memory of Sperry and collected sources that made possible the writing of the biography. In Cortland, New York, Sperry's early home, Mrs. L. B. Leach, a Cortland historian, did research in local records; Mr. Porter Bennett, also of Cortland, was one of the many Sperry admirers I encountered who independently gathered information about him and made it available to me. Eric Sparling and Perry Bigelow were members of the Sperry Gyroscope Company during the early years, and they both gave me interesting interviews. I am sorry for the inevitable distortions resulting from the biographer's inability to envisage precisely what Sparling, Bigelow, Bassett, the Sperry family, and others who knew Sperry recall in the mind's eye.

The librarians and archivists who have helped me are legion. During the three years I worked in the Eisenhower Library of the Johns Hopkins University, I found the staff professionally superb and a model of patient courtesy. I also want to thank the staffs at the Southern Methodist University libraries (especially M. Mounce and V. Templeton), the National Archives, the Library of Congress, and the Smithsonian Museum of History and Technology, who on numerous occasions provided assistance. The archivists of the National Academy of Sciences, National Research Council, the United Engineering Center Library in New York City, the American Society of Mechanical Engineers, the Engineering Foundation, and the Cortland County Historical Society were also generous in their assistance. I am especially indebted to Mr. Von Doenhoff of the National Archives, and Lee Pearson, Historian, Naval Air Systems Command, for helping solve research problems, and to Oscar Mastin of the U.S. Patent Office. Professor William K. Kaiser helped by supplying copies of Lawrence Sperry papers. At the Smithsonian Institution, the historians and curators helped me locate and better understand Sperry inventions kept there. I am especially indebted to Dr. Robert Multhauf, director of the Museum of History and Technology, Dr. Barnard Finn, and to Dr. Otto Mayr, who read several of my sections on Sperry's feedback controls.

Russell Fries, a graduate student in history at The Johns Hopkins University was nominally my research assistant, but he was in fact a colleague thinking through with me the subject matter of two sections of Chapter III, and doing much of the research and formulating a preliminary draft of these sections. Mrs. Roz Williams was a valued research assistant in the early months of the project. Mrs. Doris Stude was absolutely without peer as a typist who also took responsibility. Working with the Johns Hopkins Press convinced me that publication is a cooperative venture. The Sperry biography demanded a commitment to a complex and unusual book involving historical narrative, technical description, and many varied illustrations. I wish to thank especially Kenneth Arnold and Arlene Sheer of the press staff. I cannot adequately acknowledge the help of my wife, daughter, and son. They all made room for Elmer Sperry as a new member of the family who demanded a disproportionate amount of attention. My wife read every draft and improved every page she read. This is her book too. My daughter may be the world's best teenaged research assistant.

THOMAS PARKE HUGHES

INTRODUCTION

WHEN ELMER SPERRY BEGAN HIS CAREER AS AN INVENTOR, ENGINEER, AND ENTRE-preneur in 1880, America had embarked upon a course of rapid industrialization and urbanization. The wilderness was being transformed into an industrial society with unmatched technological power. Before Sperry's time, America's great resource had been nature; within a few decades her strength would be technology. Elmer Sperry played an outstanding role in this transformation.

In 1880, arc lights lined the streets of a few cities, but the era of electric light and power had only begun. The railroad reached across the continent and into small villages, including Cortland, New York, where Sperry grew up, but the electric streetcar and the automobile were unknown. America had drawn world-wide attention to her system of manufacturing and her civil engineering, but other technology had not yet put down roots in the new world. Water wheels turned at innumerable power sites, but more than a decade would pass before electricity harnessed the power of Niagara. Lighter-than-air craft carried observers aloft in wartime and thrilled gaping crowds at county fairs, but heavier-than-air craft remained the vision of a few. Iron-clad warships had appeared, but the steel-hulled, long-gunned, turbine-propelled *Dreadnought*, the most awesome of pre-aerial weapons, had not been launched. Industrial concerns had been founded and grown great on patents, but corporations had not yet seen the possibility of cultivating research and development—nor had governments. The next century would see the rapid rise of the industrial research and development laboratory and a heavy commitment of government funds for this activity— first in war and then in peace.

Virtually every age and man can be portrayed as transitional, but this is a remarkably accurate description of Elmer Sperry. Born near the rural village of Cortland, he died in heavily industrialized and urbanized Brooklyn He attended a small normal school offering only an introduction to science; later he was honored by great universities educating thousands of scientists and engineers each year. His first major invention was an improvement in arc lighting; the major inventions of his later years were in gyroscopic closed-loop, or feedback, controls (cybernetics). When he began as an inventor and engineer, he used a corner of a machine shop and a borrowed drafts-man's table; in the 1920's he presided over one of the world's best-equipped research and development laboratories and machine shops. His early inventions were marketed by small companies which he helped to finance; later in his career his inventions were commercialized by such rising enterprises as General Electric, Hooker Chemical Company, American Can Company, Anaconda Lead Products Company, and his own Sperry Gyroscope Company (later the technological giant, the Sperry Rand Corporation). Before 1900 he never thought of the government as a sponsor of his inventive and engineering activity; later battleships and flying fields were placed at his disposal for testing his creations. His early inventions provided goods and services for a prosperous, peaceful economy; later he made research and development available for modern war. Sperry made the transition from the era of heroic independent invention to that of organized research and development; his rich career is a microcosm of the history of recent technology.

Sperry's contemporaries recognized his importance. After 1910, news-papers and magazines featured Sperry biographical sketches; the public termed his remarkably advanced gyro devices robots. Everyone was impressed by his 350 patents and by the inventions—and adventures—of young Sperry-trained engineers, a group that included his own sons. His son Lawrence often appeared in the newspapers because of his ingenuity and daring in designing and testing Sperry aircraft instruments. Elmer Sperry himself was good copy—the *New Yorker* in 1930 made him the subject of a "profile." Energy emanated from him and he often lectured—or just talked—to fas-cinated groups of laymen eager to learn of the "wonders of science."

His peers named him, before Edison, a member of the National Academy of Sciences (Sperry was one of a handful of engineers in the academy). The venerable American Society of Mechanical Engineers elected him president, and he was an officer of the Engineering Foundation and the committee on industrial research of the National Research Council (both influential in the engineering world). The engineering societies awarded him the John Fritz Medal, their highest honor. Sperry was the most ef-fective inventor on the World War I Naval Consulting Board—itself a milestone in the history of the effort to apply research and development to modern warfare. The Japanese recognized him as the foreigner who brought the first world engineering congress to Asia; this symbolized world recogni-tion to the aspiring, rapidly industrializing nation. Many Japanese engineers and industrialists thought Sperry less idiosyncratic than Edison and vir-tually the embodiment of modern, science-based, western technology.

Although his eminence was clearly acknowledged, his personality and accomplishments were not so easily characterized. Some called him an inventor, others described him as an engineer or a scientist. Still others labelled him an industrialist. He was all of these, although the intensity of his commitment varied, and also something more—an entrepreneur, or organizer of technological change. The scope and complexity of his ac-tivity make impossible a simple portrayal of Sperry as the self-made inventor, the hard-nosed engineer, the selfless scientist, or the dynamic entrepreneur. We cannot cast him in any of the stereotyped roles of history. His twentieth-century characteristics, his work as an entrepreneur of technological change, remove him from the familiar mold.

Because Sperry is one of the new men, he demands a new approach of the biographer. Since Sperry's major mature activity was to preside over technological change, the biographer must describe and explain how he brought change through a process involving invention, development, and innovation. To portray him only as an inventor, an engineer, or an entrepreneur would be inadequate; he is best characterized and described as an inventor-entrepreneur. This implies that he conceived of new machines and processes and then brought them into use through engineering development and entrepreneurial innovation (the last term being used in the narrow sense of introducing a thing onto the market). Acting as an inventor-entrepreneur, he presided over other engineers, inventors, and industrial scientists, and also over laboratories and corporations based on research and development. If Sperry's role as an inventor-entrepreneur can be established, then other biographers and historians may find that he is a twentieth-century prototype.

Sperry was an interesting man aside from his profession, but to write of him the biographer must also write of machines, processes, and systems. Sperry expressed himself in technology and to avoid his inventions because they are complex would be like ignoring the paintings of an artist. Because mid-twentieth-century readers are not technologically naive, I have assumed that I can write of Sperry's works and—to borrow again from the world of art—follow his evolving style. This evolving style can best be studied by concentrating upon Sperry's automatic—or, precisely, feedback—controls. Throughout his career from Cortland to Brooklyn he was drawn to invent these, and the passage of time demonstrates maturity, clarity, and force in their execution. Focus upon these devices is appropriate, for history will judge Sperry one of the pioneers of cybernetics and a major figure in the history of control engineering.

Because Sperry created artifacts and was strongly visually oriented, I have made extensive use of illustrations. Not all of these will be comprehended at a glance, but seeing them is an excellent way to grasp the shape, or form, of his work. Sperry learned more from seeing than reading, and while the reader may not share this genius, he will, it is hoped, be able to understand how at least some of Sperry's devices worked through a combination of text and illustration. If the reader is willing to take the pains to do this, he will be able to follow the idea of automatic controls as it evolved in Sperry's mind and appeared in his inventions.

Some of Sperry's original devices have been preserved, and I have found the study of these rewarding. Fortunately, the inventor's family and his company were aware of the significance of these things and presented them to museums, especially the Smithsonian Institution in Washington, D.C. If the artifact has not survived, then I have relied entirely upon printed descriptions and illustrations, making extensive use of patents.[1] The patents are invaluable records for the biographer of an inventor, as are the patent files in the United States Patent Office and the National Archives in Washington, D.C. These files give a contemporary and detailed account of the process of invention and development, including much testimony, full of flavor, given by Sperry, his sons, and his associates.[2]

The chapters of Sperry's life have prevailing themes. The biographer, knowing that Sperry at twenty invented an arc-light system with feedback controls, is moved to ask—and attempt to explain—how the young man, without leaving the village of Cortland, could invent such a complex device. The effort to find an answer leads to an exploration of the rural and small-town environment, one in which many of the leading nineteenth-century American inventors grew up. The exploration reveals circumstances ignored in legends of inventors and likely to be overlooked, such as the availability of engineering magazines and patent journals in a local YMCA. Legends also overlook the powerful impact of the contrast between nature and technology upon a young, impressionable country boy for whom technology was a beneficent force.

It is also interesting to know how Sperry raised the capital that made it possible for him to found a company in Chicago to manufacture his inventions. He went to Chicago an enthusiastic inventor; he left a decade later with a profound comprehension of what Thomas Edison meant when he said that invention was 95 percent perspiration. Sperry learned that the sweat came during development and innovation; he found that one could not be a successful inventor without being an accomplished engineer and entrepreneur. He also found himself: he much preferred presiding over invention, development, and innovation—despite the risks and the insecurity —to routine engineering and management. When his first company settled into manufacturing and sales and then faltered, Sperry established a new research and development company.

Then for more than a decade, most of which was spent in Cleveland, Sperry succeeded as an inventor-entrepreneur. He turned away from an engineering staff position with a large company; he preferred to explore independently the fast moving front of advancing technology. He wanted to choose his own problems rather than have them assigned by a corporation.

[1] Where patents have been used as the source of information for describing a Sperry device, the device described might not be the same as that subsequently manufactured. Patents have been relied on in many cases because sources describing the manufactured device were not available, were not fully descriptive, or were not about the early models. It seemed reasonable to assume that later production models—especially of gyro devices—involved many modifications originating with Sperry's engineers, for Sperry tended to lose interest in his inventions once the primary problems associated with them had become production problems or minor post-innovation modifications. When an early model of a device has been preserved, and if the device is at the Smithsonian Institution in Washington, D.C., I have taken the opportunity to examine the device and have supplied the accession and catalogue numbers. With these numbers, the serious student can view the device itself.

[2] For an informative discussion of patent records as historical sources, see Nathan Reingold, "U.S. Patent Office Records as Sources for the History of Invention and Technological Property," *Technology and Culture*, I (1960), pp. 156–67.

Since he had chosen the uncertainties of independence, Sperry's major challenge was to decide what problems, if solved, would result in innovations and bring financial rewards. To survive as a free-ranging inventor-entrepreneur, he had to compare the state of technology and the needs of the market, and close the gap with a practical invention. Because he moved from electric light, to streetcars, to automobiles, and then to chemistry, his decisions, the pattern of his inventive activity, and the reasons he entered and left fields when he did, throw light on the challenges and responses of a professional inventor.

After this period as an independent wide-ranging entrepreneur, Sperry settled in the field of gyro applications. The rich correspondence and technical reports left by Sperry from these years give the biographer the opportunity to investigate in detail from conception to use the intricate history of several major inventions. The line of development was certainly not straight and autonomous, but involved personalities, politics, economics, and even foreign affairs. Chance and paradox are as much a part of this history as of, say, political history. From the well-documented Sperry innovations in the gyro field emerges a pattern showing that Sperry kept informed of the evolving system of naval warfare and invented to fulfill emerging needs.

Wars and the threat of war greatly affected Sperry and his technology. To argue that the evolution of technology is autonomous, free from external nontechnological influences, is questionable in light of the Sperry history. The armaments race before World War I directly and substantially affected his work and career; for instance, in response to the increasing number of problems the armed forces asked him to solve, he formed a company to expand his powers. In the Sperry Gyroscope Company, and especially its staff of young engineers, the biographer finds mirrored the characteristics of its founder and director. After the war, the man and the company reacted again to a momentous social transformation, for the inventiveness—the capacity to bring about technological change—of Sperry and his staff made possible the transition from the instruments of war to machines and processes of peace.

As a result of his own and his company's work as "brain mill for the military," his unquestioned world pre-eminence in the field of automatic guidance and control, and his rich experience as inventor, engineer, and industrialist, Sperry was recognized after the war by his peers, and by the nation, as a leader of the world of technology. Following him in the postwar period, the reader then gains insight into the workings of the world of technological affairs and the way in which its leaders influenced the development of the profession.

Sperry's lasting influence is easily seen. He left a group of young men, the Sperry school, who had learned his style of invention, development, and entrepreneurship. Its members have testified to his influence. His company carried on his kind of research and development, mindful of the complexity and elegance of the founder's work. Cybernetics, feedback, guidance and control, and servomechanisms have become part of our language. To think of their history without recalling Sperry's work would be, to use his metaphor, like seeing the circus and missing the elephants.

ELMER SPERRY

CHAPTER **I** The Cultivation of an Inventor

Necessity is the mother of invention. A race of inventors has sprung up in this country because they were needed. Human labor was scarce and high. A new country was to be conquered and brought under cultivation. Wide fields demanded rapid means of sowing and harvesting. A scanty population and distant markets demanded greater facilities for rapid transit. A high ideal of life demanded a thousand new elements of gratification; and to supply all these demands a thousand new machines and processes had to be invented. SCIENTIFIC AMERICAN, XXXVIII (1878), p. 192

JUDGED BY MID-TWENTIETH-CENTURY STANDARDS, ELMER AMBROSE SPERRY enjoyed few of the advantages that would prepare one for a life of invention, engineering, and innovation. He was not from a family with a background in science or engineering; he did not attend outstanding schools or graduate from a great university; he did not grow up in an urban environment permeated by the latest ideas of science and built and maintained by the most recent technology. Yet, by the age of twenty, without leaving rural Cortland County, New York, he had invented a complex arc-light system, and local businessmen had invested thousands of dollars in him and the invention. Not only had he won the confidence of others, but he had gained the self-confidence that carried him to Chicago to compete with inventors, engineers, and entrepreneurs in the risky and open field of electrical technology. The seeming paradox of the young man's achievement invites explanation. It is not enough to say, as many inventors have, that the answer is inventive genius. A more plausible explanation exists in the circumstances of Sperry's first twenty-three years.

i

The farmers who cultivated the Cortland County wilderness were sturdy stock. Many of them, like Elmer Sperry's great-grandfather, Medad Sperry, settled there early in the nineteenth century after migrating from the East.[1]

[1] Genealogy traces Medad back to Richard Sperry, first of the American Sperrys. He settled in New Haven Colony and is remembered for having secluded two of the

Medad came from Connecticut in 1811 with his wife, Elizabeth Hine, and bought one hundred acres of farm land located several miles south of Cortland village.[2] His son, Ambrose, inherited the tract and acquired 17½ adjoining acres in 1833. Ambrose and his wife, Mary Corwin, farmed the land and reared five children, four boys and one girl. In 1860, when Elmer was born to Stephen and Mary Sperry, Judson, Burdette, Stephen, and Helen were still living on the farm; the fourth son, Ezra Cortland, had moved. Burdette had a young wife, and they seem to have built another home on the property. Judson may have been helping with the farming, but within a few years he married and established himself a few miles away at Blodgett Mills, where he operated a steam sawmill and dealt in patent rights—both of which were of great interest to the growing boy. Helen, who was destined to play an important role in Elmer Sperry's life, lived with her parents while she taught in the neighboring elementary school. Stephen found various kinds of work on the countryside, but he intended to settle down and farm after he married.

Mary Burst met Stephen Sperry when she was teaching in the county elementary school, located less than a half-mile from the Sperry farm. Mary was a cultivated and sensitive girl who had graduated from Rutgers Seminary in New York City, where she established a reputation for excellence in mathematics and the sciences. Her home was in Cincinnatus, about fifteen miles from the Sperry's, where her father, John, was a farmer and general merchant. Earlier he had been a retail lumber merchant in Albany and in Jersey City. His wife, Katharine, died in 1857, not long after the family moved, and John took a second wife within a few years.

By the summer of 1859, Stephen Sperry and Mary Burst were planning to marry. Her brother then wrote to advise her, and his letter gives an intimate glimpse of Stephen and Mary. The brother had heard, he wrote, that Stephen was a sober and industrious farmer who was not addicted "to lude or nude company." He predicted that Stephen would give her a good home and a mode of life far more satisfactory than she had. Yet, he cautioned her that a home without love was barren and that she must be sure she loved "S" deeply, or she might soon long for "one educated refined & full of fun or that one full of money."[3]

Stephen Sperry and Mary Burst were married on January 1, 1860. He took the young school teacher to live with him on the Sperry farm, but she returned to the Burst home in Cincinnatus when she expected her first child. There, in a small frame house, Elmer Sperry was born on October 21, 1860, in pathetic circumstances. His mother died a few hours after a difficult delivery and before her husband, who was out of town working, could reach her side. Aunt Helen Sperry had the melancholy experience of carrying the

---

regicide judges in a cave overlooking his farm in 1661. In the last fifteen years of his life, Elmer Sperry took an interest in genealogy and once planned to erect a statue to Richard at the cave near the present town of Woodbridge.

[2] *Bulletin of the Cortland County Historical Society*, January 1955.

[3] Undated letter in the Sperry Papers. Hereafter, all references to materials in these papers will be identified by the letters SP.

2

baby to the Sperry farm. Little more is known of the birth and death except that the mother was weak, lonely, and frightened in the days before she died.

Her son would also be lonely. His mother left him only a few possessions, among them her great chart on astronomy and her books on science, with quotations written on the flyleaves. With her husband, relatives, and friends—especially Helen—she left an image of a cultured girl of imagination and sensitivity; they preserved the fragile recollection for the boy. A Burst relative believed the mother also left her image in the facial expressions of her son. These impressions, however, were pathetically little for Elmer Sperry. All of his life he felt the pain of not having known and been loved by a mother who seemed so sensitive and loving. Without Aunt Helen, however, the pain would have been sharper. Probably because a traveling job was the best one available to him, Stephen had to leave the care of his child to his parents and his sister. It is doubtful that the grandparents, who had reared five children and were both near sixty, looked forward to the additional responsibility. Helen, however, was not yet married and became Elmer's "mother." Even after she married Charles Willett six or seven years later and moved to Virgil, five miles away, the boy saw her often on visits to what he called the Willett paradise.

Stephen may have intended to farm after the marriage, but he is remembered as a dealer in lumber, a drayman, part-owner in a traveling carnival, carpenter, and employee in Cortland's wagon works. After Mary's death, little is known of his activities until 1875, when he had settled with his parents in Cortland and was working as a carpenter. A cryptic diary that he kept for several years after 1867 suggests that he was part-owner of a traveling show and may have exhibited the Cardiff Giant, one of the great hoaxes of the late nineteenth century. From 1875 until his death in 1889, Stephen lived in Cortland and, at least until Elmer left there, in his parents' house with his son. Yet, judging by Elmer Sperry's reminiscences, the presence of his father influenced him less than the absence of his mother. One vivid recollection that Elmer Sperry spoke of years later was his father's showing him a steam engine. Other than that, Elmer Sperry recalled little of his father and his influence upon him.

ii

Before Elmer was ten, the grandparents left the farm, Helen married and moved to Virgil, and Judson and Burdette established their own households. When the grandparents moved to Cortland village, young Elmer went with them.[4] The village would be a formative experience, but the farm and the countryside had made a lasting impression. Cortland County, located in the geographical center of New York, has, according to the natives, colder winters than Syracuse, thirty miles to the north, and Ithaca, twenty miles to the southeast. There the snow comes earlier and spring later.

[4] The meager information on the places in which Sperry lived as a child has been gathered by Mrs. L. B. Leach, a student of Cortland County history, from census reports, directories, and interviews.

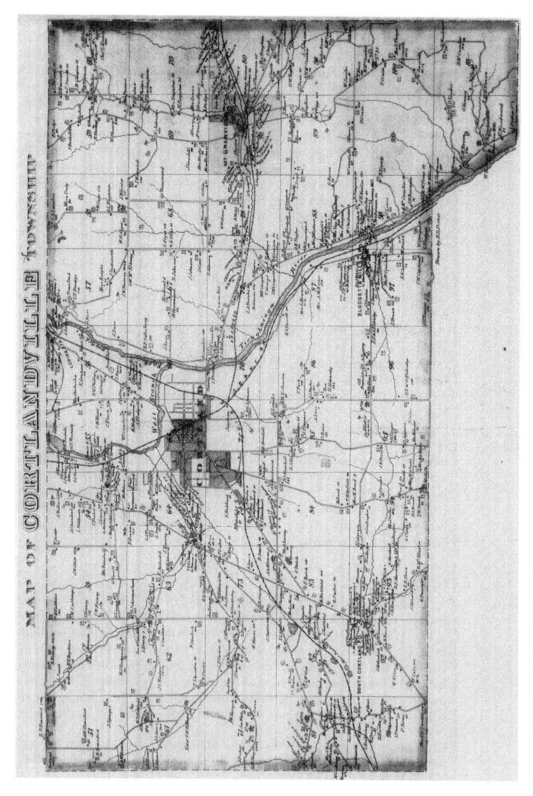

Figure 1.1 The Sperry property is in quadrant 96. From Combination Atlas Map of Cortland County, New York, Everts, Ensign, and Everts, 1876.

4

The Sperry farm stood on the slope of one of the many rolling hills to the south of the village, and these offered some shelter from the winter winds. When the rock maple and beech turned, the inhabitants justly boasted of the beauty of the countryside, but in the winter the barren trees, bare fields with outcroppings of stone, and the deforested slopes presented a cold and somber landscape. Sperry's later unsympathetic attitude toward nature would be difficult to explain unless we conclude that the general impression he carried away from the farm was one of a hostile environment. After he moved to the city, Sperry—his daughter recalled—showed no interest in the country or country life. As a child he had seen nature in terms of bitter cold, howling wind, driving rain and hard soil. Perhaps this explains why, decades later, he hung on his office wall in Brooklyn a large picture of a giant tree, half uprooted in a raging storm, clinging tenaciously to a cliff. Beneath the picture Sperry wrote, "an intrepid pioneer, conquering the sympathetic though dignified recognition of a forbidding, ever terrifying, environment."

If nature seemed hostile and disorderly, technology was, in 1860, a protective, helpful order. Unlike the urban child of the twentieth century, who looks at technology from a distance, Elmer Sperry and other farm boys of his era became thoroughly acquainted with the workings of the machines and structures of rural technology and developed thereby a keen appreciation of the problems that they were designed to solve. It is no accident that country boys like Elmer Sperry became interested in how things worked and that many inventors and engineers of his era—including Edison—came from the country. The relationship between apparent natural disorder and technological order—however primitive—was clearest in such a setting.

Two miles east of the Sperry farm lay the twenty-home community of Blodgett Mills. There, on the Tioughnioga River, young Elmer was able to study water-driven mills, which provided excellent examples of practical mechanics. Among them were a wool-carding mill, a grist mill, and, by 1869, a steam sawmill. Through his uncle, Judson Sperry, he also had access to a steam-powered mill. Uncle Judson and Blodgett Mills had additional attraction because the Syracuse, Binghamton, & New York Railroad had a station there.

Sperry was not only an observer. As a child he built small water wheels and windmills of such excellent design and construction that relatives and friends remembered them years later. These he sold as toys, but he gave his grandmother a small hand mill for grating horseradish that proved a labor-saving device. His Aunt Helen also remembered what he made for her and marveled at his pronounced interest in mechanical things at an early age.

iii

After moving to Cortland village, Sperry became familiar with a technology more complex than that of the farm. Although a village of only three or four thousand, Cortland knew industry; its boosters spoke of rapid industrialization during the decade or so Sperry lived there. Before the

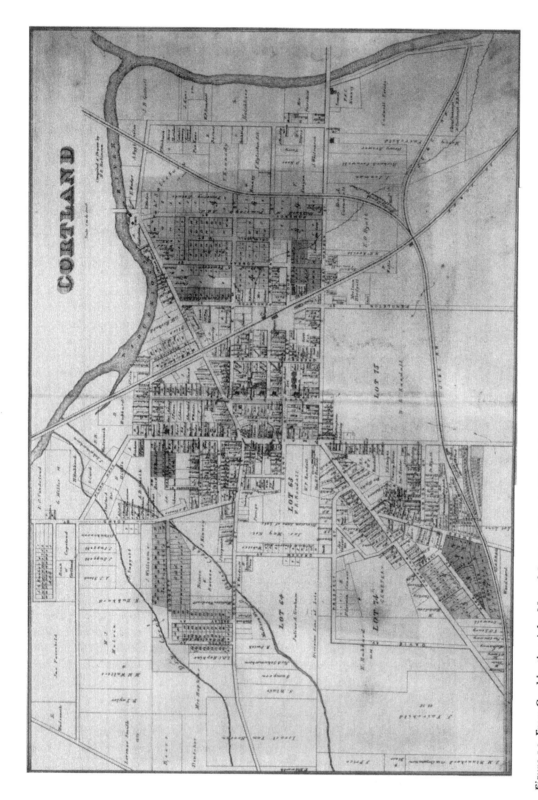

Figure 1.2 From Combination Atlas Map of Cortland County, New York, Everts, Ensign, and Everts, 1876.

6

middle of the century the village had done little manufacturing. Roads and navigable rivers coming out of the valleys brought modest commerce and settlement. Then, the Syracuse, Binghamton, & New York Railroad came to Cortland in the fifties, and early in the seventies the addition of the Utica, Ithaca, & Elmira railroad made the town a railway nexus. These two railroads involved the town in the rapidly expanding national economy, and the "almost phenomenal growth of the village of Cortland began."[5]

Not only did the number of mercantile establishments increase but also the number of manufacturing and repair shops. A foundry making agricultural implements, a machine cooperage, an oil mill, a grist mill, two planing mills, a sash, door, and blind factory, a pottery, a saw mill, and two small carriage factories were operating in Cortland in 1870.

The owners of these small enterprises did not mind a curious boy looking over their shoulders, and the railway delivery driver allowed him to ride with him to deliver new tools and machines. Sperry then watched the uncrating and observed the designs of the mechanisms. The town knew him as a frequent observer at the blacksmith shop, the machine shop, the foundry, the printing press, and the railway yard. After school and during vacations, he worked variously at a machine shop, a foundry, and a book bindery. In the Cortland book bindery, he encountered a new automatic paging machine, which was "the apple of the proprietor's eye,"[6] and on one occasion, he completely dismantled it. After he had oiled all of the parts and reassembled it, the machine worked better than before. As a result, the proprietor allowed him to maintain other machinery in the shop. At the Cortland foundry, he assisted in the construction of a hydraulic turbine, which was to be installed in a laboratory at Cornell University. Before he was sixteen he also helped insulate and wind the armature of a Gramme-type dynamo being made for Professor William A. Anthony of Cornell. The dynamo was later exhibited at the Philadelphia Centennial exposition of 1876, which Sperry attended on a week-long YMCA excursion.

The excursion to Philadelphia was a compelling experience: it greatly reinforced Sperry's interest in invention and engineering, for he was able to see the grandest achievements of technology handsomely displayed in a dramatic setting. His after-school earnings and the local YMCA, which organized the week-long trip, made the venture possible. The round-trip fare with lodging in Philadelphia for one week was only $11 but Sperry was a "parlor boarder," staying with a friend at a private residence. Although he attended a "marvelous" performance of Jules Verne's "Around the World in 80 Days"—his first theater experience—his interest seldom strayed from the exposition.[7]

Philadelphia did more than celebrate the first century of the republic

[5] *Cortland Standard* (Industrial Edition), December 1895.
[6] Gano Dunn, "The Engineering and Scientific Achievements of Elmer Ambrose Sperry," *Mechanical Engineering*, XLIX (1927), pp. 101-2. This was an address read on the occasion of the award of the John Fritz Medal to Sperry. Sperry and his secretary supplied the biographical information.
[7] *Ibid.*, p. 102.

Figure 1.3
Main Street, Cortland, New York (1865).
Courtesy Cortland County
Historical Society.

Figure 1.4
Helen Willett.

Figure 1.5 Early photograph of house believed to be Elmer Sperry's home in Cortland,
New York (located on Main Street south of Tompkins). Courtesy Cortland County
Historical Society.

in 1876; it announced to the world that the United States was growing rapidly as an industrial power. At the Great Exhibition of 1851 in London, when Britain celebrated her industrial preeminence, the ingenuity of the American republic's inventors had attracted notice. In 1876, the United States showed herself more than precocious; she appeared a rising industrial giant. Visitors from abroad were impressed by American agricultural machinery, locomotives, civil engineering, and system of practical education, as well as by the extensiveness of U.S. newspaper and book industries.[8] The most impressive symbol of this technological and industrial advance was the giant Corliss steam engine that dominated Machinery Hall.

Sperry recalled that Machinery Hall was his Mecca: "Daily he worshipped among the riot of machinery. . . . His mechanical knowledge was increased."[9] For hours he studied the Jacquard loom, an early example of punched-card automatic control, "and would not leave until he understood every movement of the intricate mechanism." He found the Corliss engine "stupendous" and, as might be expected, persuaded the engineer in charge to allow him to handle the starting wheel. He also saw the Cornell University dynamo; but no one called his attention to the exhibit of the as yet little-known Bell telephone. Altogether, Philadelphia was a memorable experience that greatly influenced Sperry's decision to become an inventor and engineer.

Besides making the trip to Philadelphia possible for Sperry, the Young Men's Christian Association influenced him in other ways. In the 1870's, before the public library was common, the Y gave many communities a library and a periodical reading room with a good selection of magazines. Many local YMCAs, following the national trend, also provided lectures and courses on nonreligious subjects.[10] The Cortland Y, for example, scheduled Professor Alexander Graham Bell to lecture on the telephone in 1877.[11] Sperry made good use of the Cortland reading room, where he studied the mechanisms described in the latest issues of the *Official Gazette* of the U.S. Patent Office. He also heard Y-sponsored lectures by the well-educated and articulate faculty of the Cortland Normal School. Many years later the inventor recalled:

Well do I remember the "Y" activities of earlier times, when I was from 15 to 18 years old; how we all used to go up on the second floor of the Taylor Building in the room where the meetings used to be held on Sunday afternoons at four, and what rousing meetings and fine uplifting times we had together there, usually under the leadership of one or another of the professors of the State Normal School.[12]

---

[8] Merle Curti, "America at the World Fairs, 1851-1893," *American Historical Review*, LV (1950), pp. 833-56.

[9] Dunn, "Sperry," p. 101.

[10] Charles N. Hopkins, *History of the Y.M.C.A. in North America* (New York: Association Press, 1951), pp. 195-200.

[11] *Cortland Standard*, November 8, 1877. The lecture had to be cancelled when Bell became ill.

[12] EAS to Edward Stillson, September 21, 1923 (SP).

Figure 1.6
Sunday school class,
Cortland Baptist Church.
Elmer Sperry is second
from the right.

Figure 1.7
The State Normal School
in Cortland.
From Cortland Standard,
Industrial Edition,
December 1895.

Figure 1.8
The Cortland Wagon
Company. Courtesy Cortland
County Historical Society.

The Y also set a moral tone that reinforced young Sperry's interest in technology and invention. By no means the least of its contributions to the Cortland community was the help it provided individuals in choosing a life work.[13] The YMCA encouraged a young man to enter a career providing an opportunity not only for hard work, but also for good works and exemplary behavior. A life given to invention and discovery fulfilled the YMCA objectives, for Americans believed that the purpose and result of technology was the creation of a material environment within which man could cultivate his higher faculties.

The First Baptist Church of Cortland also helped form Elmer Sperry's character. On November 20, 1875, church records show, he "related his Christian experience" and requested baptism in the presence of a "goodly number" at a covenant meeting. His grandfather, Ambrose, had been a leading Baptist in the community for many years, and the church records bear his name as a member who accepted many responsibilities. He saw that the grandchild left in his care had a Christian upbringing. Sundays, as a result, were full of activity. Following the "Puritanical example" of his grandparents, he felt duty-bound to attend jail meeting with other devoted souls early on Sunday mornings. He then attended morning church service, the afternoon Y meeting, and the regular evening church service—all of which made "his Sundays nothing short of Puritan."[14] After he became seventeen, he taught Sunday school.

iv

The Cortland Normal School also influenced Sperry's career. Opened in 1869, the institution provided teacher-training and offered local students tuition-free education from the elementary level through two years of college. Cortland had competed with other New York communities to obtain one of the four normal schools authorized by the legislature in 1866, and its success created a lasting transformation in the intellectual environment. The faculty first numbered only eleven, but this small group brought with it a level of education and a cosmopolitan viewpoint infrequently encountered in rural communities. If he had not attended the school, Sperry's interest in technology might have led no further than the confines of Cortland. He might have become a foreman in a local industry or perhaps the owner of a machine shop. The faculty and curriculum of the school, especially in science, raised his sights beyond tinkering ingenuity and proficient mechanics. Through the school laboratory and library, he had an opportunity to explore things and ideas that led beyond Cortland. The concept of achievement stimulated by Cortland Normal reinforced the image of intellectual quality left by his well-educated mother and fostered by his Aunt Helen.

The Normal School had three major divisions: the primary of six years, the intermediate of five, and an upper level, which offered a "normal school" program of four years and, as an alternative, an "academic" program of three or four years. The primary and intermediate levels provided a

[13] Sperry left the YMCA $1 million. See below, p. 309.
[14] Dunn, "Sperry," p. 101.

teacher-training school for the students in the normal-school program. Local students enrolled in the academic department and took some of the same courses as the normal students but did not have teacher-training or the obligation to teach. Elmer Sperry was a student in the primary, intermediate, and academic departments. Students in the academic department had a choice of either a three-year course preparing them for a business or trade career, or a four-year "classical" course designed as a preliminary to further college work. The school offered no baccalaureate but awarded to normal students a diploma that served as a license to teach. To gain admission to the academic department, students had to be at least sixteen, of "average ability," of "good moral character," able to pass a fair examination in reading, spelling, geography and arithmetic (as far as the roots), and able to analyze and parse simple sentences.[15] The "good moral character" would be strengthened by the school's location between two of the four churches on Church Street, thought a "suitable environment and a right kind of atmosphere."[16]

In the academic department for two and one-half semesters, Sperry completed about two years of the three-year course preparing the student for business or trade. He enrolled in the college-preparatory "classical" course for one term—probably a special arrangement to allow a local boy to take subjects which he desired. His course of study included mathematics through advanced algebra, geometry, and spherical trigonometry; English grammar, composition, and literature; and zoology, botany, physics (natural philosophy), chemistry, and astronomy. He also studied perspective drawing.

His grades suggest the nature of his intellectual interests. His first year in the academic department, September 1877, to June 1878, was difficult; in fact, judging from a cryptic record of his work, he failed the first term. He redeemed himself in the second term, however (probably repeating some of the work of the first), and subsequently maintained a passing level. His record was above average in advanced algebra (although he failed elementary algebra the first term), unusually good in perspective drawing, well above average in natural philosophy, and average or better in zoology, chemistry, and astronomy. He later recalled a particular fondness and aptitude for geometry; and his school notebooks suggest that he spent considerable time in mastering advanced trigonometry. In sum, he was a good science student with some strength in mathematics. He was not, however, outstanding in class, and it is doubtful that he was encouraged to seek admittance to a university. He probably had no interest in further formal education.

Although his scholastic record was not exceptional, Sperry had found his strengths and his weaknesses. Throughout his life his spelling remained atrocious (in 1879, he wrote in his notebook, "hydrochloric acid out of sault by electroleciss"), and later he depended on his wife, secretary, and editors to correct his letters, business letters, and professional articles.[17] Yet he be-

[15] *Circular of the State Normal and Training School at Cortland, New York*, Albany, 1885.
[16] *Cortland County Sesquicentennial, 1808–1958: Souvenir Book*, July 20–28, 1958, p. 96.
[17] Notebook No. 9, c. 1879 (SP).

came known as a dynamic, even compelling, speaker whose language, although not always the "King's English," was vivid and persuasive. A story he recounted some years later reveals his youthful ambition and fear of failure, and his mature satisfaction in achievement. He recalled a "terrible fear of being a total failure owing to my difficulty with spelling, grammar and languages." His teacher had warned that a poor speller "might be able to attain the dignity of a grocer's clerk . . . but nothing more." Sperry, as a distinguished leader of his profession, noted the inadequacy of the prediction. "Now that I have lived it down," he said, "I can see that the whole world seemingly does not revolve around spelling."[18]

Certainly not Elmer Sperry's world. As a youth at Cortland Normal, he found that his talents were, as he put it, "piled up in a single spot" and expressed through the conception and construction of things.[19] Fortunately, the demands placed upon him by his education allowed him to minimize his weakness and emphasize his strengths. Early successes and experiences in technology and science sustained his self-confidence and ambition, and further increased his commitment. His grandparents "found the greatest difficulty in controlling his passion for mechanics and inducing him to study books, as well as tools and machinery."[20]

By pursuing his interest in technology in the framework of other subjects, he made the distasteful more palatable. In his English composition, "England and America as Manufacturing Competitors," which was published in the school periodical, Sperry acclaimed America's increasing industrial might. He acknowledged—as did Horace Greeley, whom he quoted—that Britain had excelled in the age of generals and statesmen, but asserted that America would be supreme in the era of industry and invention. "By the aid of the infallible industry and true Yankee genius of the American," Sperry wrote, "we can truthfully say that the manufacturing supremacy of the United Kingdom is already numbered among the things that were." He condemned English conservatism because it stifled invention and attributed the success of American invention to her freedom from deference to precedent and to her excellence in mechanical manipulation.[21]

His professors encouraged his strong mechanical bent. He attracted the attention of the science faculty and won their admiration for his extracurricular activities. Shy and reserved unless with close friends, young Sperry did not excel in the more conventional extracurricular affairs, such as school politics and clubs. He felt at ease and capable in a world where things, not people, were to be organized. For example, he demonstrated a telephone system that he rigged for the Normal Debating Club, and received the public acclamation of the principal for a bell system he designed and built to maintain the school's schedule of classes. Whenever a professor was willing to display the instruments and models in the school's physics cabinet outside of regular class hours, Sperry was an enthusiastic audience. His professors not

18 EAS to a cousin, May 26, 1920 (SP).
19 EAS to Harvey Underhill, January 3, 1929 (SP).
20 "Electrical World Portraits: Elmer A. Sperry," *Electrical World*, XIII (June 15, 1889), p. 340.
21 *Normal News*, December 5, 1879.

only cultivated his interest in applied science, they also befriended him. Professor Thomas B. Stowell, Cortland's first professor of natural science, who later became Dean of the School of Education at the University of Southern California, knew the boy well enough to give him "fine" advice when Sperry became distracted by one of Cortland's most popular girls. He later recalled that Professor Stowell and his wife were "leaders of scientific thought and culture in old Cortland" and that they instilled in him a vision "of the great and thrilling unknown." Their dignified way of life and their cultured home impressed him deeply.[22]

Stowell found Sperry different from other students because of his inclination for natural science and his remarkable ability to apply scientific knowledge to the invention of devices which, the professor believed, illustrated the "practical value of scientific study."[23] The hard work Sperry expended upon his extracurricular science and technology impressed Stowell; he was especially impressed by the boy's electric-bell system for the school and his invention of a device to measure at a distance, with an electrical signal, the contents of a gas or liquid container. He commended Sperry for his independent studies in electricity and for his understanding of the generators and other electrical inventions of such pioneers as Zenobe T. Gramme and Werner von Siemens. Sperry also saw problems arising from the work of these men, problems to be solved by inventors who wished to go beyond them.

Professor James Shults, who taught physics at Cortland, also took an interest in Sperry. Shults noticed the boy when he was still in the intermediate department of the training school, for he had never encountered a "scholar" at the school who had shown greater interest in the apparatus of the laboratory and its experimental use. In the evenings, the professor provided special demonstrations for the young enthusiast.[24] Another of the faculty whom Sperry had the good fortune to know was Samuel J. Sornberger, who began teaching physics the year after Sperry entered the academic department. An outstanding graduate of the school's "classical course," Sornberger returned after receiving a doctorate, with honors, from Syracuse University. This young professor also took a liking to the boy and allowed him to experiment independently with laboratory equipment. In sum—despite Sperry's commonplace academic record—he received attention and encouragement from men to whom he looked for guidance and inspiration, men who gave him a glimpse of how to gain "dignified recognition" in an "ever terrifying environment."

Even though Sperry found his contacts with science professors outside of class most influential in cultivating his talents, the course work he did at Cortland later proved essential in his career. Although he would be handicapped by not having had the calculus, he depended on his algebra and

---

[22] EAS to Thomas B. Stowell, April 30, 1920 (SP).
[23] Stowell testifying, October 1881, in patent interference, *Sperry* v. *H. Sample and F. Rabl,* "Electric Current Governor for Dynamo Electric Machine," National Archives, Box 921.
[24] J. H. Shults to EAS, January 4, 1920 (SP).

14

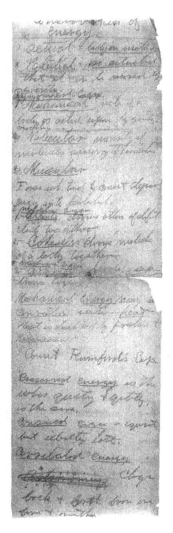

Figures 1.9
Pages on the conservation
of energy from Elmer Sperry's
Normal School notebook.

trigonometry. His knowledge of physics (including electricity) and chemistry was neither profound nor detailed, but he acquired broad principles that provided a framework within which he could pursue the solution to a technological problem. His class notebooks shows that he studied fundamental laws of physics, such as the conservation of energy, Faraday's laws of electromagnetism, and Newtonian mechanics. He wrote four pages of notes on the conservation of energy and sketched the apparatus used by James Joule to demonstrate the law.[25] The fundamentals of science learned at Cortland were simple but powerful in their ability—along with his mathematics—to explain and predict the behavior of the machines and processes which interested him. For most of the students at Cortland, science was information to be memorized; for young Sperry it provided a key to understanding the world of nature and technology—a world over which he wanted to exercise some control.

v

By the time Sperry ended his studies at Cortland Normal in January 1880, his interest in science and technology was focused on electricity. That he should have been attracted to electrical technology in 1879 is not difficult to explain. During his last term at the school, interest in electricity greatly increased with the publicity given to Thomas Edison's invention of a practical incandescent bulb. The arc lamp of Charles Brush of Cleveland, Ohio, had been famous for several years. Young and impressionable, Sperry was attracted by this new and exciting field where there was need and opportunity for inventors and inventions, judging by the success of Edison, Brush, Edward Weston, Werner Siemens, Z. T. Gramme, Paul Jablochkov, and others. Sperry's interest was fortunate because he would not be handicapped by a lack of formal education in engineering: in 1880 there were no courses in electrical engineering, and the field was wide open for bright young men.

Sperry kept abreast of the state of the art by studying technical periodicals and patents available at the Cortland Normal School library, the YMCA reading room, through his science professors, or by his own purchases. He subsequently testified, in a patent-interference hearing, that before 1880, "I was a constant student of electrical inventions, as patented in the U.S. and English patent offices . . . [and] took scientific and electrical papers by means of all of which I was enabled to keep posted as to the advances made in the art. . . ."[26]

The *Official Gazette* of the Patent Office, which he read at the YMCA, printed the claims and some drawings from all the patents issued, and other technical periodicals elaborated upon some of the most important of these. If it is assumed, for example, that he also read the *Scientific American* in 1879, he would have found over a six-month period at least a dozen helpful articles on electricity. Between July and December, there were articles on electricity as motive power, the electric light, a current regulator, magnetic

[25] Notebook No. 9, c. 1879 (SP).
[26] Sperry testifying, 1886, in patent interference 10,426, *Van Depoele* v. *Henry* v. *Sperry*, "Electric Railway," National Archives, Box 1262.

fields, and the commercial potential of electricity. Of considerable interest and help could have been the series of articles, with engineering drawings, on Edison's generator for his lighting system, written for the *Scientific American* by Edison and his assistant, Francis Upton. A young enthusiast such as Sperry could have obtained from the articles the precise dimensions of the generator's parts and even learned of the kind of wire used in the field winding. After Edison and Upton publicized their claim of greater than ninety percent efficiency for their generator, the *Scientific American* carried letters and essays debating the validity of the claim. Such an exchange would have provided Sperry with a discussion of the theory of generator design by practical experts whose ideas and information were not beyond his education. If Sperry wanted more detail on electrical machines, he had only to send one dollar to the *Scientific American* to obtain a copy of any patent issued since 1867. In this way he could have become intimately acquainted with many types of generators.

Filled with information and spurred by youthful ambition, Sperry wanted to invent something himself. He had read enough about generators and arc lights to perceive the problems that the inventors were trying to solve; the question was: upon which of these critical problems should he concentrate? The young man had the wisdom to consult a man experienced in the field before making the decision.

Knowing that Professor Anthony was a pioneer and an expert—and nearby—Sperry made the twenty-mile train trip to Cornell to ask a few questions.[27] In arranging the interview, he had the help of Edward Cleaves, a Cortland resident who attended Baptist prayer meetings with Sperry and taught free-hand drawing at Cornell. Anthony kindly saw the youth and, in the course of their conversation, noted two major problems to be solved to improve existing generators. One was to design a generator that produced a constant current despite variations in the speed of the steam engine driving it. Anthony hoped to substitute such a constant-current generator for a battery in making practical laboratory demonstrations; others wanted current regulation for arc-light systems. The second problem was to design an automatically regulated generator that would supply constant current despite load variations (as when arc lights were cut out of a circuit).[28] The problems defined by Anthony were generally recognized and attacked, but hearing of them directly from the distinguished professor gave them a dramatic clarity and urgency. On the way home Sperry sketched a governor and a current regulator which he believed would solve the problems. He showed these to Cleaves on the train and that night made additional sketches on a brown manila book cover. Sperry had conceived his first major invention.

How would he develop his invention and introduce it on the market? Since the Normal School no longer filled his days, he had the time, but he did not have the money. He continued to live with his grandparents in Cortland. His father was also there and working at the Cortland Wagon

[27] Sperry testifying, November 4, 1881, in *Sperry* v. *H. Sample and F. Rabl.*
[28] *Ibid.*

16

Company, but the total income of the family was small.[29] Sperry's resourcefulness under these conditions is indicative of his determination. He found part-time work that left him free to commute to Cortland to audit Professor Anthony's physics lectures, which regularly included discussions of electricity and magnetism. Sperry later described these as amounting practically to a course in electrical engineering. He also took the opportunity to study the electrical apparatus in the laboratories at the university, thus "keeping himself *au courant* with all the special researches and experiments going forward" at Cornell.[30] Sperry found Anthony's lectures using a generator for demonstration purposes especially informative[31] and relevant to the problems of designing his own generator.

He also needed to have engineering drawings, to make a patent application, and to construct a test generator. Part-time work and his resources would not support him, hire a draftsman, pay a lawyer, and rent a machine shop. Determined, enthusiastic, and self-confident, he approached business men in Cortland in an effort to raise capital. After many disappointments, in July 1880, he interested the managers of Cortland's most progressive industry, the Cortland Wagon Company—a company large enough to give Elmer Sperry extensive technical support. The business began building horse-drawn wooden wagons for the general market in 1872; thereafter, production and plant size increased rapidly. In 1877 the output was 4500 vehicles; it rose to 8000 in 1880 and reached 12,000 in 1884. The production equipment dispersed in various buildings throughout the town was consolidated in 1881 and supplemented in a large new factory building that was capable of producing thirty thousand vehicles annually. The factory brought the latest advances in machine shop and factory practices to Cortland and became known as a model in its field. [32]

Sperry's arrangement with the company provided him a salary, tools and supplies, assistance from skilled workers, and the advice of the company's executives and patent lawyer.[33] The company not only helped him develop his invention, but provided him practical experience in design techniques and shopwork. Among the company executives advising Sperry was Hugh Duffey, who had invented widely used wagon-manufacturing machines and had accumulated considerable mechanical engineering ex-

[29] In April 1880, grandfather Ambrose died, leaving his wife real and personal property; his son, Burdette, the 117½ acre farm; and his son, Stephen, "his gold-headed cane and relinquished to him all charges for the care and expense of rearing and maintaining his son, Elmer Sperry, up to that time." Ambrose had advanced their shares to his two other sons, Cortland E. and Judson A., and he provided household goods and furniture for his daughter Helen upon the death of her mother. The next spring, the grandmother, Mary Brown, died, leaving all of her real estate to Stephen and, after his death, to Elmer. The legacy, however, did not help when young Sperry needed money to develop his invention in the spring of 1880. D. N. Elder, "Elmer Ambrose Sperry's Ancestry," *Bulletin of the Cortland County Historical Society*, January 1955.

[30] "Electrical World Portraits: Elmer A. Sperry," p. 340.

[31] *Cornell University Catalogue, 1875*, pp. 47, 69.

[32] *Cortland Standard* (Industrial Edition), December 1895, p. 12.

[33] Sperry testifying, August 2, 1886, in *Van Depoele* v. *Henry* v. *Sperry*.

perience before becoming superintendent of the works.[34] Duffey allowed Sperry to work near him so that they could discuss mechanical problems as they arose. The wagon company also assigned a skilled machinist, Jesse Vandenburgh, to assist Sperry in his construction of the dynamo [35] (the two applied, incidentally, for a joint patent on a vehicle wheel in September 1881). Two wagon works employees made the patterns for the generator castings. The company's machine shop and forge cast and finished parts in both brass and iron for the Sperry generator. From these skilled workmen he learned the properties of materials and the uses of various machine tools. If he had been content to invent improvements in machine tools—as one of his early patents showed him capable of doing[36]—he might have remained in Cortland with the wagon works and found a satisfying challenge.

Wayland Tisdale, treasurer of the company, took a particular interest in the progress of the young inventor. From Tisdale, Sperry learned more of the financing and management of invention and development. Almost every day from August 1880, until February 1881, when Sperry was building his generator and arc light, Tisdale discussed the problems with him. As the generator moved from drawing board to successful operation, their talk turned from development problems to those of marketing, or innovation, for Tisdale believed that a company manufacturing electrical equipment based on Sperry patents would be a commercial success.

These plans depended on Sperry's preliminary success in obtaining patents. To help him the company turned to Cortland's patent lawyer, John W. Suggett, noted for his comprehension of science and technology as well as the intricacies of patent procedure. During the two years Sperry worked in Cortland developing his inventions with the support of the wagon company, he learned much about the patent process. He learned to keep a record of his inventive activity, dating, and obtaining witnesses for important verbal disclosures, sketches and drawings—a record that would be essential in case of interference. Sperry also learned how to draft an application that described his invention in a patentable way. In 1881 the company sent Sperry to Washington for six weeks to work in the office of J. R. Nottingham, the solicitor of patents who handled Suggett's patent affairs at the Patent Office. In Washington, Sperry became intimately acquainted with the complexities of shepherding a patent application through the Patent Office. From Suggett —and Nottingham—he was learning a critical part of the invention-development process.

Why did the wagon company invest so extensively in a young local inventor in a foreign field? Probably because the company was dynamic and a sanguine risk taker. Inventors also received considerable publicity, and the outstanding successes of Edison demonstrated the profitability of invention and innovation. Since Edison was one more example of an inventor's rising from humble circumstances—and the small town—to world renown, Sperry's

[34] "Grip's" Historical Souvenir of Cortland (1899), pp. 118–19.
[35] Hugh Duffey testifying, June 1883, in patent interference 8596, Daft v. Thomas v. Sperry, "Electric Arc Lamp," National Archives, Box 1005.
[36] Filed September 30, 1882, "Machine for Cutting Screw Threads," issued February 26, 1884 (Patent No. 294,092).

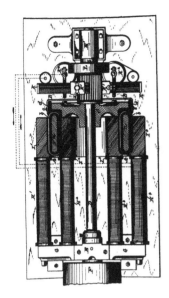

Figure 1.10
Dynamo Electric Machine
(Patent No. 260,132).
Below: front elevation.
Top: plan view.
(Key: governor, L; armature,
a; armature shaft, D; drive
pulley, F; field magnet
helices, H; field magnet
castings, I; commutator, J;
commutator brushes, K.)

chances seemed good. Furthermore, Sperry had a local reputation as an inventor, and the company had received Professor Anthony's favorable opinion of his work. Sperry's exciting, new, well-publicized field must have been an attractive diversification move for a company building wagons and carriages. An immediate motive was the company's need for arc lighting in its new buildings so that longer hours could be worked during the short winter days. In the long run, the company could use its Chicago branch to establish a foothold if it should decide to compete for the expanding arc-light market in a large city. Nor should the personality of the young inventor be overlooked, for he was already known for his sincerity, enthusiasm, and persuasiveness. Whatever its reasons, the company gave Sperry a great boost in his career.

vi

With the support of the wagon company, Sperry developed a complete arc-light system between the summer of 1880 and the fall of 1882. Initially, he concentrated his efforts on the generator, the construction of which was completed in December 1880. He borrowed a single arc lamp from Cornell to display his creation in the Cortland First Baptist Church on Christmas Eve. An account of the scene survives:

As one entered the Church on Friday Evening last the objects which first met the eye were two graceful evergreens, one on each side of the rostrum loaded with presents and brilliantly illuminated with waxen tapers. The rostrum beneath the trees was filled with packages of various shapes and sizes and in the immediate neighborhood were sleds, chairs, pictures &c.

The exercises began promptly at seven o'clock before a crowded house. The singing by the double quartet was very fine and the little folks did themselves great credit in all their parts. . . . At the conclusion of the literary part of the exercises old Santa Claus made his appearance. . . .

An extra feature of the evening was the exhibition of Elmer Sperry's electric lights. Owing to a defect in the engine which was located in the O'Neil wagon works at the rear of the Church the full capacity of the lights was not brought out, and yet all were permitted to see the light and make a comparison between it and gas—a comparison which was decidedly unfavorable to gas.[37]

A few weeks later he again displayed the light, which this time illuminated several blocks of Cortland from the tower to the lower shops of the wagon company. For a village whose streets had been dimly lit by gas, the brilliance of the display was dramatic. Wherever the arc light first appeared it caused a sensation, whether in Cortland, New York City, or Berlin.

Assured by its performance and supported by the wagon company, Sperry applied for a patent on his generator in December 1880—his first patent application.[38] The next problem for the young man was to build

---

[37] *Cortland Standard*, December 30, 1880.
[38] Filed December 22, 1880, "Dynamo Electric Machine," issued June 27, 1882 (Patent No. 260,132). When several claims in the application were found in interference by the Patent Office examiner, Suggett divided it, leaving only the uncon-

an arc light that especially suited the characteristics of his generator. Sperry had conceived of a self-regulating arc the previous August, but Suggett had advised him to delay its development until the generator proved itself.[39] After the December display, Sperry completed the construction of several lamps of his design and installed them with his generator in the wagon works in February 1881. A steam engine drove the generator and two of his lamps connected in series with the generator. To combine several lights on a single circuit successfully was a notable, if not an original, achievement. Each light had to perform satisfactorily despite variations in the performance of the others. If one of the lights burned out, then it had to be short-circuited automatically to avoid breaking the entire circuit. The lamps provided adequate light for working areas during the short winter days; gaslight could not provide the necessary intensity of illumination. Nine months later the display of the Sperry system during the Firemen's Festival brought newspaper comment:

The electric light Mr. Sperry exhibited in the evening through the courtesy of the Cortland Wagon Company, under the massive double arch on the corner of Main and Court Streets, was fairly dazzling in its brilliance, making lighter than day the space about it, and giving to hundreds who had heard of, but never seen it, an opportunity of viewing its beauty and brightness. The mass of dark evergreens, the sharply defined shadows cast by every object upon which the rays of light fell, the great crowd which gathered about with upturned faces, and the blackness of the night which furnished a background to all, made a picture which will not soon be forgotten by those who beheld it.

The Concert and Ball: under the great arch, illuminated by the rays of the electric light, the band concert was given, and a most delightful one it was.[40]

These successes were accompanied by mounting development costs, arising in part from patent-interference cases. The Patent Office in 1881 declared interferences in connection with a patent application on the generator[41] and also on Sperry's arc-lamp application.[42] Suggett had to arrange for testimony in Cortland and for preparation of briefs; Nottingham, his associate in Washington, D.C., had to follow the case through hearings and appeals at the Patent Office. Sperry and the wagon company won these cases, but only after many months. The inventors who opposed him also had financial backers and were able to withstand the sustained expense.

The wagon company also invested in the construction of Sperry's first major arc-light system. This system had twenty lamps and required addi-

tested claims in the first application. A second application which included the contested claims was filed April 28, 1881, "Dynamo Electric Machine"; issued December 12, 1882 (Patent No. 268,956).

[39] *Daft* v. *Thomas* v. *Sperry.*

[40] *Cortland Standard*, September 8, 1881.

[41] *Sperry* v. *H. Sample and F. Rabl.* The patent application at issue was the division filed April 28, 1881, taken from the application made December 22, 1880, "Dynamo Electric Machine."

[42] *Daft* v. *Thomas* v. *Sperry.*

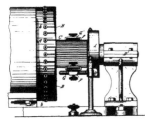

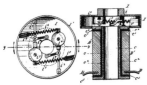

Figures 1.11
Mechanical governor for
Dynamo Electric Machine
(Patent No. 268,956),
side elevation, plan view,
and vertical section.
(Key: governor, L; armature
shaft, D'; generator
commutator, C; commutator
brushes, G, G';
governor weights, i3 i4;
governor springs, i5 i6.)

tional manufacturing facilities not available at the wagon works; therefore, the company subcontracted work to the Phoenix Iron Foundry of Syracuse, New York, and Sperry went there, in April 1882, to supervise construction. Work continued until September, when he made a public display of the system during the democratic convention held in Syracuse's Globe Hotel. The Syracuse *Standard*, in September 1882, described the demonstration as "a success from the moment it was started." Hundreds witnessed the exhibition and remarked upon the steadiness and peculiar softness of the light. "Mr. Sperry's associates," the *Standard* added, "are highly gratified with the results of the first test, and it is their purpose to push the manufacture and bring the light into general use as soon as possible."

vii

Sperry's earliest patents suggest the characteristics of the electrical generators and arc lights he and his backers hoped to sell in Chicago? Sperry patents described three major improvements in the generator: the positioning of the armature in relation to the poles of the field magnets; a mechanical governor to maintain uniform output automatically, despite variations in the speed of rotation of the prime mover; and an electromagnetic control to adjust current output with load variations. The objective of his armature arrangement was to "subject a greater proportion of the wire wound on the armature to an intense magnetic field."[43] Many inventors strove to maximize the effect in various ingenious ways. Sperry designed an annular armature that revolved between the pole pieces of the generator's electromagnet (see figure 1.10) and thereby exposed both the internal and external surfaces of the armature coils to the surfaces of the pole pieces. The pole pieces of most generators faced only the external surface of the armature. As late as 1888, a leading scientific expert on the generator cited Sperry's as an unusual form of armature worthy of note.[44]

To control output current automatically, despite variations in the rotation speed of the armature shaft driven by a steam engine, Sperry resorted to the mechanical governor (see figure 1.11). Driven by the armature shaft, the governor consisted of rotating weights under adjustable spring tension. The levers carrying the weights were linked to the commutator, which was free to rotate around the armature shaft and under the brushes. Sperry designed the mechanism to maintain the brushes in a position giving a uniform current at various armature or prime-mover speeds. In essence, the operation of the centrifugal governor was similar to the conventional feedback device used on steam engines: instead of using the governor to control steam flow by positioning a valve, Sperry's governor positioned the commutator in relationship to the brushes and thereby controlled the amount of current collected by the brushes. For every position of the brushes

[43] Sperry patent, "Dynamo Electric Machine," June 27, 1882 (Patent No. 260,132).
[44] Silvanus P. Thompson, *Dynamo-Electric Machinery* (London: E. and F. N. Spon, 1888), p. 135, places it with three other examples in a category under the general heading, "ring armatures."

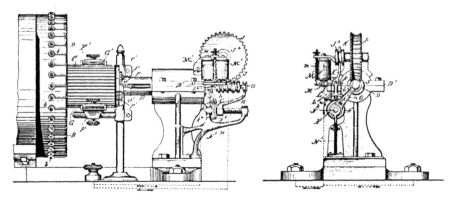

Figures 1.12 Electromagnetic current regulator for Dynamo Electric Machine (Patent No. 268,956).

there was a related current output. Sperry had located by experiment the brush position giving constant output at each armature speed; he adjusted the governor to move the commutator under the brushes to these positions as the speed varied.[45]

The third Sperry invention for the generator was an automatic electromagnetic regulator. It was placed in the lamp circuit so that variation in the electrical resistance of the circuit, resulting from lamps switching in or out, caused a change in the relative position of the brushes and the commutator. The change of position, as in the case of the Sperry mechanical governor, adjusted the current of the generator in proportion to the variation of resistance. Later, the Sperry company claimed that 75 percent of the total number of arc lamps in a given system might be instantly extinguished without damaging the generator.[46]

The functioning of the electromagnetic regulator can best be understood from Sperry's description of a similar regulator in a patent for which he applied April 28, 1881 (268,956). The device was for use with a generator supplying electricity to incandescent lamps connected in parallel. The problem, as Sperry saw it, was to reduce automatically the amount of current in the main circuit when a lamp or lamps in the circuit were extinguished. Otherwise the intensity of the lamps would have varied. Sperry had in mind a direct-current generator with field coils excited by a small direct-current generator. His solution was to use a very high-resistance electromagnet in parallel with the main circuit; therefore, current passing through the magnet, although less than that passing through the external circuit, was proportional to the voltage across the circuit. Opposite the face of the magnet, Sperry placed an iron armature which was attracted by the magnet but restrained by a spring. He adjusted tension on the spring to increase or decrease the armature's resistance to the magnet. The tension was set to hold the armature in a neutral position when the current flow was at the desired level. If the current increased, the armature approached

[45] *Ibid.,* pp. 64ff, discusses the distribution of potential around the collector (commutator). (Sperry Patent Nos. 260,132 and 268,956).
[46] *Electrical Review,* February 16, 1889, p.13.

the magnet, and vice versa. The armature's movement triggered a mechanism driven by the generator shaft (see figure 1.12) which altered the commutator's relationship to the brushes to raise or lower the current output. This increase or decrease in current output then fed back to the electromagnet, and the magnet armature returned to the neutral position. In the regulator, the tension of the spring was a function of the desired current, and the force of the electromagnet a function of the actual current.

In the electromagnetic regulator, Sperry used a patentable combination of well-known elements and principles. Elihu Thomson, the electrical inventor, received the best known of the automatic current regulator patents in 1881; it was similar to Sperry's.[47] Sperry's use of the counterforce of a spring to create a differential was a favorite device of early arc-light designers, although instead of a spring they often used the force of gravity or the buoyancy of a float as a counterpoise.[48]

Sperry's arc lamp also had distinguishing and patentable features.[49] Since Humphry Davy, the British scientist, had publicly described an arc light in 1808, inventors could not patent the striking of an arc between carbon rods, but they could patent their ingenious devices for regulating the distance between the carbons as these burned down. It was necessary to maintain a certain distance because a gap too large for the system extinguished the arc. Fortunately for the inventors, variations in the distance between the carbons changed the flow of current through the circuit (the greater the gap, the smaller the current flow); and Sperry, as many inventors before him, used these variations of current flow to vary the force of an electromagnet, and so to move the carbons and control the gap. To do this he used two coils, or solenoids. The first was a high-resistance coil wound on a core, which carried the upper carbon (see figure 1.13). The second was a low-resistance coil within which the other moved up and down. The first coil was in parallel with the circuit passing through the arc carbons and creating the arc. The second was in series with this circuit.

When the arc was struck, the current normally passed from the generator through the low-resistance coil, into the upper carbon, across the arc (gap), through the other carbon, and then returned to the main circuit and the generator. Only a small amount of current normally passed through the high-resistance shunt circuit. The two magnets were so designed that with the normal current flow their respective magnetic forces were such that the exterior fixed coil held the interior movable coil (with the movable carbon rod) in a position a fraction of an inch above the tip of the lower rod. When the carbons burned down, the gap then opened, the resistance

[47] The Thomson device is singled out for description in Edwin Houston, *A Dictionary of Electrical Words, Terms and Phrases* (New York: W. J. Johnston, 1889), pp. 53-54.
[48] W. James King, *The Development of Electrical Technology in the 19th Century* (Washington, D.C.: Contributions from the Museum of History and Technology, Paper 30, 1962) describes early arc-light regulators.
[49] Filed April 15, 1881 (30,916), "Electric Arc Lamp," issued September 9, 1884 (Patent No. 304,966).

across the gap increased, and more current shunted through the high-resistance coil and increased the force of the magnet. At the same time, the force of the exterior magnet decreased with the decrease in current across the arc gap. This change in the relative force of the magnets—or the differential—allowed the upper carbon to descend (under the force of gravity) until the gap was the normal width; the magnetic forces were restored to their normal values with the return of normal current flow.

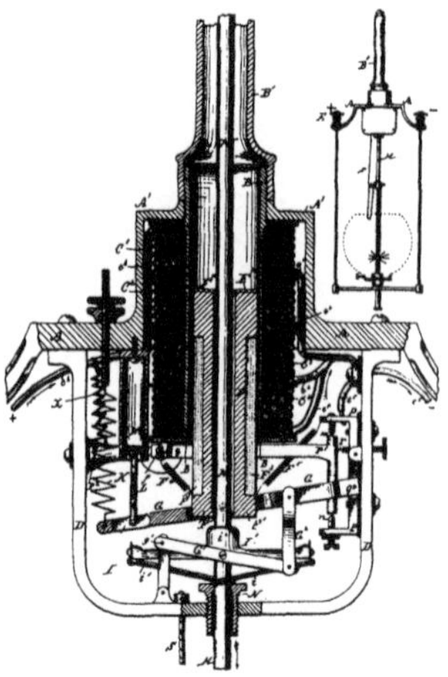

viii

Sperry probably incorporated automatic regulation in both the dynamo and arc light shown at Syracuse in September 1882. Made confident by the success of the display, the wagon company and Sperry then proceeded with plans to manufacture the system and introduce it on the market. The major problems of invention and development seemed past, and the problem was now to innovate or to market the system. They chose bustling, expanding Chicago as the scene for this new phase in their activities.

Electric light was not unknown in Chicago. In 1878 the North Side Water-Works had been illuminated briefly, and the next year arc lights were installed in the Grand Pacific Hotel. Other isolated plants were operating before Sperry arrived. The large manufacturers outside Chicago, such as the Brush Electric Company of Cleveland, were establishing Chicago branches and two electrical manufacturing companies selling lighting equipment originated in Chicago before the Sperry Company was incorporated: the Van Depoele Electric Light Company in 1880, and the Western Edison Company in 1882. Within a year of the founding of the Sperry Company, there were 11 other electrical firms doing a total annual business of over $1 million.[50]

To break into this competitive environment, Sperry shipped to Chicago the lighting system shown in Syracuse and prepared to demonstrate it to "capitalists contemplating the formation of a company."[51] He arrived on October 20, first visiting the Chicago branch of the Cortland wagon works to discuss plans with L. J. Fitzgerald, president of the company, who was visiting Chicago. The next day, Wayland Tisdale arrived in Chicago to help Sperry promote the new electrical company. The two worked through the county commissioner, who was much interested in the system and arranged a demonstration at the centrally located Chicago court house. Sperry attended to all the details of setting up the steam engine, the dynamo, and the arc lights. On November 2, 1882, the system performed successfully in the presence of the commissioner. Local investors were then invited to inspect the system and discuss the founding of a company with Tisdale.

In February 1883, twenty-three-year-old Elmer Sperry saw a company founded bearing his name and capitalized for $1 million. The principal assets of The Sperry Electric Light, Motor, and Car Brake Company of

Figure 1.13
Electric Arc Lamp
(Patent No. 304,966), general view (upper right) and vertical sectional view (left). (Key: compound solenoid or helices, C1 C2 C3 C4; movable core or armature, F; high resistance coil, F3; carbon holder, H.)

[50] Bessie Louise Pierce, *A History of Chicago* (New York: Alfred Knopf, 1957), p. 224.
[51] Sperry testifying, August 2, 1886, *Van Depoele* v. *Henry* v. *Sperry*.

Chicago were the patents, patent applications, and inventive potential of Elmer Sperry. The objective of the corporation, as described in its Illinois charter, was to manufacture and sell materials, machinery, apparatus, and appliances for the generation, storage, and measurement of electricity to be used for electric lighting and power.[52] The corporation owned no plant, no equipment, no goods in process, little good will, and only a small amount of operating capital.

Sperry and the wagon company received $350,000 of the capital stock for his patents and applications. The corporation acquired the three dynamo patents on his regulators, the design of his armature and field coils, and the patent application for the arc light.[53] The company also acquired half interest—Sperry retained half—in other inventions which Sperry had disclosed to his associates, but for which applications had not yet been filed. These included an electric train brake, a system of electric lighting for cars or trains of cars, a combination steam engine and dynamo, a locomotive headlight apparatus, a car lighting apparatus, and a street-car motor.[54] Sperry and the wagon company shared the $350,000 in capital stock equally. The *Cortland Democrat* predicted that "no one will begrudge the Wagon Company its share in the fruits of success, for to its wise liberality Mr. Sperry has been indebted for the financial backing which has enabled him to perfect and defend his inventions." The editor also anticipated that "Mr. Sperry, who has a host of friends and admirers here, will receive sincere congratulations on the well won and amply merited returns for his genius and industry."[55]

Chicago was involved as well. With Sperry's court house exhibition to support them, Tisdale and Edwin Palmer, manager of the Chicago branch, found local promoters to organize the company. To the promoters went $350,000 in capital stock, leaving $300,000 in the treasury to be sold as working capital. For the stock, the promoters put up a large sum of money "as a guaranty of the fullfillment of their contracts." They were also to provide financial and managerial advice. All of the "leading spirits" in the new company were described as having been interested in prominent electric-light companies, and as undertaking the new enterprise with the conviction that "Mr. Sperry's inventions are far in advance of all competitors."[56] The Reverend Galusha Anderson, Charles S. Cleaver, Loren Greene, and Professor Allen A. Griffith of Chicago, along with Tisdale, Sperry, and Palmer, were the seven original stockholders.[57] Anderson was President of the University of Chicago. He had previously achieved a reputation for strong convictions and forceful leadership in several very large and influential Baptist congregations in various parts of the country. His supporters hoped that he

[52] State of Illinois, Office of the Secretary of State, Records of the State of Illinois, Box 183, #8733.
[53] EAS to J. W. Suggett, July 2, 1883 (SP).
[54] *Cortland Democrat*, February 2, 1883.
[55] *Ibid.*
[56] *Ibid.*
[57] Records of the State of Illinois, Box 183, #8733. According to this record, the

would be able to save the University of Chicago from impending financial disaster resulting from a heavy burden of debt. He held on until 1885, shortly after which the lands and buildings of the old university fell to an insurance company holding its mortgage. A few years later John D. Rockefeller helped found the new University of Chicago. Anderson became a faculty member in the Divinity School.[58]

Anderson was president of the Sperry Electric Light, Motor, and Car Brake Company. Fitzgerald, president of the wagon company, became vice-president of the new corporation; Palmer, treasurer; Griffith, secretary; Cleaver, superintendent; and Tisdale and Loren Greene, directors. With two such prominent citizens as Anderson and Fitzgerald as officers, the new enterprise could reasonably expect to raise operating capital in Chicago and in Cortland by the sale of some of the $300,000 capital stock held in the treasury. Elmer Sperry was named company "electrician, inventor, and superintendent of its mechanical department," and he negotiated a separate contract with the company. He formally agreed to remain in the service of the company as long as needed at a starting salary of $30 per week to increase to $3000 annually when the sale of stock, or manufactured articles, allowed. Sperry anticipated, beyond the salary, a share of the company's net earnings and dividends on his stock. The company agreed to manufacture his inventions and to give Sperry half of the net proceeds from inventions that he and the company owned jointly. Of crucial importance to the young inventor was the agreement that the company supply him with all the machinery, tools, materials, and assistance he needed for inventing, improving, and developing electrical apparatus.[59]

At the age of twenty-three, Elmer Sperry had become an inventor, financed by an enterprise which promised to support his creativity and to manufacture his creations. Three years earlier, when he conceived of his first patentable invention, could he have also conceived of such good fortune?

---

original subscribers received the following shares of capital stock valued at $100 per share: Tisdale, 3000; Sperry, 3000; Palmer 500; Anderson, 1000; Cleaver, 1000; Greene, 500; and Griffith, 1000. On the face of it, this information does not agree with the apportioning of stock as described by Sperry to Suggett in his letter of July 2, 1883. Since Suggett was well informed, and because Sperry was seeking advice that depended upon the accuracy of information supplied by Suggett, this letter has been taken as authoritative. There is no record of any of the subscribers having paid for their stock, and it seems reasonable to assume that most of it was given in return for patents or promotion. As noted, the Chicago promoters, whom I have assumed to be the subscribers, provided some cash.

[58] Richard J. Storr, *Harper's University: The Beginnings* (Chicago: University of Chicago Press, 1966), p. 75n.

[59] Notarized agreement, February 12, 1883, signed by Sperry, Anderson, and Griffith (SP).

CHAPTER II Chicago: The Maturing of an Inventor

I am young yet. . . . We have the best electric generator that has ever been made; why just think of giving 92 net per cent of the total energy. . . . Do you suppose for one moment that I am going to fail in the great Chicago where one can turn a hundred ways—no sir. ELMER SPERRY, 1883

WHEN SPERRY LEFT CORTLAND, HE WAS UNDOUBTEDLY THE TOWN'S OUTSTANDing inventor, but Cortland numbered a few thousand people and Chicago several hundred thousand; Cortland had no other electric-light manufacturer, but Chicago had several, most of whom were backed by powerful eastern capitalists. In Cortland, his success was judged by patents and a model arc-light system. He had not, however, shown that he could develop his inventions to meet competition, to function in varied environments, and to survive in the market. This challenge demanded the response of an engineer and a professional keenly sensitive to the economics of production and to the requirements of the market. His career in Chicago would depend to a large extent on the fortunes of the Sperry Electric Light, Motor, and Car Brake Company, but it would also depend on his own creative development. Sperry was already an inventor; he would have to become an innovative engineer.

i

The objectives of the Sperry company were similar to those of other early electrical enterprises, including the Edison, Brush, and Thomson-Houston companies: to carry on the invention and development, manufacturing, design and construction, and operation of electric-light plants. Thomas Edison and his backers had founded the Edison Electric Light Company to support him in invention and development and to license his patents; the Edison Electric Illuminating Company in New York to build and operate a public electric-light station; subsidiaries to manufacture

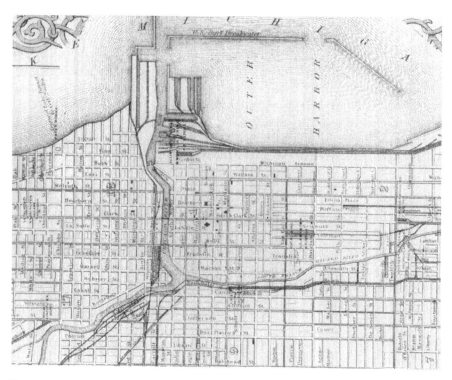

Figure 2.1 Map of Chicago, Illinois (1885), showing area in which Sperry offices, shops, and arc lights were located. Courtesy Chicago Historical Society.

components; and a sales organization to sell and install privately owned plants. Sperry had similar goals, but he was only able to form, besides the parent company, the Sperry Electric Illuminating and Power Company, "to sell or otherwise utilize and market" the Sperry patents abroad.[1]

The company factory was at 40–42 West Quincy Street,[2] about a block west of Chicago's Union Station at Canal and Adams Streets. (This was about a mile on the horse-car line from Briggs House, where Sperry lived.) The three-story, yellow-brick factory building measured 50 feet by 80 feet. It had a Buckeye steam engine and boiler in the basement, machine tools located on the first floor, and light machinery on the second. On the third floor were the finishing room and the small laboratory. By 1885 about fifty men worked at the factory, where generators, arc lights, and related components were made.[3] In the laboratory, Sperry tested new products and old ones in need of repairs. He recalled that in the early days the instruments

[1] Sperry and the Cortland Wagon Company each received one quarter of the stock of the Sperry Electric Illuminating and Power Company, and Cyrenius M. Greene, manager of the European company, who became general manager of the parent company in 1884, acquired the other half. The company was capitalized for $2 million; there is no further record after the founding. Records of the State of Illinois, "Sperry Electric Illuminating and Power Company," Box 187, #8853.
[2] Sperry testifying, June 2, 1887, in patent interference, *Sperry* v. *Wooley,* "Steam Dynamo," National Archives, Box 1339.
[3] *Electrical World,* VI (1885), p. 73.

for measuring voltage and amperage were exceedingly crude, but, using these, the company specified, as best it could, the characteristics of its products and analyzed their performances.

Besides manufacturing, the major work done by the company was installing its equipment. During the life of the Sperry Electric Light, Motor, and Car Brake Company (1883 to 1887), the company installed arc-light systems in Chicago and throughout the Midwest. Most of its installations were small plants of 20 to 40 arc lights. The largest central station was in Chicago, where generators in the basement of the Commercial Bank Building supplied light to buildings in the vicinity. By August 1885, more than 1000 Sperry arc lamps were burning nightly, with the largest concentration in Chicago (400) and Omaha (200).[4] A year later the company claimed more than 2000 of its lamps throughout the country, but this number was not large in comparison to several other companies. In Chicago there were about 2000 arc lights in late 1884. In 1885 there were estimated to be 96,000 arc lamps in service in the United States; in 1886, 140,000. Of the 96,000 in use in 1885, 30,000 were manufactured by the Brush Company, 8000 by Thomson-Houston, 5000 by Fort Wayne Electric, and the same number by Weston. The remainder were the products of approximately twenty smaller manufacturers who, like Sperry, had each installed a few thousand lamps.[5]

Sperry recalled that he was "lighting up the entire Northwest in an effort to put out the stars."[6] Although his installations were not collectively as brilliant as some of his competitors', his plants were widespread. Among their locations were Ohio City and Mason City, Iowa; North Platte and Grand Island, Nebraska; Carthage, Carrollton, and St. Louis, Missouri; Neilsville, Wisconsin; Coffeyville, Independence, Ottawa, and Paola, Kansas. These were mostly privately owned installations involving one or two generators supplying twenty to forty arc lamps used by the owner. Perhaps half of all the lamps the company installed were in small cities or towns. Sperry also helped light up the Midwest by installations on Mississippi steamboats.

Cortland was one of the small towns in which Sperry installed an arc light system. In 1884 he constructed a central station supplying three lights to the village for its streets, seven lights to a skating rink, three or four to hotels, one to a saloon, and the others to various commercial establishments. The lighting plant was located at, and took steam power from, the Cortland Machine Company. Sperry came down from Chicago in September, almost two years after leaving the village, to supervise the installation. Tisdale handled the business affairs before Sperry's arrival. A year later the village, with the approving vote of its citizens, installed another Sperry plant to

[4] *Ibid.*
[5] *Electrical World,* VIII (1886), advertisement; *Electrical World,* IX (1887), p. 50; Harold Passer, *The Electrical Manufacturers, 1875–1900* (Cambridge, Mass.: Harvard University Press, 1953), pp. 56, 70; W. H. Preece, "Electric Lighting in America," *Journal of the Society of Arts,* XXXIII (1884), p. 66.
[6] Dunn, "Sperry," p. 103.

Figure 2.2 Sperry Electric Light Station, Chicago (1889). From Electrical World, XIII, 1889.

supply twenty more street lights. Cortland depended on the moon to light the streets for ten days each month and contracted for Sperry light only twenty days a month at a fixed fee. The arcs were to burn from dusk to 11 P.M. Inhabitants found that the lights were a considerable improvement over the 120 gas lamps previously lighting the streets ("It is not necessary to carry a lantern to determine if they [the arcs] are burning").[7] Yet they found the moon proved unaccommodating and, spoiled by technology, they criticized the city for its parsimonious contract.

A typical Sperry installation in a small community consisted of a steam engine supplying about 20 horsepower, a single generator, and 20 arc lights connected in series. Each arc light was rated at 2000 candlepower and used 29 amperes. The company claimed that the carbons burned 15 hours before needing replacement by the "trimmer." Besides the automatic feed, the more expensive lamps had two sets of rods, the second set cutting in automatically when the first had been consumed. The Sperry generator, which weighed a ton, was shunt wound and equipped with his current regulator, which maintained a constant current throughout the circuit despite lamps burning out or being switched out.

In addition to installations in the small towns, Sperry and the company secured contracts for some spectacular festive lighting in Chicago. This kind of installation was deemed desirable by the lighting companies because of the publicity. For several years Sperry equipment lighted the annual Chicago Exposition and the summer orchestra concerts held in the exposition building.[8] In 1886 he installed 100 arc lights and 400 incandescent lights for the exposition, and 80 Sperry arcs illuminated the Chicago Stock

[7] *Cortland News,* September 25, 1885; *Cortland Democrat,* August 15, 1884, September 12, 1884, and May 15, 1885.
[8] EAS to Jean Goodman, December 24, 1926 (SP); and the *Electrician and Electrical Engineer,* V (1886), p. 437.

Figure 2.3 Sperry arc lights atop Chicago Board of Trade building (1889).

Show in November 1886. His most spectacular display in Chicago was a brilliant corona of lights on the 300-foot tower of the new Board of Trade Building located on Jackson and La Salle Streets. Sperry asserted that the corona was the greatest concentration of light in the world, a crowning jewel for the edifice, a major tourist attraction, a beacon for mariners on Lake Michigan, and light for the streets. It was also excellent publicity for the Sperry system. He persuaded the Board of Trade to furnish the steam power and to buy one of his generators; the owners of surrounding property agreed to pay part of the maintenance costs. Among these were the Union League Club, the Police Department, and the Grand Pacific Hotel. The Sperry company furnished the lights, installed them, and provided a man to maintain the plant and trim the arc lamps. Sperry had hoped to light the lamps at the inauguration ball planned for the dedication of the building, but he became ill and the work was delayed some months until the end of 1885. Then two weeks of sleepless nights were sufficient for him and several assistants to install the lights for illuminating on New Year's Eve.

At dusk on that day, Miss Zula Goodman, whom Sperry had recently met, seized the switch, "and saying playfully 'let there be light' turned the current upon the circuit."[9] And there was light—visible as far as Michigan City, 60 miles distant. The Chicago streetcar conductors and drivers were enthusiastic, reporting all of the different viaducts in the city well lighted. Even in the outskirts of the city there appeared to be a bright moonlight.

[9] *Chicago Tribune,* January 1, 1886, p. 3.

"The atmosphere," the *Tribune* observed, "was very luminous, and as far away as Douglas Park houses cast shadows from the light. . . . [Even] various railroad yards through the city, particularly those in the vicinity of Sixteenth Street, were fairly lighted. . . ."[10]

Sperry had designed the entire installation. He fastened four fourteen-foot arms to the top of the tower to suspend a ring. Pulleys and a windlass lowered the twenty lights, attached to the ring, to the level of a balcony where the carbons were accessible to the trimmer. The lamps, the standard manufactured by Sperry, together produced 40,000 candlepower. About one-third of the output from a sixty-horsepower steam engine drove the single generator. The lights remained on the tower for several years, burning every day from dusk to dawn—despite the hosts of birds that regularly burned to death. Subsequently it was removed because of the overload on the tower and the expense of maintenance. The Sperry company thought the installation worthy of illustration on a new letterhead.

Although Sperry central stations in downtown Chicago attracted far less publicity than the Board of Trade installation, they demanded more arc-light technology and were more challenging to Sperry and the company. Although called central stations, which now connotes size, the plants then often had only one dynamo and twenty lights. They were designated central stations, rather than isolated plants, because the owner sold electricity to others in the vicinity. When the designation of isolated plant was applied, it meant that the owner used all of the current. Sperry constructed one of Chicago's first central stations, the Sperry Electric Illuminating Station, in 1883 or 1884, in the basement of the old Commercial Bank Building on the northeast corner of Dearborn and Washington Streets. It was built and owned by Sperry,[11] who acted not only as the electrical engineer in charge of its operation and maintenance, but also as financial manager. When the building burned "over our heads," he quickly re-established the station in the basement of the Hale Building at the southeast corner of Washington and State Streets. There it expanded and, by 1885, was one of the largest central stations in Chicago,[12] supplying lamps located over an area of several blocks and connected by two miles of underground conductors. Among the customers were the Chicago *Tribune*, the Inter Ocean building, Carson Pirie Scott, the Kranze Candy Co., Kinsley's Restaurant, and a large number of small stores and restaurants in the vicinity.[13]

The station had perhaps 20 or more generators while Sperry owned and operated it.[14] It had a boiler room where a fireman tended the furnace and an engine room where an electrician tended the generators. Each reciprocating high-speed horizontal steam engine operated at least two, and sometimes four, belt-driven generators. Sperry and others found that one power trans-

[10] *Ibid.*

[11] Power of Attorney document, June 1886, names Elmer Sperry the owner (SP).

[12] *Electrical World*, XIII (1889), p. 340.

[13] EAS to William S. Keily, January 20, 1917 (SP).

[14] James M. Kent to EAS, January 17, 1909 (SP). Other evidence suggests as few as six generators in the station; *Western Electrician*, II (1888), p. 217.

mission belt could be placed on top of another to drive generators set in line at different distances from the steam engine.[15] The electrician manually adjusted current and voltage if the regulators malfunctioned. Commutator brushes wore down quickly and had to be replaced often, but if serious trouble developed, as was not infrequent, Sperry had to be called in.

The lighting of the *Tribune* was of unusual interest because Sperry placed arc lamps and incandescent lamps in the same circuit, a combination that tested his ingenuity. The generators supplied electricity at high voltage (700 volts) and low amperage (29 amps). By contrast, the incandescent lamps made by Edison, Swan, Maxim, Weston, Stanley-Thompson, and others, required a generator of low voltage (110 volts in the Edison circuits; 44 volts in the Stanley-Thompson). In order to combine the two kinds of lights on a circuit energized by an arc-light generator, Sperry wired the circuit according to a series-parallel plan, which had sufficient incandescent bulbs wired in parallel on a sub-circuit so that their combined amperage equalled the amperage of the circuit (about 15 bulbs for a 29-ampere dynamo). The power consumed was therefore the same as that across one arc lamp. For each sub-circuit of incandescent lights used, he removed one arc from the circuit. He recalled the venture with pride: "The [incandescent] lamps themselves were far from uniform and we found we could only use from 25 to 30 percent highly selected lamps for the series-parallel work, owing to one lamp robbing from another and being more brilliant because of this unequal distribution in the small multiple groups."[16] He also found that the compositors at the *Tribune* unscrewed the best lamps and took them home. Although problems of combination installation were great, customer demand for the newer incandescent light for interior spaces pressured the arc-light companies into resorting to such improvisations.

As the engineer for the company, Sperry had to adapt the system to varied needs and conditions, such as those encountered at the Chicago *Tribune*. Customers in small towns needed plants that operated without experienced supervision; or a local telephone company complained if the arc lights interfered with telephone transmission; city ordinances imposed severe constraints upon the layout; and some customers demanded the installation of a plant in inaccessible and crowded spaces. All of these engineering problems, and many more like them, were the responsibility of Elmer Sperry, the "electrician," for the company had no other engineer of reputation on its staff. Despite the presence of a Professor Allen Griffith among its directors and a Charles Cleaver listed as "superintendent" in 1883, the surviving record does not show that Sperry had substantial assistance in solving the multiple problems of system design and construction. Consequently, during the life of the company, he had to devote much of his time and energy to work that required ingenuity and knowledge of the state of the art, but work which was not that of the inventor.

One problem troubling Sperry and other engineers resulted from constraints imposed on their system design by legislation. The Chicago city council, for example, regulated the methods of running wires, constructing

15 Typescript of three pages of Sperry reminiscences (no date), p. 1 (SP).
16 *Ibid.*, p. 3.

plants and appliances, and so on, within the city limits. The most immediate and pressing cause of the legislation was the danger to life resulting from the poorly insulated, high-voltage wires that ran above the city streets. Serious accidents and death by electrocution were reported in cities throughout the country, but Chicago was the first to see the danger and adopt remedial legislation.[17] The code forbade the erection of any telegraph wire or electric conductor above the streets within the corporate limits after December 1883, and required that "from and after the 1st day of May 1883, every telegraph line or wire, or electric conductor, used and operated with the corporate limits of the city of Chicago, shall be laid under the streets. . . ."[18] Although the law provided substantial fines for violations, the difficulties and costs of rerunning existing conductors under the streets resulted in evasions: "Every now and again bright new wire and insulators grew to their places in the dusk of night or gray of the early morning."[19] In other cities, such as New York, where legislation came later, the arc companies had a considerable advantage. Leniency in Chicago allowed delays in placing existing lines underground. If some leniency had not been shown them, the light companies would have been severely penalized by additional costs during a period of sharp competition with the gas light. When the limits of concession had been reached, however, and a company with overhead wires remained obdurate, the mayor himself often appeared to sever the lines. By 1885, as a result, Chicago had a reputation as a hard market for the arc light.

Since Sperry erected his major station after the ordinance went into effect, he had to comply immediately. This caused him many difficulties. At a meeting of fellow station engineers and owners in 1885, he aired his problems:

Now we are operating about two miles of underground conduit from our central station in Chicago. We are compelled to go underground all over. We cannot cross even an alley overhead, and I would say that my experience is that the underground system is a very expensive one to maintain. We have tried every cable that money can buy. We commenced with a lead-covered cable which had to be abandoned entirely. We bought about two miles of lead-covered cable. Then we tried the various other kinds, which at first seemed to be better. I might say, however, that iron conduits are used in crossing streets. . . . We have to take our wires under the area-ways, and I may say that the cost of insulation and maintenance of these underground lines, especially where they go under area-ways, is very great. People occupying area-ways are constantly putting boxes against them, or throwing something against them. On the whole, I do not think that it is a very satisfactory system, although it is in its infancy, and things may improve.[20]

Another problem that Sperry and other early arc-light engineers encountered arose from electrical interference between their lines and the

[17] *Electrician and Electrical Engineer*, IV (1885), p. 111.
[18] *Ibid.*, III (1884), pp. 89–90.
[19] *Ibid.*, IV (1885), p. 329.
[20] EAS comment at meeting; *Proceedings of the National Electric Light Association*, I (1886), pp. 224–25. Hereafter cited as *NELA*, I.

single-wire telephone distribution system then used. Communities, large and small, had been introducing the telephone for almost half a decade when Sperry entered the arc-light field, and he and the other arc-light pioneers often found a telephone system in operation when they installed the newer arc light. As Sperry recalled:

The [telephone] switchboards were often in the kitchen of a house in a moderate-sized town. They were a struggling industry, as were our arc light systems. Just here a difficulty arose. The moment we started the arc light system at night, every telephone in town 'went on the blink.' "[21]

ii

Faced by the demands of the cities, the complaints of the telephone companies, and other common problems, the arc-light station owners and engineers decided to organize. Sperry took a major part in this effort, which resulted in the founding of the National Electric Light Association (NELA) —an organization that became the spokesman for American electric light and power companies. In 1933 the name was changed to the Edison Electric Institute.

In the winter of 1884–85, the arc-light station people in Chicago and the Midwest forgot their differences and began to discuss common problems. Sperry recalled that the talks took place in Chicago rather than New York, not because Chicago was the center of arc lighting, but because she had "earned some title to distinction as . . . the first to adopt and also to enforce the horizontal rule of buried wires."[22] Sperry was a leading participant in these conversations. After a meeting on February 3, he, with F. S. Terry, operator of Chicago's largest electric-supply house, and George S. Bowen, who operated a Van Depoele station in Elgin, Illinois, drew up a circular inviting central-station operators from all parts of the country to gather in Chicago on February 25, 1885, at the Grand Pacific Hotel for the purpose of organizing, and discussing common problems.

Sperry and his committee of call were pleased when almost 100 representatives from local companies as far away as Boston and Baltimore attended the convention. The mayor of Chicago welcomed the group, observing that even though a visitor from an eastern metropolis had said that Chicago was the darkest town he had seen, Chicago was determined to forego street lighting until it was made safe by underground conductors. The mayor, who knew of accidents in his own city from poorly insulated high-voltage lines, said that "as the father of over 600,000 people, all looking to me for protection [laughter] (I can prove it by these reporters), I say we want electricity, but we do not want death dashing like a horrid monster through our streets."[23] (His most cynical listeners wondered how much influence the gas-light interests had in Chicago.)

[21] Typescript of a Sperry address before a National Electric Light Association meeting in 1924 (SP).
[22] Typescript of a Sperry address before the NELA Annual Convention in Kansas City, 1910 (SP).
[23] *NELA*, I, p. 15.

Since a prime purpose of the meeting was to define common problems, a program committee composed of representatives from the companies using the major systems of arc lighting, including the Sperry system, prepared an agenda of topics to be discussed. The list gives an excellent summary of the problems, including the two basic concerns, underground wire and telephone interference.[24] The committee also suggested discussing the possibility of adding incandescent lights and motors to arc-light systems because incandescent lights and motors would increase the load and help the arc-light stations counter the competition from the Edison incandescent stations. Motors connected to the system would also distribute the load more evenly, for motors operated during daylight hours when the lighting load was small. Other subjects thought worthy of discussion were electric lighting by water power, rates and rebates, and special insulation for conductors at dangerous points.[25]

During the convention, Sperry often participated in the discussions, and his comments show that he anticipated major developments in electric light and power. He foresaw the need for economic transmission of electricity over long distances, arguing that "the matter, with regard to the transmission of the electric light to long distance, can only be, in my mind, solved by some method of carrying a high tension current to the point where the incandescent light is to be used. Then there should be a storage cell, or some means, or motor, to transfer the current to a low force, suitable to the operation of incandescent light." The convention reporter noted that Sperry went into a lengthy discussion, "which, being all technical language, and being uttered by him very rapidly, was not understood by the reporter."[26]

The problem of lowering load peaks and raising load valleys to even out the daily load was also discussed. A few years later it was analyzed and identified as the "load factor" problem, a major determinant of the pattern of central-station growth. Sperry suggested to the NELA convention in 1885 that customers be persuaded to use electric motors during the summer months when the arc-light load was relatively small. Later during the convention he instigated a lengthy discussion by suggesting the practicability and safety of connecting a central station to a larger load than the capacity of the station, on the assumption that all customers never used all their equipment simultaneously.[27] Subsequent practice supported Sperry's contention. Besides entering the discussions, he was named a member of the committee of invitation, the committee on the organization of the NELA, and the committee on program. In view of his activity at the first NELA

[24] With seeming ease, the telephone interference problem was quickly solved. During the meeting, a message from Theodore Vail of the Bell telephone system lessened tension between telephone and arc-light stations by suggesting that the telephone companies use two wires (a twisted pair) instead of one, thus eliminating the ground circuit. Elmer Sperry, typescript of an address before NELA convention, June 1929 (SP).
[25] *NELA*, I, pp. 21–22.
[26] *Ibid.*, p. 79.
[27] *Ibid.*, pp. 63, 66.

convention, it is clear why one of the leading trade journals, the *Electrical Review*, called him "probably the youngest of the Chicago electricians . . . one of the brightest, and full of enterprise and resources as he is of electricity."[28]

Earlier Sperry had also played a part, although a less important one, in the founding of the American Institute of Electrical Engineers. In the spring of 1884, N. S. Keith, a consulting engineer, C. O. Mailloux, technical editor of *The Electrical World*, and T. C. Martin, news editor of *The Electrical World*, all of New York, circulated a call for the organization of a national association of electrical engineers. Although Sperry did not sign the call, which attracted the signatures of many leading figures in the electrical field, he attended the first meeting. On April 15, he joined others at the American Society of Civil Engineers in New York, where they debated the scheme of organization and elected Keith secretary. In May, another organizational meeting was held, and in October 1884, the newly formed association met at the International Electrical Exhibition sponsored by the Franklin Institute in Philadelphia. Because of his participation in the first meeting, Sperry was named a charter member of the American Institute of Electrical Engineers.[29]

iii

In his contract with the Sperry Electric Light, Motor, and Car Brake Company, Elmer Sperry agreed to be its electrician, or electrical engineer. Engineering problems, such as those discussed at the NELA meeting, absorbed much of his time. Added to this burden was an increasing load of responsibility arising from the financial problems of the company. Together these demands at times strained him physically and exhausted him emotionally.

A few months after the founding of the company, for example, he fell ill while on business in Cincinnati. The attendant physician found him "a little delirious" and with an extremely high temperature. Not long afterwards he "got desperately sick and was really miserable in a perfect dry-goods box of a hotel room [in Chicago] . . . and had sunk to the lowest point in depression and got so I didn't care whether school kept or not. . . ."[30] Then, he later recalled in a letter, his Aunt Helen miraculously appeared, assumed command, took him to visit relatives in Wisconsin, and "the whole face of nature changed." In August 1883, when the company was unable to meet its payroll, the young man feared he would "break down, myself, between work and worry."[31] Nor did his work and worry let up the next year. In the spring he thought himself, as he wrote his older friend and lawyer, J. W. Suggett, "a fellow . . . crushed to the earth with multitudinous

[28] *Electrical Review*, VI (1885), p. 5.
[29] C. O. Mailloux to EAS, July 27, 1923 (SP); T. C. Martin to EAS, July 28, 1923 (SP); *Electrical World*, III (1884), pp. 100, 111, and 131.
[30] EAS to Helen Willett, February 11, 1919 (SP).
[31] EAS to J. W. Suggett, August 21, 1883 (SP). Many of the quotations in this section

burdens and so crippled as to appear both hateful and helpless."[32] Often he closed his letters with an excuse for their appearance or content, explaining that the hour was late and that the day had been exhausting. Yet his spirit was resilient, his self-confidence unflagging, and his outlook fundamentally optimistic. Although writing that he was afraid of breaking down, he hastened to add that he still had strength to work. "I am young yet," he wrote Suggett, and "we have the best electric generator that has ever been made; why just think of giving 92 *net* per cent of the total energy. . . . Do you suppose for one moment that I am going to fail in the great Chicago where one can turn a hundred ways—no sir."[33] Although he appeared crushed by his burdens for a time in 1884, he asserted that "there may be worth discovered in . . . [me] yet. . . . There is certainly more experience."[34] Chicago was a hard school, but the lessons he learned were to remain with him.

Financial problems and internal dissension, which seriously impeded the company's operation, caused much of his worry.[35] Asked in a patent hearing about the financial condition of the company since its organization, Sperry testified that "we have always been hard up. They have not even had money enough to pay salaries of the men, and have had to borrow money and mortgage our machinery and chattels for security."[36] Dissension between the wagon company and Sperry, on one hand, and an unidentified "other party" on the other, developed in 1883. The wagon company feared that the other party would gain control of the treasury stock. To forestall this, the company officers urged Sperry to agree to turn over the half-right he held in his future patents to the electric-light company. In return, he would be given more stock—stock created by increased capitalization. If Sperry acquired the $100,000 new stock, he would then own $275,000 of stock, or a quarter of the total issue. This block, combined with that owned by the wagon company, would guarantee control of the company to Sperry and his Cortland colleagues.[37] Sperry, with the advice of Suggett, hoped to reach some *modus vivendi* because Professor Elisha Gray, best known for his telephone patent litigation with Alexander Graham Bell, but also a gifted inventor and electrician, was considering joining the company if its problems were resolved. Subsequent evidence indicates that the stock scheme was carried out, although Gray did not join the company.

By early 1884, Sperry thought that business affairs were "looking decidedly up." He anticipated that at the forthcoming annual meeting the "objectionable" parties among the stockholders, "the ones to whom most of

are taken from a series of letters written by Sperry to his lawyer, John W. Suggett of Cortland, between 1883 and 1886. The originals are in the files of the Cortland Historical Society.

[32] EAS to Suggett, March 6, 1884 (SP).

[33] EAS to Suggett, August 21, 1883 (SP).

[34] EAS to Suggett, March 6, 1884 (SP).

[35] *Electrical World*, XIII (1889), p. 340.

[36] *Sperry* v. *Wooley*, p. 14.

[37] EAS to Suggett, July 2, 1883 (SP). Edwin Palmer of Cortland voted $60,000 of stock; in addition, the wagon company retained its original $175,000 in stock.

The Sperry Electric Light System.

During the past three years there has been developed in Chicago a system of electric lighting by the Sperry Electric Light, Motor and Ar-brake Company, which possesses some new and interesting features.

Although the system was organized as late as 1883, there

are at the present time over 1,000 Sperry arc lights in nightly use in different parts of the country. Of these some 400 are in Chicago; 200 in Omaha, Neb.; 40 in North Platte, Neb.; 40 in Grand Island, Neb.; 40 in Ottumwa,

development of a commercial electric motor, and with good results, quite a large number of their motors being now in use.

Certain other electrical appliances developed by the company will receive attention in these columns at some future time.

The company now occupy the works situated at Nos. 40 and 42 West Quincy street, Chicago, a three-story building 50 × 80 feet, and employ more than 30 men. Most of the heavy machinery is located on the first floor, and most of the lighter machinery on the second floor. The third floor is the finishing room, the laboratory being also located here.

The Sperry dynamo, which is illustrated in Fig. 1, is of

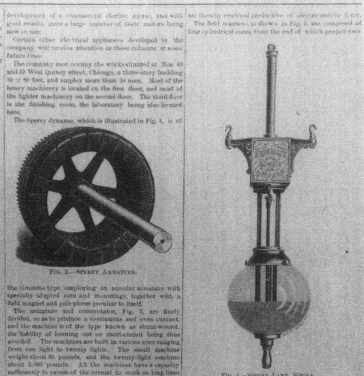

FIG. 2.—SPERRY ARMATURE.

the Gramme type, employing an annular armature with specially adapted core and mountings, together with a field magnet and pole-pieces peculiar to itself.

The armature and commutator, Fig. 2, are finely divided, so as to produce a continuous and even current, and the machine is of the type known as shunt-wound, the liability of burning out on short-circuit being thus avoided. The machines are built in various sizes ranging from one light to twenty lights. The small machine weighs about 80 pounds, and the twenty-light machine about 2,000 pounds. All the machines have a capacity sufficiently in excess of the normal to work on long lines and still not fall below their respective number of lights.

A glance at the illustration will show the departure made by the Sperry Company from the construction adopted by other manufacturers of dynamo machines.

The armature is built up of thin sheets stamped from

are thereby rendered productive of electro-motive force.

The field magnets, as shown in Fig. 3, are composed of four cylindrical cores, from the end of which project two

concentrically disposed pole-pieces of equal length, between which the armature revolves. The cylindrical part supports a small helix of fine wire, as shown. The peculiar disposition of the masses of metal forming these pole-

FIG. 4.—SPERRY LAMP, DOUBLE.

FIG. 5.—SPERRY LAMP, SINGLE.

FIG. 1.—SPERRY DYNAMO.

FIG. 3.—SPERRY DYNAMO FIELD MAGNETS.

Figure 2.4

the disastrous results of the recent past are attributable," would be removed.[38] What these parties did the record does not show, but Sperry makes reference to officers' salaries and dividends preventing net earnings. Without earnings, improvements needed in the company's product could not be made; these were of prime concern to the inventor. In March the company appointed a new general manager, Cyrenius Greene, who did not, however, greatly improve its fortunes. Despite a "big boom" in central station construction in the fall of 1884, and an increase in the number of arc lights in Chicago (from 1000 to 2000 during the twelve-month period),[39] the company was unable on occasion to meet the payroll or to pay the fees of Suggett and his Washington associate, Nottingham. For a time, Suggett threatened to drop the company's affairs, but Sperry and Greene persuaded him to continue. Suggett was of vital importance because he conducted interference cases for the company, including the complex and important arc-light case.[40] These stresses strained relations within the company: Palmer once wrote Suggett that Sperry required plain talk "to keep him within bounds. . . . He has imagined that others are to work for him without consideration or compensation and that he runs the world and has it fenced in."[41]

A pressing need for operating capital in 1885 caused the company to assign about half of its total assets in patents, tools, and machinery to T. J. Foley, a capitalist in North Platte, Nebraska, in return for cash and collateral. The Sperry company also turned over to him the stocks, bonds, and securities it had received from central stations in payment for equipment. It appears that Foley bought out the original Chicago promoters, including Galusha Anderson.[42] Foley, who held an interest in the Sperry local plant at North Platte, became manager and treasurer of the parent company. Only thirty-four years old, he had already made a fortune conducting a general mercantile business. *Electrical World* reported that the change put the Sperry company on a solid financial basis and predicted that "with Mr. Sperry in the laboratory, and Mr. Foley in the business management, a large increase in the company's already very satisfactory business may be confidently looked for."[43] Foley not only acquired existing patents, but in all probability a half-share in the company's rights to Sperry's future patents. If Foley did not live up to promise, then Sperry's inventiveness would remain in frustrating bondage.

Foley, who came in November 1885, lasted until spring, when L. Everingham bought him out and became president. Everingham was a wealthy and prominent Chicago commission merchant "known all over the West and Northwest."[44] George H. Bliss, general superintendent of the Edison Light Company in Chicago until 1885, became the secretary and treasurer. Sperry thought Bliss a grand acquisition. Everingham, Bliss, and Sperry

[38] EAS to Suggett, January 22, 1884 (SP).
[39] *Electrician and Electrical Engineer*, III (1884), p. 201; Preece, "Electric Lighting in America," pp. 66–73.
[40] *Daft* v. *Thomas* v. *Sperry*.
[41] Palmer to Suggett, February 20, 1884 (SP).
[42] The evidence on the transaction is fragmentary.
[43] *Electrical World*, VI (1885), p. 205.
[44] *Electrical World*, VII (1886), p. 170.

became the new committee of management.[45] In general, Sperry, who had had days of "depression and anxious work," was decidedly pleased, for he believed that Everingham would protect the inventor's interest in "each, any, every and all ways." Sperry now predicted that "Sperry light" would "win"; he wrote his fiancée, Zula Goodman—the young lady who lit the Board of Trade arcs:

I worked nearly all last night on an invention and got it this morning, too. I knew it was to come out correctly if I but could find the time and could have the heart to work upon it until I mastered the problem. It is one of the most valuable I have yet been able—under His guidance, Dear—to bring out, that is, I think it is and trust it will prove to be. I say "have heart". You see, my patents and inventions are tied up in this Company and so long as it was upon an insecure basis and hands in whom I had not the utmost confidence, I did not have the heart to push my work in this line as my position should seem to demand. . . . *Just* won't you and I turn out the inventions, though.[46]

But the troubles of the Sperry Electric Light, Motor, and Car Brake Company were not over, and Sperry could not yet dedicate an unfettered mind to "turn out the inventions." In August 1887, the company finally went out of business, selling its property to some friends of Sperry's, "together with whom I organized the 'Sperry Electric Company.' "[47] Sperry, however, was not a shareholder.

Sperry's arc-light patents were a major asset of the new company, which, with the arc-light patents, continued to manufacture the Sperry system and to sell replacement parts for existing plants. Sperry contracted to work part time for the Sperry Electric Company as an engineering consultant and later negotiated the assignment of several of his new inventions to the company; most of his inventions, however, were disposed of elsewhere. The system manufactured and sold by the Sperry Electric Company, therefore, remained essentially the original Sperry system of 1883.[48] Reorganized in 1888, the company continued to advertise Sperry products, but

[45] These negotiations resulting in the reorganization of the company early in 1886 are described in a series of letters from Elmer Sperry to his fiancée, Zula Goodman. The financial records of the company do not exist. Everingham and his lawyer acquired Foley's half of the company by paying him about $10,000, assuming his collateral of about $40,000 and giving an indemnifying bond against certain action being brought against Foley amounting to about $12,000. EAS to Zula Goodman, January 29, 1886, and February 6, 1886 (SP). Remarks in Sperry's letters to his fiancée suggest that he may have acquired earlier the stock owned by the Cortland Wagon Company. This would leave Sperry and Everingham each owning about half of the company. Sperry may have sold his stock in the Chicago central station to buy out the wagon company.

[46] EAS to Zula Goodman, January 29, 1886 (SP).

[47] EAS to E. M. Bentley of the patent office of Thomson-Houston, May 23, 1892 (Thomson-Houston letter book, SP). L. Everingham became president of Sperry Electric Company in 1888. In 1889 the capital stock was increased from $250,000 to $500,000. A Henry G. Dickenson owned 2,495 of the first 2,500 shares issued. Records of the State of Illinois, Box 423, #14295.

[48] Descriptions in trade journals in 1889 show no notable changes except an increase in generator capacity. *Electrical World*, XIII (1889), pp. 343-45; and XIII (1889),

by 1893 it seems to have been absorbed by the Standard Electric Company of Chicago,[49] and Sperry was no longer associated with it. In 1894, the stock of the Sperry Electric Company was devalued from $500,000 to $10,000, and shortly thereafter the company disappeared.

About the time that the first Sperry company went out of business, the Sperry central station on Washington and State Streets was also sold to the newly formed Chicago Arc Light and Power Company, which bought nine independent Chicago central stations and began consolidating them physically. An inventory of the generators owned by the company in 1888 showed six Sperrys among the total of sixty. At least one Sperry was used in the new consolidated station built at Washington and Market Streets in 1888.[50] In 1893, the Chicago Arc Light and Power Company was acquired by Samuel Insull's Chicago Edison Company; the combination was the basis for the Commonwealth Edison of Chicago later formed by Insull.[51]

iv

Little evidence survives that could completely explain the demise of the Sperry Electric Light, Motor, and Car Brake Company. Financial weakness and managerial inadequacy, and the company's failure to support invention and development, were major factors. A comparison of the company's innovations with those of the leading electric-light companies, and an analysis of Elmer Sperry's inventive activity, shed some light on the question.

Between 1883 and 1887, the technology of electric light and power changed rapidly, and the competitor in the field had to be innovative to keep abreast of the state of the art. A system of electric light and power that responded to the needs and fulfilled the potential of 1883 was obsolete by 1887. During these formative years, radical changes occurred with greater ease than in later years when the momentum of a going system offered strong resistance to fundamental change. To remain competitive in the electric light and power field, therefore, the Sperry company needed to support invention and development. Instead, the company used its talented young inventor in routine engineering and could not supply the capital needed to support his inventions and their development. Not only did Sperry lack time and capital, but he was harassed by the company's financial troubles. Under such trying circumstances, he and the company could not compete with heavily capitalized and rapidly expanding manufacturers such as Thomson-Houston and Westinghouse, which supported

Figure 2.5
The Sperry Double Arc Lamp.

---

p. 89. The new company advertised dynamos ranging from 3 to 35 arc-light capacities and stressed a Sperry automatic regulator. *Catalogue of the Sperry Electric Company,* 194–198 Clinton Street, Chicago, Ill.

[49] EAS to Robert A. Clapp of the legal department of the General Electric Company, October 3, 1893 (Thomson-Houston letter book, SP). The reorganization of 1888 was noted in the *Western Electrician,* III (1888), p. 52.

[50] *Western Electrician,* II (1888), p. 217.

[51] Forest McDonald, *Insull* (Chicago: University of Chicago Press, 1962), pp. 59, 124.

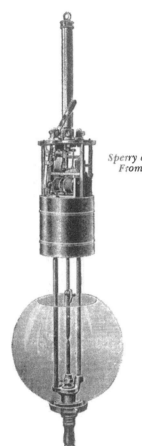

Figures 2.6
The improved
Sperry arc-lighting system.
From Electrical Review,
February 16, 1889.

such pioneer inventors and engineers as Elihu Thomson, William Stanley, and Nicola Tesla.[52]

Developments between 1883 and 1888 determined the future characteristics of electric light and power. Incandescent lighting spread rapidly. The Edison company without first entering the arc-light field, pioneered in the manufacture of a direct-current incandescent light system. Leading arc-light companies later established footholds in the incandescent field. The Brush company acquired rights to manufacture and sell the incandescent light invented and patented by the English inventor, Joseph Swan. The Thomson-Houston company, also an arc-light manufacturer, bought rights to manufacture the Sawyer-Man lamp in 1883. In 1886 the Westinghouse Electric Company introduced a system of incandescent lighting.

Perhaps the outstanding innovation of the mid-eighties was the introduction of alternating current for the transmission of electricity by George Westinghouse in 1886. The system he used was invented abroad, but he acquired the rights and employed William Stanley, the young American inventor and engineer, to improve the system. Not long after Westinghouse established a.c. stations, Thomson-Houston followed suit using the inventions of its founder, Elihu Thomson. Initially there was no practical alternating-current motor, but Nicola Tesla sold his patents for a polyphase

[52] David Woodbury, *Elihu Thomson, Beloved Scientist: 1853–1937* (Boston: The Museum of Science, 1960), pp. 168–70.

motor to Westinghouse and by the end of the decade long-distance transmission for light and power was being rapidly developed.

The Sperry company was left behind as incandescent light and alternating current seized the market. Sperry saw the future of, and was informed about, the incandescent light and alternating current. He did not, like Edison, adamantly refuse to recognize the merit of alternating current. In 1888, during the "battle of the currents," the Chicago Electric Club sponsored a series of lectures to debate relative merits of the alternating and the direct-current system. One of the six engineers and inventors who read papers, Sperry advocated alternating-current transmission, the incandescent light, and the alternating-current motor.[53] At times during the discussions he believed himself "to be the only advocate of the alternating current system."[54]

Although grasping the importance of incandescent light and alternating current, Sperry did not contribute noteworthy inventions. The 19 patent applications he filed during the life of his company (1883–87) focused on improving the company's arc-light system rather than on breaking into the new incandescent and alternating-current fields. Of the 19, none were for alternating-current systems and only one was for incandescent lighting. The incandescent-light patent covered a device that automatically introduced a resistance into a circuit to bypass a burned-out lamp.[55] This device was needed when incandescent lamps were wired in series on an arc-light system, for if one lamp burned out, the entire series followed. The Edison and other practical incandescent systems, however, placed the incandescent lamps in parallel, so that if one burned out it would not affect the others. Of Sperry's 18 remaining applications, 7 were for arc-light generator improvements and 2 for improvements in an electric motor. Four others were for improvements in an arc-light system (regulator for arc lamp, switch, lighting arrester, and use of rails for transmitting electricity). The 5 remaining defy categorization (alarm system, electric regulator to control flow of a liquid, vitrified asbestos, an electric car brake, and a gas engine).

These were fallow years for Sperry. Nineteen patent applications in five years (1883–87) amounted to half his *annual* average during his years of invention (1880 to 1930), and was less than half the average of nine per year that he would file during the next five years (1888–92). The pattern of his applications mirrors the fortunes—or more precisely, the misfortunes —of his company. In 1883, when the company was new and he had a backlog of ideas from his fruitful Cortland period, he made eight applications. Six of the eight were made before the summer, when the company's history of financial troubles began. In 1884, a year of serious managerial and

---

[53] E. A. Sperry, "Motors and Alternating Currents," *Western Electrician*, II (1888), pp. 264–65. Read before the Chicago Electric Club, May 21, 1888.
[54] Report of discussion at the Chicago Electric Club of a paper read by H. Ward Leonard on alternating versus direct current, on March 5, 1888. *Western Electrician*, II (1888), pp. 129–30.
[55] Filed May 22, 1886 (203,030), "Safety Device for Incandescent Lamps," issued April 26, 1887 (Patent No. 361,844).

financial problems, he applied for only four patents. He did not make a single application in 1885—an event rare in his entire career—and this was the year Foley was brought into the company. By April 1886, Everingham and Bliss came into a company Sperry thought well-reorganized; in the remaining eight months of that year he applied for six patents (although before April, none). The next year, he made only one application—the year the company went out of business. As Sperry said, he needed an unfettered mind "to turn out the inventions."

v

Although Sperry was not inventing devices of consequence for incandescent lighting and alternating current, he was, probably without plan, gaining experience with complex contrivances that later played a major role in his career and in the history of technology. In his superficially barren period from 1883 to 1887, 11 of his 19 patent applications disclose, upon closer examination, automatic controls, and half of these were closed-loop feedback controls. Sperry never described himself, nor was called by his contemporaries, a control engineer or a pioneer in cybernetics, but his stress on control devices in the eighties and his great work in gyro controls several decades later justify his being so characterized. His consistent stress upon control devices as he shifted from one field of technology to another —from electric light, to electric streetcars, and to gyro applications—suggests that concern with control was his prime characteristic as inventor. Although it would be interesting to search for the roots of this trait, only a few tentative suggestions, particularly applicable to his electric-light period, can be offered here. His first two important inventions, the generator and arc light, featured automatic controls; their success may have caught his interest and imagination. Or he may have been hoping to increase the sale of his lighting system by lowering its cost of operation; his automatic devices performed some of the complex tasks of the station attendant. Perhaps the idea of controlling large forces with delicate devices attracted him, for often in his lifetime he referred to getting control of the "brute," meaning a ponderous machine or vehicle.

In Sperry's day the science of controls, particularly closed-loop feedback controls, was not articulated and there is no scientific description of his controls by Sperry. Later analysis by others, however, of devices for which Sperry was in large measure responsible, helps one to understand his creations.[56] Feedback, an important characteristic of Sperry controls, is now defined as "a method of controlling a system by reinserting into it the result of its past performance."[57] An automatic control with feedback

[56] For a discussion of Sperry's place in the long-term evolution of controls, see below, pp. 283–85.

[57] Norbert Wiener, *The Human Use of Human Beings: Cybernetics and Society* (Garden City, N. Y.: Doubleday, 1954), p. 61. I am indebted to my colleague in the history of technology, Otto Mayr, for criticism of my discussion of feedback controls. Further elaboration can be found in his *Zur Frühgeschichte de Technischen Regelungen* (München: Oldenbourg, 1969).

is now called a closed-loop system to contrast it with an open-loop in which the output is not compared with the input to generate an error signal. The "error signal" is the difference between the desired performance of a system, mechanism, or process, and the actual performance.[58] Taking, for example, Sperry's electromagnetic control for his generator, the desired current intensity was set into the control and was compared to the actual flow of current through the circuit. Sperry's device had an adjustable spring for setting the desired current and an electromagnet with a movable armature to represent the actual current. The force of the electromagnet tended to move the armature in one direction and the tension of the spring, attached to the armature, in the other. The forces when balanced represented the no-error state; movement in either direction showed an error. Movement activated a servomechanism that rotated the brushes of the generator until output matched desired current. Thus, the past performance of the generator was reinserted to correct the performance.

Servomechanism, another term contributing to an understanding of Sperry's automatic controls, is defined as a control system in which the controlled variable is a mechanical position.[59] It often involves a train of linkages that carry an error signal to a motor which performs the necessary corrective action. In the generator regulator, Sperry had the generator perform the remedial action: the control engaged the generator shaft with a mechanical train that positioned the brushes or the commutator.

In the mid-eighties, when he was devising controls for electric-light systems, Sperry relied on the electromagnet as the prime component. In various of his patents of the period, he used an electromagnet to move a rheostat to increase or decrease the external resistance of the field of a motor or generator, thus regulating the output; he also used it to vary the length of an arc gap in the field circuit of a generator and thereby the resistance (used to regulate current output), and to move an armature into and out of the field of a generator to vary the output. For his linkages in his servomechanisms, he used the motion of the armature of the magnet in various ways—to close, for example, an electrical contact activating a motor; to open a valve in a hydraulic circuit; to engage gear teeth; or to throw a pawl into a ratchet wheel.[60]

vi

Sperry's early resourcefulness and expertise in automatic control matured years later. In the meantime, in 1887, he faced a discouraging situation. He had given six years of unstinting effort to a company that was in pressing financial difficulty, had never paid his salary, and to which

[58] These terms are defined more precisely in the American Institute of Electrical Engineers Committee Report, "Proposed Symbols and Terms for Feedback Control Systems," *Electrical Engineering*, LXX (1951), p. 909.
[59] *Ibid.*
[60] These inventions were not unique. In 1884 there were twenty-two U.S. patents issued for generator regulators alone. *Electrician and Electrical Engineer*, IV (1885), p. 45.

46

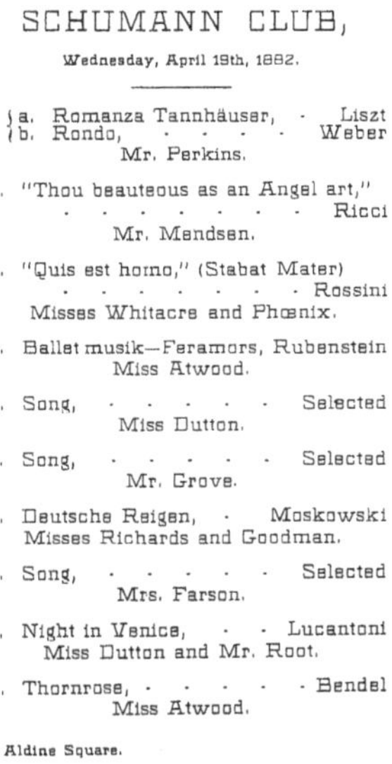
SCHUMANN CLUB,

Wednesday, April 19th, 1882.

1. a. Romanza Tannhäuser, · Liszt
   b. Rondo, · · · · Weber
   Mr. Perkins.
2. "Thou beauteous as an Angel art," · · · · · · · · Ricci
   Mr. Mendsen.
3. "Quis est homo," (Stabat Mater) · · · · · · · · Rossini
   Misses Whitacre and Phœnix.
4. Ballet musik—Feramors, Rubenstein
   Miss Atwood.
5. Song, · · · · · Selected
   Miss Dutton.
6. Song, · · · · · Selected
   Mr. Grove.
7. Deutsche Reigen, · Moskowski
   Misses Richards and Goodman.
8. Song, · · · · · Selected
   Mrs. Farson.
9. Night in Venice, · · Lucantoni
   Miss Dutton and Mr. Root.
10. Thornrose, · · · · · Bendel
   Miss Atwood.

36 Aldine Square.

Figure 2.7
Zula Goodman plays for the Schumann club.

he had made a loan that was never repaid.[61] Nor did he profit from the sale of his first company's assets to the new Sperry Electric Company. His years with the first company were filled with engineering work, much of it routine. It qualified him as an electrical engineer at a time when construction, maintenance, and central station operation were rapidly developing fields. Another man might have then chosen a career in electrical engineering, but Sperry found his way back into independent invention and development. He had the enthusiasm, self-confidence, and optimism to take the risk, but the strength and support he derived from a fortunate marriage undoubtedly reinforced his determination and sustained his effort.

Elmer Sperry was often lonely when he was not absorbed by his work. A Cortland teacher remembered him as a lonely boy, and Sperry himself recalled his loneliness as a child. In Cortland, his curiosity about, and interest in, technology helped fill the void as did the church, Sunday School, and YMCA. The pattern was extended in Chicago. He filled long days with engineering and invention; but he also transferred his church affiliation, taught Sunday School, and gave lectures on technology and science to YMCA groups. Yet there were interludes such as the desperately lonely one when his Aunt Helen found him sick at the Briggs House in Chicago.

He met Zula Goodman in January 1885. Earlier he had seen the slender, blue-eyed girl, with dark brown curly hair, playing the organ at his church, the First Baptist Church of Chicago, and in the role of Priscilla in the story of John and Priscilla Alden, dramatized by the church drama group. When they met, there were mutual friends to discuss because both young people were well acquainted with the large Baptist community of Chicago. Galusha Anderson, president of Sperry's company, was well connected in the Baptist business world, and Zula's father was senior deacon of the First Baptist Church and publisher of a leading Baptist newspaper. More than mutual acquaintances attracted them as they came to know each other during the following months. Her friends throughout her life described her as a person of strength but of gentle demeanor. She was alert, with broad interests, but at the same time stable, even tranquil. These traits complemented Sperry's restless energy and more mercurial temper.[62] While his cultural horizons were bounded by his work, she taught piano and gave recitals at local music groups. She had, as well, an interest in history and art and was a competent writer. (That she also spelled well was not the least of her cultural distinctions in the opinion of young Sperry).

Zula had a secure family life, and her home near the University of Chicago may have reminded Sperry of the professors' homes that had

[61] EAS to Robert A. Clapp of the legal department of General Electric, October 3, 1893 (Thomson-Houston letter book, SP). Sperry cited Galusha Anderson as the supporting authority for the statement that he had never received a salary and had made the loan.

[62] Zula later described her husband as extremely irritable when annoyed. Zula to Herbert Goodman, c. 1915 (SP).

captured his imagination in Cortland. Both her mother and father were living. Her father, Edward Goodman, was a native of England who came to America at mid-century and settled in Chicago, where he became part owner of the *Chicago Standard,* an official organ of the Baptist Church in America. As publisher, he developed the paper into one of the foremost denominational journals in the country.[63] Goodman and the *Standard* helped raise funds for the new University of Chicago, and he was named to its first board of trustees, serving in that capacity until his death in 1911.[64]

Zula's brother, Herbert, became an influential part of Sperry's life. As a business associate and adviser in a number of enterprises, Herbert freed Sperry for invention and development. Zula and her brother were—and remained—quite close. With support from Dr. Galusha Anderson, who was a family friend, she had helped persuade Herbert to attend the University of Chicago after attending business school.[65] Herbert left the university because of poor health in 1884, but after a trip West to restore his energy, he gained business experience by selling advertising for the *Standard* and the *U. S. Official Postal Guide* in Boston. He would use this experience in business with his brother-in-law.

By New Year's Eve, when Sperry chose Zula to throw the switch to light the Board of Trade Tower, they were in love. Their letters, written when she and her mother visited Herbert in Boston, provide another perspective on Sperry. They document the profound effect she had on his life—an effort that he acknowledged on numerous occasions.[66] Having known each other for nearly a year, Sperry was able to write that Zula's ability to understand him and the kind of understanding and friendship that existed between them guaranteed that they would "be happy."[67]

He found himself "a little lonely" during her absence. After calling on her father, he wrote that he "missed a bright happy face and when I saw the piano I wished I could carry it to Boston for you. . . . It really longs for you, one could not help but notice. . . . It gave me a dark silent look and never even spoke."[68] These are the letters of a young man in love, but one not so carried away by passion that he has not labored to write well, spell well, and introduce subjects congenial to the ear of a cultivated young lady.

Zula developed facets of his personality foreign to the engineering world in which Sperry spent his working hours. He often wrote her of reactions he would not have shared with another. While measuring an art

---

[63] *A History of the City of Chicago* (Chicago: Inter Ocean, 1900), p. 39.
[64] Article on Herbert Edward Goodman in *The Cyclopedia of American Biography* (New York: Press Association Compilers, 1918), VIII, pp. 563–64; Storr, *Harper's University,* pp. 42–43.
[65] Zula to Herbert, September 17, 1880 (SP). On Herbert Goodman, see Bernard Drell, "The Role of the Goodman Manufacturing Company in the Mechanization of Coal Mining" (Unpublished dissertation, University of Chicago, 1939); see also *In Memoriam: Herbert E. Goodman* (Chicago: privately published, 1918), 31 pp.
[66] EAS to Zula, January 20, 1886 (SP).
[67] *Ibid.,* January 26, 1886 (SP).
[68] *Ibid.,* January 25, 1886 (SP).

Figure 2.8 Elmer Sperry, about the time of his wedding, age 27.

museum for arc lights in New York, he found himself studying the paintings and later writing that the pictures affected him so much he almost wanted to paint himself. In Washington he scrutinized the architecture of new buildings, told her of them, and also shared his reaction to the country-side as his train left the capital:

But, Dear, you should see the wild flowers. I hope we will be obliged to stop in the woods here so I can get you some—a large white flower like this with three petals, then a delicate pink, smaller, the shade of which I cannot get as we pass. The dogwood (I presume the "dog rose" of our "Mill on the Floss") is in blossom—delicate white flowers. Something like a rose, only smaller with very many more petals, and then a bush with pink flowers, the last two make a very beautiful contrast with green on the hillside across the valley. . . .[69]

Zula was awakening new interests in him, but he also extended her horizons. She became interested enough in his work so that he could write her of the scientific books he bought in New York at Van Nostrand's—"some $35 worth, some on thermo dynamics [sic], one on ice machinery, and the others on electricity, the very latest one marked 1887."[70] She wanted to know his business affairs in detail, and during the reorganization of the company early in 1886, his letters kept her abreast of the complex negotiations involving Foley, Everingham, and himself. She had some stock in the Sperry company, perhaps only a few shares he had given her as a present, or perhaps purchased by the Goodmans to help their future son-in-law. Zula never lost interest in Sperry's business affairs, and if she found her husband troubled by purely business problems, she sometimes turned to her brother Herbert for advice which she then passed tactfully on to her husband.

Zula Goodman and Elmer Sperry were married on June 28, 1887. After a wedding trip to Lake Delavan, Wisconsin, they received friends and relatives during several "at homes." The extensive list of wedding presents and the long guest list show how widely the Goodmans were connected in Chicago. The guest list also contained the names of Sperry's friends—mostly his business associates. Aunt Helen, and Uncle Judson Sperry and his wife visited the newlyweds. Sperry's father did not attend the wedding because his health had deteriorated (he concealed this in his letters to his son for fear of causing him anxiety).

The young couple first lived with her parents at 3343 Vernon Avenue. This arrangement proved agreeable because her mother was in poor health and needed Zula; and Zula had company during Elmer's many business trips. The young couple entertained often. After one affair, Zula predicted to her brother that "a few more receptions and the like will transform Elmer into quite a society man. He seems to take quite naturally to it."[71] Zula continued to cultivate his interests and tastes. Within a year she

[69] *Ibid.,* April 30, 1886 (SP).
[70] *Ibid.,* December 3, 1886 (SP).
[71] Zula to Herbert, August 4, 1887 (SP).

wrote Herbert that "Elmer is wild over French art and history" and was planning a business trip to Paris with one of his and Zula's mutual friends, the patent attorney, Francis Parker.[72] Zula also had her husband choosing his wardrobe carefully, and was of the opinion that his new suit and ulster overcoat fit as well as anything the tailor made and "were much less trouble and expense."[73] Not by chance, Sperry bought his suit at Willoughby Hills, where he had installed arc lights.

By the fall of 1888, their life had assumed a cast that lasted a lifetime. Birth of their first child, named for Aunt Helen, on October 17, was followed by those of three boys: Edward on January 2, 1891, Lawrence on December 21, 1892, and Elmer on May 9, 1894. After a profitable business transaction,[74] Sperry bought a home for his family at 4406 Ellis Avenue, a short walk from the Goodman home. Zula still found time to care for her children and to keep closely informed about her husband's work. She continued to play the piano which she had studied as a girl in Florence Ziegfield's Chicago Academy of Music, and her husband found this soothing after his long work days. She also began reading to him from novels and popular non-fiction. This not only relaxed him but fulfilled her objective of extending his interests. She, in turn, heard out his descriptions of his inventions, placing her name as a witness to early conceptions in his notebook. Both had reason to consider themselves fortunate in their marriage; and many years later his friends found that his zest for life seemed to pass with Zula's death.

At this time, Sperry wore a full beard that made the twenty-seven-year-old inventor, engineer, and entrepreneur look older and dignified. Short, wiry, and quick in his movements, he had ears well-set and eyes wide-spaced and "honest." A distant cousin and business associate remembered him as "exceedingly smart in appearance, with a pleasant expression, and almost without an equal in repartee."[75] Throughout life he retained his "smart" appearance and his gift for impromptu speaking. In the Chicago days his quick wit and engaging presentation brought him many requests to speak on electricity before YMCA and fraternal organizations. He was especially effective before a group of newsboys who met regularly to hear popular science lectures. His style was informal, his witticisms good-natured and shrewd, and his metaphors, used in explaining science to lay audiences, imaginative and vivid.

The quality of his mind is suggested by a characteristic recalled by his bookkeeper during the Chicago period:

I have many times seen him seemingly just looking into the air, when all at once he would pick up a pad and hold it at arms length, then with

[72] *Ibid.,* fall or winter, 1888 or 1889 (SP).
[73] *Ibid.,* December 1, 1888.
[74] Sale of the Elmer A. Sperry Company for $12,000. See below, p. 61.
[75] William R. Goodman, "Recollections of Mr. Elmer A. Sperry's Early Years in Chicago" (thirty-one-page typescript dated 1933, SP). Goodman joined Sperry as bookkeeper in 1888; in the 1920's he was treasurer of the Sperry Gyroscope Company.

a pencil in the other hand he would begin to draw. This habit aroused my wonder to such an extent that one day I asked him why he held his pad up in the air instead of putting it on the table. In his answer he seemed to disregard me and the question entirely but said with one of his quick motions, 'It's there! Don't you see it! Just draw a line around what you see.' Whatever he saw he saw 100% perfect, there in the air but it took a long time and many changes to reproduce in wheels the thing which he saw. He had infinite patience of mind and persistence of will in working out a detail with his engineers. It is true physical limitations and the comparative slowness of other people's minds, and the fact that he would sometimes start an explanation right in the middle, interfered now and then and he was thought to be temperamental and various other things. Be that as it may, I have heard it said that judging his activities compared to ordinary people, he had lived hundreds of years.[76]

He certainly possessed remarkable powers of concentration and reflection. His associates never ceased to marvel at his ability to resume a technical conversation precisely at the point of its interruption hours, or even several days, before. In Chicago he often broke off a conference to receive visitors and then resumed the complex discussion without a word of resumé. He was known to hold his mind concentrated upon a single invention for days and was sometimes unable to disengage from a problem to sleep. Such extreme concentration, coupled with irregular working hours, "laid him on his back every few years."[77]

His typical workday after he formed the Elmer A. Sperry Company in 1888 was arduous and self-imposed. He arrived at the factory shortly after 8:00 A.M., threw off his coat, disconnected his cuffs, and worked with the machinists until noon. Indifferent about food, his tastes were simple, running to soup, bread, roast beef, and milk. He never drank alcohol and only occasionally indulged himself in tea or coffee. Often he had lunch with engineering and business associates, probably promoting his most recent inventions. From lunch until the factory closed at 6:00 P.M., he worked on inventions using the experimental and shop facilities in his "lab." Then he dictated letters until 8:00 or 9:00, sometimes forgetting not only his own supper but that his secretary had not eaten. Determined to fill every waking minute with work, he sometimes had his secretary accompany him as he rode to catch a train, or even go along to the first stop taking dictation. He also liked to have some congenial associate, engineer, or technician, join him on a brief day's excursion to give a lecture. There is no evidence that he engaged in small talk, unless engineering details fall within that category.

vii

Perhaps Zula's support and the spirit of change engendered by his marriage encouraged Sperry to take a bold step that proved a turning point in his career. Whatever the reason, the year after his marriage, he

[76] *Ibid.*
[77] *Ibid.*

broke free from the restraints of routine engineering and business affairs to establish his own research and development enterprise, the Elmer A. Sperry Company. Obligated to give only one-third of his time to the new Sperry Electric Company, he used the remainder to invent and to develop his own inventions and those of others. He hoped that the new company would not hamper him with routine matters or harass him with anxieties, for as he wrote in his notebook about the time of the founding, "no man can work and worry too."[78] (The concerns and the frustrations of the Sperry Electric Light, Motor, and Car Brake Company were still fresh in his mind.)

The Elmer A. Sperry Company of Chicago was founded in October 1888, "for the purpose of general manufacture, buying, selling, owning, leasing and dealing in machinery, goods, apparatus and fixtures, inventions, patents, licenses and territorial rights."[79] The company issued 500 shares of stock listed at $100 per share; Sperry owned 498 of these (Zula had one and Arthur Dana of Chicago, who became a life-long friend and business associate, another). Thus, the company could not be plagued by dissension among stockholders, unless there were family spats over the dinner table. Nor would Sperry have to worry about others owning and failing to develop the patents he assigned to the company. The board of directors, incidentally, consisted of Elmer Sperry, Zula Sperry, and Arthur Dana.

This was the first of several invention and development companies with which Sperry became associated during his career. Institutionalization of invention (research) and development is not as recent a phenomenon as is popularly believed. Even before Sperry formed his company, the pages of the *Electrical World* carried an advertisement for the "Electrical Development and Manufacturing Company" of Boston, Massachusetts, which promised to assist inventors and owners of electrical inventions, discoveries, and devices to "develop, patent and render commercially valuable their inventions." Early in 1888, a Chicago Electrical Development Company was organized with $100,000 capital stock to deal in patents and securities, and to promote electrical inventions.[80] There were undoubtedly other such companies existing at the time Sperry organized his.

The Elmer A. Sperry Company office and shop at 215 South Clinton Street was only a few steps from the Sperry Electric Company factory at 194–198. Sperry ran back and forth as occasion demanded, but D. P. Perry, manager of the Sperry Electric Company, felt at times that the invention and development shop "across the way" took more than the apportioned share of the inventor's time.[81] Undoubtedly, Sperry's ability to compartmentalize his thoughts, store and retrieve them as needed, proved invaluable in his dual capacity on South Clinton Street.

Unfortunately there is little information about the way in which Sperry outfitted his invention and development shop. His bookkeeper,

[78] Sperry Notebooks No. 17 (July 14, 1888 to ?, SP).
[79] Records of the State of Illinois, "Elmer A. Sperry Company," Box 470, #16411.
[80] *Western Electrician*, II (1888), p. 84.
[81] Goodman, "Recollections."

William Goodman, a distant cousin of Zula's, recalled that the payroll never rose over $500 a week, so perhaps there were thirty or more workmen in the shop. Equipment probably consisted of general purpose machine tools capable of building small machines such as electric motors and fine appliances such as switches. The lab of the shop undoubtedly had testing and measuring instruments. A single steam engine of moderate size could have powered the shop. With his years of experience working with shop mechanics—beginning with the Cortland Wagon Works—he could see his inventions through the shop to their emergence as test or production models.

As head of the development firm he drew upon his special competence in patent affairs, also hard won by experience. His patent attorney was his friend Francis Parker, a former examiner in the U.S. Patent Office and a specialist in electrical inventions. Sperry patented and developed his own inventions, but there were also a large number of "jobs" working out schemes and patents brought to him "from day to day" by others.[82] Not all of these were worth development. The inventions brought to him were diverse, for he did not limit himself to electrical light and power. For example, he brought onto the market a method for electrically charging prairie fence wires, and he also secured for an inventor a patent on a cartridge to be put into wolf bait to blow up the animal (this may not have been marketed).

His own inventions, developed through his company, were generally in electric power and light. Those for which patents were applied during the active life of the company, 1888 to 1892, can be categorized as mostly electric mining machinery, electric traction (streetcar and railway), and improvements for electric-light systems. He continued to provide his devices with automatic regulation or control. One invention which cannot be categorized to which he devoted great energy and resources, was his gas engine. His emphasis in invention shifted, during the life of this company, from electric lighting to electric mining machinery, and then to electric railway and streetcars.

His notebooks, in which he jotted down his ideas, provide an indication of his ingenuity and of inventions never patented or developed. In 1888, for example, he conceived of, and hoped to develop, an elevator annunciator, an automatic crossing alarm to operate a whistle or bell in the locomotive before each crossing, an automatic gate for bridges, an automatic gate for railroad crossing, automatic switches, an automatic register for chemical processes, an automatic temperature regulator, "automatic return current for typewriter," an automatic indicator of grounds in a fire-alarm circuit, an automatic gas burner, and "electric temperature regulator—latent heat."[83] Although his electric mining machinery and electric traction took precedence over these devices in his development work, they show his continuing interest in automatic control.

From a broad array of inventive ideas, Sperry selected those most worthy of development. Patenting, constructing the first model, and testing

[82] *Ibid.*
[83] Sperry Notebooks No. 16 (December 1887 to April 1888) and No. 18 (July 1888 to ?, SP).

were expensive. In the face of these expenses Sperry resorted to a practical means of deciding which of his inventive ideas to develop—he chose those for which he could find financial support at the early development stage. This sometimes meant that an invention more ingenious or more original was neglected while another was brought onto the market. Thus, controllers of private capital were a powerful factor influencing Sperry's decision as to which of his many technologically feasible inventions would be developed. He had insufficient capital of his own to finance his developmental activity, and at that time it would not have occurred to him to seek governmental development funds.

Each of Sperry's principal enterprises had an "outstanding citizen" as backer.[84] His gas engine was supported by Joseph Medill, mayor of Chicago and editor of the *Tribune* after 1874. Medill, who carried the *Tribune* into the Republican party believing it represented the educated classes of moral worth and business enterprise,[85] had much confidence in young Sperry. W. R. Goodman recalled a visit to Medill to present the monthly bill for development costs. Medill declared in paying the bill, that he had too many irons in the fire for a man of his years, but that he liked Sperry ("He is a very wonderful young man").[86]

The development of the gas engine continued long after Sperry left Chicago and with funds other than Medill's, but several other invention and development projects of the Elmer Sperry Company reached the market or innovation stage in a comparatively short time. He applied for his first electric mining machinery patent in October 1888, with the backing of A. L. Sweet, head of the Chicago, Wilmington, and Vermillion Coal Company. Sweet regularly paid a monthly bill for labor and material used in developing mining machinery, and when the machinery and the market seemed exceptionally promising, he financially backed the formation of a new company based on Sperry's patents. In 1890 Sperry became interested in electric streetcar and railway systems and, when he found capitalists interested in this facet of his inventiveness, he concentrated his efforts there and soon founded a company on his patents in this field.

When Sperry founded his own invention and development company, he probably did not foresee how new companies would evolve from it. He undoubtedly thought that he would develop his, and others', inventions and then sell the patent rights for large-scale manufacture and marketing. Two new Sperry companies, however, were formed within four years on the basis of his inventions developed first within the Elmer Sperry Company: the Sperry Electric Mining Machine Company, formed on April 17, 1889, and the Sperry Electric Railway Company, established in 1892.

viii

Sperry was on his wedding trip at Lake Delavan, Wisconsin, when he met Albert L. Sweet, whose company was the second largest producer

---

[84] Goodman, "Recollections."
[85] Pierce, *A History of Chicago,* p. 411.
[86] Goodman, "Recollections."

of coal in Illinois. When their conversation turned to problems of mechanizing coal mining, out of the exchange came Sperry's idea for an electrified coal puncher for undercutting coal seams. Sperry, "who was always ready to design a new machine overnight," submitted a sketch, and Sweet agreed to finance the inventor's enthusiastic proposal for developing the machine.[87]

Machinery had long been used to pump water and to raise coal from the mines, but the miner still used a hand pick, bored the hole for the explosive charge with a hand drill, and manually loaded the coal into cars hauled to the foot of the lifting shaft. These operations were not mechanized more completely because power could not be transmitted economically from the surface down into and through the mine (steam engines below made working conditions intolerable). In some mines, steam-driven, compressed-air machines at the surface supplied air to coal punchers below, but this required an expensive and usually leaky transmission system. The major obstacle to mechanization, then, was power transmission, not the tools or machines to apply the power. The availability of cheap electricity in the late eighties changed the situation. Coal mining was an obvious market for electrical power—a potential early recognized by Sweet. The small inexpensive wires of the electric transmission system could solve the major problem of getting the power economically to the machines underground. Furthermore, electricity was not noxious and transmission losses were small.

At Lake Delavan, Sperry sketched his idea for electrifying a coal puncher. He realized that the main problem in using electricity was not transmitting power to the puncher, but driving the puncher's reciprocating cutting tool, or pick, with a rotating electric motor. Another problem was to prevent the pounding of the pick against the seam from disrupting the motor. Sperry thought he had the solution, but the pressure of other affairs at the Sperry company and the problem of expressing the idea as working drawings and as a full-scale model took him more than a year.

On October 8, 1888, Sperry applied for his patent on an "electric mining-machine," which he described as "a cheap and simple automatic mining machine or pick."[88] The most ingenious component of the machine was a clutch for engaging and disengaging the electric motor from the reciprocating pick (see figure 2.9). The clutch consisted of a compound cam rotating within, and concentric with, the hollow crank driving the reciprocating pick. The compound cam with its four eccentric exterior curves was keyed to a shaft geared to the electric motor. As the motor began to turn, the compound cam rotated within the bore of the crank until the rollers, moving between the eccentric curves of the cam and the

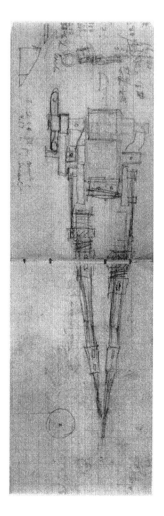

Figure 2.9
Sperry sketch for mining machine.

---

[87] The quotation is from Herbert Goodman. This section on the electric mining machine company draws on Drell's "Mechanization of Coal Mining." Drell's dissertation was written from research he did for a biography of Herbert E. Goodman and a history of the Goodman Manufacturing Company (coal mining machinery). Drell used the Goodman papers which include business correspondence and company records as well as letters of the Goodman family. The Goodman papers are now on microfilm and in the Sperry Papers.

[88] Filed October 8, 1888 (287,481), "Electric Mining-Machine," issued May 23, 1893 (Patent No. 497,832).

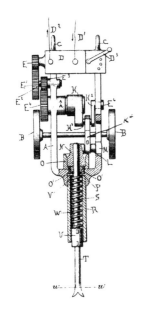

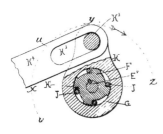

Figure 2.10
Electric Mining Machine
(Patent No. 497,832).
Below: crank release
mechanism.
Top: plan view. (Key:
wheels, B; electric motor, D;
cam, F; eccentric curves, G;
rollers, J; crank, K; pitman,
K⁴; cross head, L; drill
point, T.)

concentric interior curve of the crank, jammed where the space between the curves narrowed. Then the powered shaft turned the crank which, by means of a pitman (connecting rod), crosshead, and slide, compressed a heavy spring that drew back the pick. When the crank reached the turning point (the pitman began to move forward instead of back), the spring then moved the crank forward faster than the cam, the rollers moved toward the wider space, and the clutch disengaged.

The motor, gear train, clutch, crank system, and pick were all mounted on a two-wheeled frame that could be pushed about by the operator. In the patent claims Sperry stressed that his machine had "a continuously acting motor, . . . a cutter-holder reciprocating longitudinally," and "a power storing device, . . . a spring receiving and accumulating tension from the continuously acting motor and discharging the same at regular intervals." The puncher also had a lost-motion device to absorb recoil. The ingeniousness of the device and the ease with which Sperry conceived of it demonstrated his mastery of both mechanisms and electrical systems.

Sperry built the first model in the winter of 1888–89. The newly formed Elmer A. Sperry Company contracted to construct the machine, subcontracting parts that could not be cast or machined in the shop. Sweet and Sperry displayed the first machine at the meeting of the National Electric Light Association in Chicago in February 1889. They then shipped it to the Paris Exposition, where it won a gold medal, although the foreign rights could not be sold. In Chicago, Sweet announced to interested mine owners and operators that production would not begin until the machine was thoroughly mine-tested.[89] Sperry and Sweet modified the machine on the basis of limited tests, but time revealed that they should have tested the first machine in many mines under various conditions before assuming the basic design sound.

Their initial optimism also led them to form a company before the machine was proven. They incorporated the Sperry Electric Mining Machine Company on April 17, 1899, with Sweet as president, Sperry as "electrician," and Herbert Goodman, secretary and treasurer. Goodman knew nothing more of coal mining than what he had read in the publicity for the new Sperry machine, but when he opened the new company office at 175 Dearborn Street, he was convinced that "immediate success was assured!"[90] The original subscribers to the 1000 shares of $100 stock were Sweet (500 shares), Elmer A. Sperry (499 shares), and Herbert Goodman (one share). On the books of the company, Goodman wrote, "the Capital Stock nor any part of it not being actually paid up, the estimated value of the patents and services . . . [are] considered an equivalent in value to the full amount of the Capital Stock."[91] Once again Sperry patents formed the basis of a corporation with few other assets.

89 Drell, "Mechanization of Coal Mining," p. 20.
90 Herbert Goodman, "History of . . . the Goodman Manufacturing Company," an address given at the twenty-fifth anniversary of the concern, February 2, 1914 (quoted in Drell, "Mechanization of Coal Mining," p. 23).
91 The Sperry Electric Mining Machine Company "Journal," June 1, 1889 (quoted in Drell, "Mechanization of Coal Mining," p. 24).

The company placed the coal puncher on the market in June 1889, guaranteeing the machines for six months. It was priced at $300, but the company recommended a system composed of a Sperry dynamo (between $300 and $400), driven by a five horsepower steam engine ($250), and powering four punchers and an arc lamp. Also in June, *Electrical World* carried a well-illustrated story on the Sperry mining machine. The writer stated, "miners experienced with mining machines . . . are loud in its praise." In June, not one machine was at work in a coal mine.[92]

The first order for a mining machine came from Sweet's coal company, which ordered a battery of ten, and then the Watson, Little & Company with a mine at Brazil, Indiana, ordered six machines in August. Because the Indiana order took precedence, the first major test of the Sperry system occurred there. Unfortunately, the Sperry company had not tested the machines on bituminous coal of the hardness found at the Brazil mine, and the punchers repeatedly failed. Because iron had been used sparingly to reduce the weight of the machine, the severe jolts of the reciprocating pick broke the brittle cast-iron parts and the motor components. Clutches, springs, gears, bearings, and armatures failed. The design of replacement parts was changed, which resulted in each machine's having unique components that complicated the replacement and spare parts problem of the Sperry company. Finally, the company recalled the whole system and placed it in one of the Sweet mines. There, with softer coal and sympathetic operators, the machines began operation early in 1890. When operating, the machines more than fulfilled specifications in undercutting coal—but, as at Brazil, there were repeated breakdowns. One of the pit bosses remarked that "we needed bushels of them armatures to keep the machines going." On one occasion a machine puncher and a mule collided, and "it was found that the mule was all right, but that the machine was 'out of business.'" Sweet became discouraged in the spring of 1890, and, although he continued to believe in the ultimate success of the coal puncher, he was ready to give up his stake in the company to make way for a new capitalist willing to support further development.[93] He lost the $5000 he had advanced the company for "experimentation."

Sperry did not want to abandon the mine machine despite the difficulties. He was determined to maintain the momentum he had built up and to apply the acquired experience. He knew from prior invention and development that once an inventor immersed himself in a field new horizons and opportunities for invention opened if one pushed forward. He had seen himself and other inventors go into electricity with only an arc-light system and then invent substantial improvements and entirely new applications.

The need for a new capitalist was imperative and Sperry's enthusiasm and persuasiveness helped him here. His close friend, and a board member of the Elmer A. Sperry Company, Arthur Dana, suggested William D.

[92] Drell, "Mechanization of Coal Mining," p. 26.
[93] Goodman, "The Goodman Manufacturing Company" (cited in Drell, "Mechanization of Coal Mining," pp. 31–32).

Ewart, founder and president of the Link-Belt Machinery Company, a Chicago firm of which Dana was treasurer. Ewart agreed to advance the Sperry Electric Mining Machine Company the money necessary to continue the development of coal mining machinery. After stock was redistributed the stockholders were: Sperry, 400 shares; Ewart, 333 shares; Dana, 250 shares; and H. E. Goodman, 17 shares. Ewart became president of the company.

With the new capital, Sperry invented and developed a coal cutter of basically new design and then an entire system of electric mining equipment. He continued to test the puncher and improve the design, but his tests showed more conclusively the limited applicability of the reciprocating machine. He then changed from reciprocating motion, or punching, to rotary cutting. In so doing, he departed further from the motions used by the workmen with his tool—a progression not unusual in the history of mechanization. Sperry applied for a patent on his chain-type, rotary coal cutter in May 1891, and tried out his first model the same year.[94] The basic principle of the machine was not new, for there were British patents on a similar cutter driven by compressed air. His problem, as in the case of the punch, was to design a mechanism for transmitting the power from an electric motor to the cutting tool. With this machine, however, the rotating motion of the cutter chain suited the rotating motion of the electric motor drive. He also had to design a machine whose parts could withstand the great strain of cutting, although in the new machine the problem was less severe because the work formerly done by one cutter was distributed among a whole series of cutting surfaces.[95]

The rotary-chain cutter had a design better suited to cutting coal than the puncher, but its acceptance in the mines was delayed because it was a radical departure from the customary way of cutting coal in American mines and demanded a change in related techniques. To use the rotary-chain cutter, mines had to change to longwall mining which in America had been resisted before mechanization because labor costs were higher. In Europe, where labor was cheaper and coal less plentiful, the owners had used the more thorough longwall system.

Sperry further expanded his system of electric mining in 1891. This resulted from the possibility of creating a system of electrical devices dependent on the electrical generating plant. To use the coal puncher or the rotary cutter the mine needed an electric generator; this initial investment having been made, the mine owner was more easily persuaded to buy electric light and other appliances. The company offered a patented means for distributing electricity through the mines to the various appliances.[96] To expand this system further Sperry invented and developed an electric locomotive for mine haulage.

[94] Filed May 4, 1891 (391,469), "Mining Machine," issued July 5, 1892 (Patent No. 478,141).
[95] Drell, "Mechanization of Coal Mining," p. 41.
[96] Filed April 1, 1889 (305,582), "System of Electrical Distribution for Mines," issued February 10, 1891 (Patent No. 446,030).

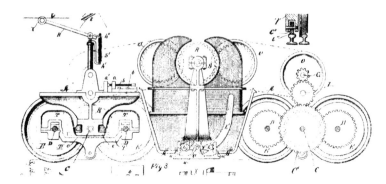

Figure 2.11 Electric Mining Car (Patent No. 478,138), side view.

Before he applied for his patent on a mine locomotive, several locomotives had already been used in American mines. The first, the design of a German engineer, W. M. Schlesinger, began operation in July 1887, in Pennsylvania. Sperry designed a more practical and economical machine. After analyzing the environment within which a mine locomotive functioned, he foresaw that the locomotive had to turn sharply, ride on light, irregular rails and an uneven roadbed, and still exert a strong pull. This meant a heavy locomotive on light rails. To satisfy these conditions Sperry invented a locomotive with two four-wheel trucks and a single driving motor centrally mounted on the locomotive frame. By this he distributed the weight and the load on the rails over eight wheels and provided flexibility for turns. Furthermore—and Sperry was proud of this—if one truck lost contact with the rails, the other truck took the power. If there had been a motor for each truck then the truck that lost contact would spin ineffectively. His design was an ingenious solution that nonetheless created the challenging problem of driving all wheels from a single motor. The problem of power transmission was further complicated because the trucks moved independently and the motor was mounted on the frame.

To solve the problem, he modified one of his earlier ideas for power transmission on an electric streetcar.[97] He transmitted power from the motor by pulley and belt through the gears on shafts G, I, and C—which were rigidly mounted to the frame of the locomotive—to the gear wheel (M) driving the four wheels on the swiveled truck (figs. 2.10, 2.11). He used an ingenious mechanism for transmitting the power of rotation from shaft C to gear wheel M. To do this he formed a spherical portion on the shaft (C) and gave the gear wheel (M) a hollow spherical hub adapted to fit around the spherical portion of the shaft. A pin passed loosely through the spherical portion of the shaft and rode in the meridional slots in the concave surface of the hollow hub. This resulted in the shaft rotating the gear wheel but allowed variation in the angle between the plane of rotation of the gear wheel and the axis of the shaft.

Sperry built his first haulage locomotive in the winter of 1890–91,

[97] Filed March 17, 1890 (344,219), "Power-Gearing for Vehicles," issued August 12, 1890 (Patent No. 434,097).

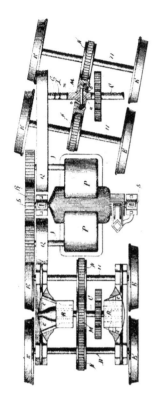

Figure 2.12
Electric Mining Car
(Patent No. 478,138).
Below: axle to traction
wheel drive.
Top: plan view.
(Key: shaft, C; spherical
portion formed on shaft,
m; pin, n; meridional slots
in the hub, o.)

and "compared with the coal puncher, the haulage car worked well from the very first."[98] After some modification in the design of the first full-scale test model, the company manufactured locomotives weighing eight and twelve tons and having fifty and 125 horsepower motors.[99] The company then advertised its electrified system of mining equipment and obtained large orders in 1891 and 1892. The mining machine company was so successful that it increased its manufacturing facilities by buying the Elmer A. Sperry Company from Sperry in 1891 for $12,000, acquiring as well the developmental work and other manufacturing activities of that company. With sales reaching $137,402 in 1892, the company employed more than fifty men and rented a part of the Link-Belt factory at 39th and Stewart Avenue for manufacturing. The expanding enterprise was sold, however, in 1892, after the Thomson-Houston Electric Company, continuing its policy of buying competitors, patents, and enterprises in fields into which it wished to expand, made an attractive offer to the Sperry Mining Machine Company. On May 10, Thomson-Houston bought the mining machine patents, jigs, and special tools for $50,000. Sperry agreed to act as consultant when Thomson-Houston took over.

His association with mining machinery did not end here, however. Herbert Goodman proceeded to buy patents for the mining machinery of other inventors and began manufacturing under the name of the Independent Electric Company in 1893. Sperry, with Goodman and Dana, had equal interests in this company. It also manufactured other products such as fuse wire, which the Elmer A. Sperry Company had first manufactured. In 1900, Goodman reorganized the company as the Goodman Manufacturing Company in which Sperry also acquired a large block of stock. Sperry became vice president of this company, and throughout the remainder of his career he helped make policy decisions and contributed much on the technical side as its consulting engineer. The company became a leader in its field.

ix

The purchase of the mining machine company brought an end to Sperry's Chicago years, for he decided to move to Cleveland, Ohio, in order to develop and manufacture his streetcar with the aid of a syndicate of Cleveland businessmen. Sperry had arrived in Chicago a promising young inventor with an arc-light system that performed well in limited use. He had the good fortune to help found a company intended to support his inventive and developmental activity and to market his patented inventions. A decade later he left Chicago a mature inventor and an engineer who had not only developed and manufactured a system of electric mining machinery, but had developed, in addition, a streetcar financed by a wealthy syndicate. Both the mining machinery and the streetcar were known by his name. Yet the intervening ten years had also brought him

[98] Drell, "Mechanization of Coal Mining," p. 39.
[99] One of the first of seven made was repurchased after fifteen years of service and is now displayed at the Museum of Science and Industry in Chicago.

many disappointments and frustrations. His first company demanded routine engineering of the inventor and harassed him with financial and managerial problems. He watched other inventors, who had been no further advanced in electric lighting than he in 1883, forge ahead of him. He saw his company fail before the competition of more heavily capitalized and innovative enterprises. For five of his ten years in Chicago he had to accept conditions that frustrated his inventiveness.

During the Chicago years, however, his enthusiam, self-confidence, and determination triumphed over frustration and disappointment. Neither he, nor a sizeable number of supporters, lost faith in his ability as an inventor. After his first company dissolved, he recovered and helped establish two more Sperry companies, one of which, an invention and development firm, allowed him to concentrate where his genius lay. It was not altogther chance that he found the way beyond disappointment and handicapping circumstances shortly after he married Zula Goodman. Her sophisticated taste and refined interests enriched his personality; her strength of character and faith in him reinforced his determination and sustained his effort.

Chicago was a maturing experience. He not only sounded the depths of his character, but he learned more of his capabilities and limitations as an inventor. He also became a capable engineer through his experience in developing inventions for the market, and he grew wise about the innovation process. He learned that successful innovation involved financing, manufacturing, and marketing of the developed invention. In addition, he realized that as an inventor and engineer his success, or failure, might depend upon the businessmen and managers with whom he associated.

CHAPTER III Entrepreneur of Invention and Development

If I spend a life-time on a dynamo I can probably make my little contribution toward increasing the efficiency of that machine six or seven percent. Now then, there are a whole lot of arts that need electricity, about four or five hundred per cent, let me tackle one of those. ELMER SPERRY, ADDRESS TO BROOKLYN YMCA, 1922

WITH THE FORMATION OF THE ELMER A. SPERRY COMPANY IN 1888, SPERRY became an independent inventor-entrepreneur, a role in which he chose to remain for two decades. As an independent, not salaried employee, he was free to select his own problems and to determine his own approach to their solution, but his position also meant that he could not depend on regular work and income. Independence, however, allowed him to devote himself to major innovations instead of to marginal refinements. The excitement of such a pursuit more than compensated for its insecurity. As an inventor-entrepreneur, Sperry presided over the process of invention, development, and innovation; he had to be, therefore, intimately acquainted with the state of technological development in his field and he had to identify a market for particular inventions made possible by the state of the art. His concern, in other words, was the transformation of an idea into a practical device that could perform well in a competitive environment. After a patented invention was ready to be marketed, Sperry then might use the patents as the principal capital asset of a company formed to manufacture and sell the invention. He left the management of such new enterprises to others. It was not that management bored him, however, as some of his contemporaries believed: managing a business would have forced Sperry to abandon his profession—invention and engineering.

Presiding over invention, development, and innovation required organizational ability as well as technical skill. How expert a patent lawyer, how many assistants, how excellent a laboratory and testing facilities, and how large a company the inventor-entrepreneur acquired depended on his ability to raise money. Financing was not invention, but as long as Sperry remained free of corporate ties, he assumed that burden. "So to get these things going, inasmuch as I never could borrow any money from banks to help me on these queer dreams of mine," Sperry recalled, "I had to pitch in and earn money."[1] The money he earned from one invention he invested in the next "queer dream." The assertion of Sperry, Edison, and other inventors that they wanted money only to pour it into more inventions seems entirely understandable. If the inventor could not raise sufficient funds from his own savings, then he had to appeal to financiers or corporations to enter into *ad hoc* agreements, by which he sold them a share of his invention or patent.

During his decades as an inventor-entrepreneur, Sperry partially financed his inventions with income from Chicago companies that stemmed directly or indirectly from his inventions of the Chicago period. The company that began as the Sperry Electric Mining Machine Company and metamorphosed over ten years into the Goodman Manufacturing Company was most notably successful. Herbert Goodman proved an excellent manager, and Sperry's substantial stock in the company, originally payment for his inventions, increased in value. Goodman and Arthur D. Dana, Sperry's friend and business associate, directed the Independent Electric Company (later the Chicago Fuse Wire and Manufacturing Company), which manufactured under Sperry patents and in which he held an interest. William R. Goodman, one of Zula's cousins, also became a business manager in the fuse wire company and in other enterprises in which Sperry was involved. Sperry often acted as an engineering consultant to these businesses and sometimes served as an officer or director. In a 1902 financial statement, he estimated his worth "as something over $100,000"; most of the assets he named were joint Dana, H. Goodman, and Sperry enterprises.[2] These holdings, however, did not provide enough cash for his entrepreneurial activities; he supplemented them with the sale of shares in his patents to new financial backers or companies. On occasion, a company that bought his patents retained him as a consultant, and such fees helped him over lean periods. For a time in Cleveland, Ohio, in fact, he styled himself, "ELMER A. SPERRY: Consulting Electrical & Mechanical E.N.G.I.N.E.E.R., Cleveland, Ohio."

i

Sperry's success as an inventor-entrepreneur undoubtedly depended upon his ability to choose fields of endeavor in which his particular characteristics as an inventor were needed. The environment in which he as an inventor-entrepreneur had to survive was complex, shaped, as has been noted, by the state of the technology, the market potential, and the avail-

---

[1] Elmer A. Sperry, "The Creative Spirit—Pioneer Work in Industry," *Industrial Management Lecture Course* (Brooklyn: YMCA, 1922), p. 8.
[2] Typescript dated May 6, 1902, giving a three-paragraph description, in general terms, of his worth (SP).

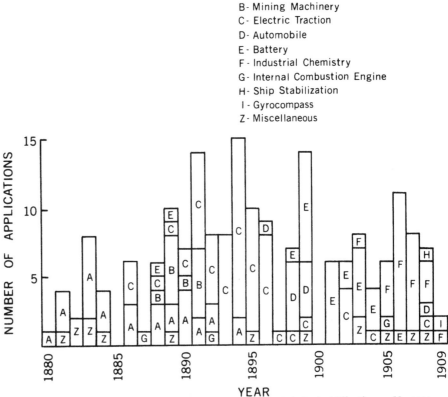

A- Electric Light and Stationary Power
B- Mining Machinery
C- Electric Traction
D- Automobile
E- Battery
F- Industrial Chemistry
G- Internal Combustion Engine
H- Ship Stabilization
I - Gyrocompass
Z- Miscellaneous

Figure 3.1 Elmer Sperry's patent applications, 1880-1909.

ability of capital. There is no evidence that Sperry systematically analyzed these and other factors before making a move into a field, but it seems reasonable to assume that if he did not make this analysis, he had other ways of sensing circumstances conducive to his inventive endeavors. His efforts to keep abreast of the activity in technology and industry through professional journals and meetings, the close contacts he maintained with capitalists, and his determination to develop practical devices were ways in which he sensed new possibilities for invention and development.

His shifts from one field to another can be narrated on a level of immediate cause and effect; an effort will be made, however, before narrating the inventive activity of the Cleveland period, to find more general explanations for Sperry's decisions. The effort will be exploratory and by no means conclusive, but it might suggest explanations for his seemingly uncanny knack, as his associates put it, to be in the right place at the right time with the right invention.

During the three decades from 1880 to 1910, Sperry concentrated his inventive activity in a sequence of industrial fields (see figure 3.1). These were electric light (mostly arc lights) and power (stationary), mining machinery, electric traction (mostly streetcar), automobiles (mostly electric), batteries, and industrial chemistry (especially electrochemistry).

A comparison of the pattern of Sperry's applications with general industrial activity helps explain his tactics. When he applied for his first patent in arc light and electric power in 1880, capital invested in the electrical machinery and supplies industry was $1.5 million; the number of electrical machinery and supply firms was 76, and only two new firms entered the field during the year. Five years later, in 1885, capital invested was about $7 million, the number of firms totaled about 100, and 24 new firms were established. Another indication of the rapid development in the field, one in which Sperry was particularly interested, was the rapid rise in the number of central electric stations beginning operation each year; in 1881, 8 began operation; in 1885, there were 55, and in 1886, 100 new stations.[3]

Sperry began sustained concentration in the streetcar field about 1890, when capital employed in streetcar production was about $3 million and the number of producers was around 17; the miles of electric street railways in Massachusetts, Ohio, and New York (the states for which information is available) totaled about 480. Five years later, the capital employed had risen to $5 million, although the number of producers remained around 20; the total mileage in the three states totaled 2,500.[4] Sperry shifted his emphasis to electric automobiles in 1898. One year later —the first in which census figures are available—the capital invested totaled $5.8 million; there were 57 companies producing 3,723 cars. Five years later, in 1904, the capital invested was $23 million, the number of producers totaled 178, and the number of automobiles produced totaled 21,692.[5] Statistics on the production of batteries are not available.

Sperry shifted from electric automobiles and batteries about 1905, when he applied for four industrial chemistry patents. In 1904 the value of chemical products produced with and of electricity was $7 million; five years later, the value had risen to $18 million.[6] Electrochemistry and

---

[3] U.S. Bureau of the Census, *Thirteenth Census of the United States: 1910: Manufactures*, X, p. 283; and *Abstract of the Twelfth Census of the United States: 1900*, p. 409. The 76 firms included a large proportion of manufacturers of telegraph equipment. According to the census of 1880, there were only three producers of electric-lighting equipment in 1879; capital invested totaled $425,000; and the annual product was $458,000. There is no information on electric-lighting equipment manufacturers for 1885; for 1890, information was limited to several states.
[4] The three states are used because electric street railway statistics for the entire country have not been compiled in census and other available reports for the years under scrutiny. The mileage for the three states was compiled from items in annual volumes (1890–95) of H. V. and H. W. Poor, *Poor's Manual of Railroads* (New York: H. V. and H. W. Poor, 1868–1924). The other statistics on street railways are from the U.S. Bureau of the Census, *Census Bulletin 84* (1907), pp. 67–68. Russell Fries, the author's research assistant, compiled and organized the statistical information used in this section.
[5] U.S. Bureau of the Census, *Fourteenth Census of the United States: 1920: Manufactures*, X, pp. 867, 874; U.S. Bureau of the Census, *Special Bulletin 66* (1907); and U.S. Bureau of the Census, *Twelfth Census of the United States: 1900: Manufactures: Automobiles and Bicycles and Tricycles*, X, pp. 10, 12.
[6] U.S. Bureau of the Census, *Thirteenth Census of the United States: 1910: Manufactures*, X, p. 545.

industrial chemistry received his attention until he turned to gyroscopes. This ended his inventor-entrepreneur decades, for he formed a company to develop and manufacture gyro devices and directed its activities for his remaining years.

Statistics suggest, then, that Sperry moved into a field when it was new and developing rapidly.[7] The statistical evidence is reinforced by the qualitative evidence in histories of the fields in which Sperry concentrated. The

[7] Other inventors also concentrated on areas in which Sperry was working. Often, he was among the leaders in a field; in other areas, he was a part of the second wave. This is shown by comparing Sperry's patents with those of other inventors. To count the number of patents from other inventors in fields in which Sperry concentrated, I used the *Annual Report of the Commissioner of Patents,* which lists all patents by title. Since the patents are not classified by subject, it was necessary to judge from the title alone whether the patent covered arc lighting, electric-power generation, street railways, or another field. (Field has been defined somewhat differently in this discussion of patent correlation than it was in the discussion of industrial development above, where census categories were used.) If the nature of the patent could not be established with reasonable certitude from its title, the summary of the patent in the *Patent Office Gazette* was used. (I am particularly indebted to Russell Fries for this research and compilation.)

A tabulation of arc-light patents shows that the number issued by the U.S. Patent Office rose dramatically from 8 in 1878 to 62 in 1882. The next year the level was maintained, but 1884 witnessed a precipitous drop. Sperry received his first arc-light patent in 1884. In 1888, after the number of all U.S. arc-light patents had reached the lowest point since 1882, the number rose to a new peak in 1892. After receiving two arc-light patents in 1884, Sperry's activity in the field was slight until 1892, when he again received two patents. Consideration of Sperry's patent activity in other fields also shows a correlation with the pattern of activity of other inventors. Classifying Sperry's early patents in the broader category of light and stationary-power systems permits another comparison. A compilation of patents with titles specifying alternator, armature, current motor, electrical distribution system, electric generator, electric machine, or electric motor—all major components in electric light and power systems—shows that the number of patents increased dramatically from 25 in 1880 to 158 in 1883 and then maintained a plateau of about 150 patents per year until a general decline beginning in 1892 terminated in 1898. Although the curve might result from changing usage in titles, the correlation of all patents with Sperry's patents is nevertheless notable. His first patents in this general category came in 1882 and his last in 1895. (The comparison is best made by subtracting the number of arc-lamp patents from Sperry's stationary power and light patents, for arc-lamp patents were not included in the general category for all inventors.)

Compilation of patent titles specifying electric-railway and railway systems, streetcars, and trolley equipment reveals a general trend upward from 5 in 1883 to 80 in 1896. The sharpest rise occurred from 1887 to 1890. Sperry's first electric streetcar patents were issued in 1887; in 1896 he received a total of 16 streetcar patents, the peak of his activity in the field.

Patents titled automobile electric, vehicle controller, vehicle brake, electric, and electric vehicle indicate the area in which Sperry concentrated from 1899 until 1900. The total number of patents was small, and the correspondence with Sperry's may be distorted by the inclusion of the general title "vehicle controller"; the compilation, however, shows a general cluster from 1888 to 1891 and another from 1898 through 1914. In the case of the battery, the correlation is of more interest and more substantial. Patents on the battery grid, battery plates, the electrical battery, the storage battery, and the secondary battery rose, after remaining near 30 annually

beginnings of the electric light and power industry are taken to be around 1880; the electric streetcar industry had its substanial origins about 1887; the automobile industry emerged between 1895 and 1900; and the electrochemical industry was established by the availability of cheap power at Niagara after 1895.[8] The major exception to Sperry's general pattern was his entry into the mining machinery field, for this industrial activity was well established then. If, however, a distinct industrial category, electric-mining machinery, could be identified, then Sperry's patents in electric-mining machinery would undoubtedly correlate with the origins of this particular subdivision of the industry.

The circumstances under which Sperry was attracted to a new field partially explain his reasons for leaving the old one. More can be said about his abandoning fields, however. He appears, judging by his patent applications, to have lost interest in a field after about five years. This suggests that an in-rush of inventors, engineers, managers, and corporations, brought by capitalization and size, convinced Sperry that his talents could best be applied elsewhere.[9] Desiring to remain independent, he preferred fields not occupied by corporations with staffs of dependent inventors and engineers. The dependent inventors often concentrated upon refining the technological systems in which the corporations hiring them had heavy investments. Sperry was wise not to compete against them, for their number reduced what he could contribute.

Hindsight makes Sperry's decisions to move into young and expanding

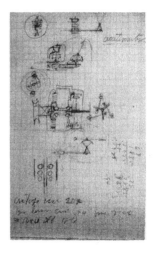

from about 1891 to 1899, to a peak of 61 in 1903. Sperry's period of patents on batteries extended from 1900 to 1907. (He had also concentrated on battery patents between 1888 and 1890, but it seems reasonable to assume that these batteries were for streetcars and central stations; Sperry's major battery work focused upon electric automobiles.)

Mining machinery patents and electrochemistry patents do not show the clearly defined clustering that allows close correlation with Sperry's clusters. Assuming that patents on coal cutting, mining machinery, and rock-drilling machinery cover Sperry's own mining machinery, the general pattern of patenting shows that one of the high plateaus of activity between 1880 and 1908 extended from 1890 to 1892 (about 40 patents annually). This three-year period corresponds to Sperry's own activity (1889 to 1893) and both might have resulted from increased opportunities to apply electricity to mining machinery. In electrochemistry—patents bearing the title "electrochemistry and electrochemical apparatus"—there was a general trend upward from 1888 to 1907. Sperry's activity came at the end of this period. The patent statistics for chlorine detinning, the branch of industrial chemistry into which Sperry moved after venturing into electrochemistry, show the first significant number (7 to 15) between 1906 and 1910. Since Sperry and his associates produced a large number of these, however, the correlation is not significant.

[8] Major histories that establish the origins of the industries about the time that Sperry entered them are: Passer, *Electrical Manufacturers;* John Rae, *American Automobile Manufacturers* (Philadelphia: Chilton, 1959); and William Haynes, *American Chemical Industry* (New York: D. Van Nostrand, 1954), I.

[9] It cannot be *assumed* that inventors leave a field and that patents decline as the industry becomes organized. One student of patents has argued convincingly that inventive activity in any industry is probably dependent on the absolute amount of capital being invested annually. Jacob Schmookler, *Invention and Economic Growth* (Cambridge: Harvard University Press, 1966), p. 172.

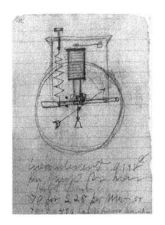

Figures 3.2
Pages from Sperry notebooks.

fields seem obvious. History records, however, numerous promising fields that did not develop or did not develop for decades after the initial promise was revealed. It also records false starts, flashes in the pan. How did Sperry, surrounded by a chaos of technological change, identify, with such impressive foresight, the promising trends? He brought some order out of the chaos and narrowed his range of choice somewhat by focusing his attention upon problems that demanded knowledge and experience in his special competence: electrical technology. Though his fields of concentration seem diverse—mining machinery, streetcars, automobiles—all involved the application of electricity.

Sperry further refined the focus of his attention by concentrating on critical problems which in his judgment could probably be solved by invention. The problem was critical if it was retarding technological or industrial development. Throughout his career, Sperry's patents show his focus on critical problems. He did not simply invent a dynamo, an arc lamp, or a streetcar, although the patent titles might indicate that he did; he invented very specific components and processes. In the case of Sperry's arc-light patents, it is clear that he considered the design of automatic carbon feed critical; his generator patents reveal his concentration on designing automatic regulation of output. When he turned to mining machinery, he directed his attention to specific machine-design problems involving the application of electrical power. When working in the streetcar field, he concentrated on power gearing and transmission, as well as on the problems of braking and speed control. In electric automobiles, he also focused on control, although battery design—especially the battery's physical characteristics—received much of his attention. In industrial chemistry, he continued to direct his attention to problems of mechanical design.

His close study of the technical journals and the records of patents, and his regular attendance at professional meetings helped him find these critical problems, for the community of inventors and engineers in their articles, their conversations, their lectures, and their patents converged upon them. Furthermore, the better informed and more experienced—the more professional—the inventor, the more likely that he, like Sperry, would concentrate on the critical problems. Only the naive inventor assumed that the challenge was to invent an arc lamp, an electrical generator, a streetcar, or an automobile.[10]

In summary, the pattern of Sperry's patenting, the level of industrial

---

10 Other inventors recognized the critical problems in arc light and electric power and further tabulation of patents might reveal this to be the case in Sperry's other areas of concentration as well. Russell Fries used the *Annual Report of the Commissioner of Patents* for the years 1878–82 to tabulate arc-light and generator patents and the *Patent Office Gazette* to determine their content, if the title was not specific enough for the purpose of his analysis. From a sample of 113 arc-light patents, he found about 78 percent focused on regulation, as did Sperry. In the case of electric generator patents, he considered 228 and found that 30 percent focused on regulation.

activity in his fields of concentration, and the critical problems in specific fields suggested the problems which could be solved by his kind of inventive experience. His choice was also influenced by his assumption that the field was attracting capital and that there would be financial reward for the inventor who solved the problems hindering technological expansion and industrial growth. He left a field, it seems, when the level of industrialization became so high that large corporations with staffs of inventors and engineers became involved in refining the technological systems.

ii

Sperry chose to live in Cleveland from 1893 to 1905, mainly because of his close association with several of its leading financiers and because of the ready availability of manufacturing facilities in what was then a rapidly industrializing metropolis. His first connection in Cleveland was with the streetcar-manufacturing company founded by the Cleveland financiers on the basis of his patents; they continued to finance him when he developed other devices. The agreement that he negotiated with the Cleveland syndicate in 1890 is an example of the means by which he maintained his status as an independent inventor and consulting engineer at a time when other inventors and engineers were being drawn onto the engineering staffs, and into the new research and development laboratories, of large corporations. On March 6, 1890, several Cleveland industrial and banking leaders, persuaded by Sperry's achievements and by the promise of his electric streetcar inventions, formed the Sperry Syndicate.[11] Sperry came to know the Cleveland men through his friend, Washington Lawrence, head of the National Carbon Company, from which Sperry had probably bought his arc carbons. Associated with Lawrence in the syndicate were Webb C. Hayes, James Parmelee, Myron T. Herrick, and Benjamin F. Miles. Parmelee, Herrick, and Hayes were all founders of the National Carbon Company. Hayes, the son of President Rutherford B. Hayes, had his father's financial backing when the company was founded. Myron T. Herrick, the secretary and treasurer of the Society for Savings (one of Cleveland's major banking institutions) later became a leading industrialist, as well as American ambassador to France; Parmelee, life-long friend of Herrick, was also an officer of the Society for Savings.[12]

The syndicate represented "western" capital, but there were "eastern" associations as well. The members of the Sperry Syndicate also held interests in the Brush Electric Company of Cleveland, a manufacturing concern that had pioneered in arc lighting. The Thomson-Houston Electric Company of Boston, later General Electric, also held stock in the Brush company. Sperry came to know both sets of capitalists well.[13]

[11] The contract and date is mentioned in an instrument executed by the syndicate, May 29, 1905 (SP).
[12] T. Bentley Mott, *Myron T. Herrick, Friend of France* (Garden City: Doubleday, 1929), pp. 92–93. Later the National Carbon Company merged with the Union Carbide Company of Chicago and the Linde Air Products to form the Union Carbide and Carbon Corporation.
[13] These relations between Cleveland and Boston are implied by the correspondence in the Sperry files but not explicitly established.

THE BRUSH ELECTRIC CO.

ELECTRIC LIGHTING APPARATUS, GENERATORS & MOTORS.

OFFICE & WORKS COR. BELDEN & MASON STREETS

Cleveland, Ohio July 9th, '95.

His initial relationship with the syndicate was similar to that he had had with the financiers who backed the Elmer A. Sperry Company. Although his 1890 contract with the syndicate is not in the Sperry records, its outline can be surmised from another Sperry contract made under similar circumstances a few years later. Sperry probably agreed to build a prototype streetcar embodying his inventions, the development costs of which would be charged to the syndicate. If this prototype proved practical in operation, the syndicate would agree to sell or license the patents, or to form a company to build the streetcar. Sperry preferred to leave "the business part" to the other parties, although he wished to keep some control over, or to supervise, early manufacture. Therefore, he probably agreed to become a consulting engineer for the new company or licensee.[14]

Sperry built a streetcar embodying his patents some time between 1890 and 1892 in Chicago. The streetcar's performance also persuaded the Thomson-Houston Electric Company, when it bought Sperry's mining machinery patents in May 1892, to purchase his streetcar patents from the syndicate and to enter into an agreement with Elmer Sperry and the Sperry Syndicate to exploit the patents. Under the arrangement, Sperry was to serve two years at an annual salary of $5000 as consulting engineer in the development and sales of the mining machinery and streetcar. Thomson-Houston and the Sperry Syndicate also agreed, judging by later events, to establish jointly, and own stock in, a new company, the Sperry Electric Railway Company, founded to manufacture the Sperry streetcar under his patents.[15] The agreement, it appears, also specified that Thomson-Houston manufacture, at its Lynn, Massachusetts, plant, motors and power transmissions for the Sperry streetcars; the Brush company factory in Cleveland would make other parts. Furthermore, the Sperry company would lease unused manufacturing facilities in its plant from the Brush company. The agreement with Thomson-Houston indicates how an independent inventor survived in an era of transition to corporate invention and development, and demonstrates the merger trends in the rapidly developing electrical manufacturing industry.

[14] Sperry proposal to a capitalist interested in his invention of a separator to take iron from clay. Outlined in EAS to J. M. Freeman, February 28, 1893 (SP).

[15] James Parmelee to EAS, May 2, 1892 (SP) suggests this. Sperry placed his streetcar patents in escrow for Thomson-Houston; EAS to E. M. Bentley, May 23, 1892 (Sperry–Thomson-Houston letter book; SP).

Figure 3.3 The Sperry streetcar at the Chicago World's Fair, 1893.
Elmer Sperry, right, has his hand on the brake.
With him were three pioneers in electrical engineering education:
Louis Duncan of Johns Hopkins, Dugald C. Jackson of Wisconsin, and H. J. Ryan of Cornell.

The agreement brought Sperry a number of advantages. Besides a substantial salary as a consultant, he was to receive, as a member of the syndicate, a share of the profits from the sale of streetcars by the Sperry Electric Railway Company. He would also have the support of the patent, or legal, department of Thomson-Houston and its engineering staff, when acquiring patents and developing streetcar inventions for the Sperry company. Furthermore, he had access to greater manufacturing facilities than ever before in his career. With such patenting, engineering, and manufacturing resources at his disposal, he hoped to make regular improvements in the Sperry streetcar to keep it competitive in a period of rapid technological change—a state in which he had not been able to keep his electric-light system. When he had only the limited facilities of his electric light and power company, Thomson-Houston, Edison General Electric, and Westinghouse left his small enterprise behind. He had learned then the futility of trying to compete with limited resources in an area of technology increasingly dominated by large corporations.

The trend toward merger and the transitional state of invention and development in the industry helps explain Thomson-Houston's willingness to negotiate with an independent inventor and the syndicate. Under the aggressive business leadership of its president, Charles Coffin, Thomson-Houston was absorbing its competition.[16] In this way the company acquired manufacturing facilities and patents, as well as inventors and engineers, for the leading personalities of the companies often joined Thomson-Houston. With the patents, Thomson-Houston synthesized the best components (patent protected) for various products; with the help of the engineers and inventors, the company improved on the products and extended its patent structure. This was a decade or so before electrical manufacturers formalized invention and development by establishing research-and-development laboratories.

By such agreements the company also acquired control of the market. Some of the firms brought into the Thomson-Houston fold were allowed to continue to operate, specializing in products that Thomson-Houston did not manufacture, or did not manufacture as well. Subsidiaries also agreed to buy components from Thomson-Houston; the Sperry Electric Railway Company agreed to do both.[17] Through standardization and coordination of the subsidiaries Thomson-Houston created a system with a synergistic effect. It controlled the manufacturing of interrelated generators, arc lights, incandescent lights, stationary and traction motors, and other components of electric light and power systems. By coordinating the characteristics of the various patented products, Thomson-Houston brought

16 The acquisitions included, besides Brush, controlling interest in the Schuyler Electric Company, about half the stock of the Fort Wayne (Indiana) Electric Company, the patents and assets of the Chicago firm of Van Depoele Electric Manufacturing Company, and also secured the Excelsior Electric Company of New York and the Indianapolis Jenney Company. From the electric traction field, Thomson-Houston also acquired Bentley-Knight Electric Railway Company in 1889. Passer, *Electrical Manufacturers*, pp. 52–53, 228.

17 John T. Broderick briefly describes this "system," as later developed at General Electric in *Forty Years with General Electric* (Albany: Fort Orange Press, 1929), p. 63ff.

pressure on a customer, once he had bought a major Thomson-Houston product, to purchase coordinated products. Depending upon Thomson-Houston to help design and manufacture streetcar components, Sperry and the Sperry Electric Railway Company were thus integrated into the Thomson-Houston system.

What began as an agreement with Thomson-Houston became one with the General Electric Company, when Thomson-Houston and the Edison General Electric Company merged on June 1, 1892. Three years later, Sperry and the syndicate made another agreement with General Electric. During these years, in his free time, Sperry concentrated on the development of an electric brake for streetcars; by 1895 the brake was so well protected by patents and so well proved by tests that General Electric and the syndicate agreed to develop it further and to market it.

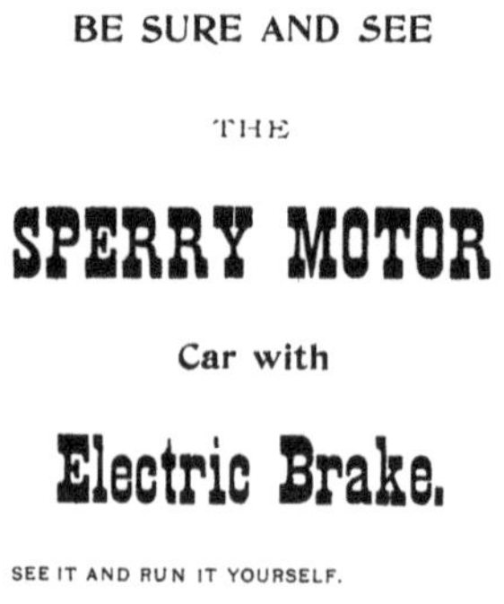

Under the new agreement (signed May 28, 1895), General Electric acquired Sperry's patent applications, patents, and future patents on the brake. The company also agreed to manufacture it. General Electric was to pay to a trustee (the Old Colony Trust Company of Massachusetts) 12 percent of the net proceeds received from the sales of the electric brake, and the trustee was to use these funds to meet all patent expenses, including litigation. Income beyond expenses was to be divided between the syndicate (20 percent) and General Electric (80 percent). After General Electric had received $200,000 and the syndicate $50,000, the income would be divided 75 percent and 25 percent. The General Electric Company could terminate the contract if a competitor introduced a competing electric brake outside of, and not controlled by, the Sperry patents. This was determined, the agreement specified, by bringing suit against a competitor assumed to have infringed. If the suit were not decided in favor of the Sperry patents, then they would be judged not to control and General Electric could terminate the contract. The agreement was to last fifteen years.[18] As with the earlier agreement, Sperry was salaried by General Electric as a consulting engineer during several years of post-innovational development. Under the new agreement his salary was $2250 annually.[19]

Through these agreements with General Electric, Sperry became well acquainted with the leaders of the electrical manufacturing industry and with the strategy and structure of a major American enterprise in its formative period—knowledge that influenced his own engineering and entrepreneurial style.[20] Sperry also developed an intimate awareness of the scientifically based engineering which was becoming more and more important in the industry in the 1890's. Otherwise, he might not have been able to compete with the young engineers coming in increasing numbers

BE SURE AND SEE

THE

SPERRY MOTOR

Car with

Electric Brake.

SEE IT AND RUN IT YOURSELF.

Figure 3.4
Broadside for the
Chicago World's Fair, 1893.

18 Copy of agreement of May 28, 1895 (SP).
19 W. W. Rice to EAS, July 23, 1895 (SP).
20 The extent of his connections with industry leaders is clear from the correspondence in Sperry letter books. These, each including several hundred letters (not all legible), are: "Elmer A. Sperry-Personal, February 29, 1892 to April 24, 1894"; "Thompson-Houston Electric Co., May 15, 1892, to July 13, 1893"; "E. A. Sperry, July 13, 1893, to November 11, 1895"; "E. A. Sperry, September 21, 1895, to September 19, 1896"; and "E. A. Sperry, November 20, 1895, to February 2, 1897" (SP).

74

from the nation's engineering schools. Around the turn of the century, these graduates were having a substantial influence on the character of engineering—especially in the electrical industry, which was hiring them in large numbers. When Sperry entered his career, there were no courses in elecrical engineering in America or abroad; in those circumstances, ingenuity and experience gave him an advantage. Fifteen years later, he might have become obsolete, had it not been for his interest in continuing his education and his broadening association with General Electric.

A good indication of the change in the character of electrical engineering was General Electric's insistence on thorough quantitative tests of the devices that Sperry wished the company to develop and market. Another indication is the emphasis upon testing and quantitative data in the papers and articles that Sperry prepared for professional meetings and trade journals. His communications and articles of the eighties did not provide reports and analyses of tests run on his equipment. They gave basic specifications and made assertions, but validation depended on the authority of the inventor or his company, not on an objective appraisal of test data. In the nineties, he sent the engineering staff at General Electric quantitative reports on his inventions. When, for instance, he began intensive development of his electric brake in 1893, he provided full reports on its performance.[21] E. W. Rice, Jr., Walter Knight, and the other General Electric engineers needed such objective reports to make decisions about the allocation of development funds in a profit-conscious corporation. General Electric also wanted reports from Sperry for the use of its sales representatives selling technically informed customers. Sperry also published the results of quantitative tests in order to persuade the technically informed of their merits.[22] Streetcars were bought by heavily capitalized traction companies in large cities, and his correspondence with the managers and engineers of these companies shows their technical comprehension. No longer was he selling to the managers of hotels or enterprising businessmen in midwestern cities.

Sperry became an advocate of thorough testing. After running an early model of his brake for 70,000 miles on a streetcar, he wrote: "It is only under such rigorous conditions of actual operation that rapid progress can be made in reduction to practice. All machinery or apparatus must pass this ordeal successfully before it can be brought into thoroughly commercial shape."[23] He had not subjected his first arc-light system to such thorough testing before his Chicago company marketed it. He learned—and General

21 EAS to Rice, July 15, 1893, and EAS to W. H. Knight, August 1, 1893 (Sperry letter book, July 13, 1893, to November 11, 1895; SP). In his letter books, additional test reports to General Electric on various devices will be found.

22 Elmer A. Sperry, "Traction and Street Railway Trucks," read before the American Street Railway Association, Milwaukee, October 18, 1893, and abstracted in *The Electrical Engineer*, XVI (1893), 391–92; and Elmer Sperry, "The Electric Brake in Practice," *Transactions of the American Institute of Electrical Engineers*, XI (1894). See tables on pp. 686, 687, 694, 702, and curve on page 696.

23 Sperry, "Electric Brake in Practice," p. 690.

Electric encouraged him to learn—that:

A plan, especially one pertaining to electrical matters, after having been proven mathematically to be feasible, is far from being realized. Many are the practical difficulties to be surmounted before a thoroughly commercial, or anything like a standard apparatus has been produced. Especially is this true in the electrical field. Many subtle influences and energies are at work which well nigh overwhelm the experimenter. In a new field but few precedents are at hand, and these are apt to be extremely unreliable. . . .[24]

iii

The contractual relationship between Sperry and General Electric—and the benefits Sperry received from that relationship—resulted from his excellent inventions in the streetcar field. To this point, however, little has been said about these. The streetcar inventions—the power transmission, the electric brake, and a related controller—echoed his earlier work in electric light and in mining machinery and reiterated his great interest in automatic control. They also demonstrate his capability as a mechanical engineer and as an electrical engineer—virtuosity that came to be a distinguishing characteristic of the man.

His streetcar was notable for its superior hill-climbing ability, which resulted from his employing a mode of power transmission similar to that incorporated in his mine locomotive (also noted for its excellent traction, or drawbar pull). In both cases, he used a single motor, while other streetcars employed a motor for each axle. Because all axles and wheels were a part of a coupled system on the Sperry streetcar, one set of wheels and one motor could not lose traction and slip when the streetcar carried an unbalanced load or was on an incline. Sperry made much of this characteristic, explaining that when the car ascended a hill and the front wheels lost contact all of the power went to the rear wheels, which maintained contact where the power was needed.[25] He also argued that with a single motor he used a larger unit, and large motors were known to be more efficient than small ones (see figure 3.5).

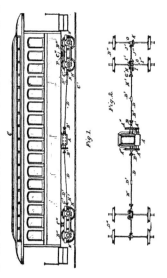

Figures 3.5
Power Gearing for Vehicles
(Patent No. 434,097).

Using a single motor, however, posed a critical power-transmission problem. Between 1890 and 1896, Sperry applied for no fewer than fifteen patents on power-transmitting systems, almost a quarter of the total number of patent applications he filed during these seven years.[26] The patents reveal that his basic problem with the streetcar was to transmit power from a flexibly mounted motor to twisting axles and wheels absorbing heavy shock from pounding on the rails. The nature of the problem and Sperry's ingenious solutions are reminders of his outstanding abilities as a mechanical engineer. (Throughout his career he returned occasionally to designing mechanical transmissions, even if concentrating at the time in another field.)

[24] *Ibid.*, p. 689.
[25] Catalogue of the Sperry Electric Railway Co. (no date).
[26] The patent numbers are (in order of date of application): 434,097; 534,676; 565,935; 587,019; 567,418; 562,498; 563,425; 560,375; 565,936; 562,499; 562,500; 561,354; 557,162; 545,920; and 565,938.

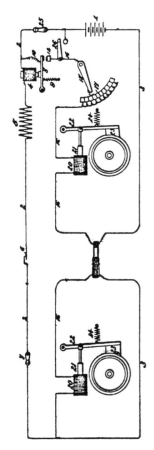

Figure 3.6
Electric Car Brake
(Patent No. 360,060).
Circuit 2-3 normally closed;
circuit 3-16 closed when
braking. (Key: battery, 1;
electromagnet, 4; resistance,
5; lever armature, 8; spring,
9; wire, 10; rheostat, 17;
electromagnet, 20;
sliding core, 21; brake
shoe, 23; spring, 24;
switches, 25, 26.)

Sperry helped the Sperry Electric Railway Company sell the streetcar by traveling widely to explain and demonstrate the equipment, especially its hill-climbing ability. One of the most memorable endeavors involved the Westinghouse Electric Company, the Sperry Electric Railway Company, and the United Columbian Electric Company. The tests were made over a 6000-foot course, on a hill rising with a maximum grade of 12 percent. Each of the competing cars made two trips with a specified load, and on one trip the cars made stops every 1000 feet. Data were compiled by the officials of the transit company. The motor completing the course with the lowest power consumption was to be accepted by the transit company, and the successful manufacturer given a letter stating the facts pertaining to the test. Sperry's car was accepted; and his company widely publicized the test data and published a booklet on the tests.[27] The Sperry streetcar also received publicity from demonstrations of hill-climbing ability at the Chicago World's Fair of 1893.[28]

Sperry built his first electric brake in 1882, included "car brake" in the name of his company founded in 1883, and applied for a patent in 1886. He began concentrated work on the brake again in 1893 and made eight more patent applications from 1893 through 1896.[29] In his 1886 brake (see figure 3.6), designed for a mainline railway, he used an electromagnet in a battery-energized circuit that was normally closed. When this circuit was opened by any one of several control switches, the electromagnet lost its force and released a spring-loaded armature that closed a second circuit normally open. When this second circuit closed, it energized a second electromagnet that pulled into its helix a sliding core; this core moved a lever which applied the brake shoe to the braking surface on the vehicle wheel.

Sperry redesigned his electric brake by 1894. Instead of using an electromagnet to pull the brake shoe against the braking surface of the wheel, he designed an annular electromagnet concentric with the axle of the wheel, but attached to the frame of the vehicle. Mounted on the spokes of the wheel and turning with it was an annular mass of iron, the face of which matched the face of the electromagnet. When the electromagnet was energized, its face and the face of the iron on the wheel were drawn together.

While experimenting with the electromagnetic brake, Sperry observed an "unexpected phenomenon" that led to a further improvement in design.[30] His calculations showed that the retarding effect of the brake upon the vehicle was much more than predicted on the basis of the magnetic attraction and the coefficient of friction of the brake shoe and the braking

[27] *An Important Railway Test.* (Cleveland, Ohio: Sperry Electric Railway Co. [no date]).

[28] *The Sperry Electric Railway Co. at the World's Fair* (Cleveland, Ohio: Sperry Electric Railway Co., 1894).

[29] Filed June 3, 1886 (204,031), "Electric Car-Brake," issued March 29, 1887 (Patent No. 360,060). His electric-brake patents, in the order of his application, were U.S. Patent Nos. 360,060; 534,974; 534,975; 534,977; 577,119; 565,937; 571,409; 574,120; and 571,409. See Appendix for titles.

[30] Sperry, "Electric Brake in Practice," p. 698. Sperry first read the paper at the meeting of the American Institute of Electrical Engineers, September 1894.

surface. Further experiment, "made to ascertain the cause," showed it "to be due to Foucault, or eddy, currents set up in the masses" of the brake.[31] Sperry then constructed the brake to maximize Foucault currents.[32]

During braking, the streetcar motor operated as a generator to supply electricity to the brake. Sperry stressed the automatic function of his brake: when the streetcar was moving fastest—and the need for braking was greatest—the electricity supply to the brake was heaviest. "The intensity of its [the brake] application," he wrote, "should be greatest when the speed is greatest, and decrease as the speed drops off. . . . This point has been fully covered in the electric brake, which is the first time that the varying application has ever been embodied in practice, and especially in such a manner as to perform its important function automatically.[33]

Sperry's talent for designing automatic devices also appeared in his controller for the streetcar. The controller not only provided the motorman control over the speed of the car, it also operated the brake. Sperry applied for his first and basic controller patent in August 1893. In the next four years, he applied for 13 additional controller patents, which, with the brake patents, came under the 1895 agreement with General Electric.[34] Sperry's controller automatically provided a fixed sequence, or program of operations (see figure 3.7).[35] One problem of streetcar-motor operation solved by the controller was the introduction, when starting the streetcar, of electrical resistance into the armature circuit to prevent current flow from burning out the coils. As the motorman used the controller to increase speed, this resistance was gradually and automatically removed. Sperry also put switches in his controller that changed circuit connections so that the motor became a generator for use with the brake. These switches were thrown when the motorman, having started the car and controlled its speed by moving the main control lever, or handle, to the left, returned it to the right. After this, when he moved the contact lever to the left, it varied the intensity of braking. Another move to the extreme righthand position released the brake and gave the motorman speed control again. In this way, Sperry combined controls on a single control lever which, he argued, increased the safety of operation. He later used a single control

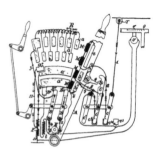

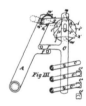

Figures 3.7
Electric Controller (Patent No. 509,776).
Description in the patent of the first of several operating conditions: "With reference to the circuit connections of the controller shown (above) . . . let a trolley be indicated at T, the regular field coils of the motor at F, an auxiliary or teaser coil for the motor at t, the armature of the motor at E with the commutator brushes indicated by r r'. It will be seen . . . that upon closing the circuit the current flows on wire 1 to contact switch c on contact button c⁴, and by wire 2 to the contact b, by way of the blow-out magnets MM; thence by contacts a and resistance R to wire 3 and contact button B⁵, switch arm B, and wire 4 to commutator brush r, armature E, commutator brush r', wire 5, switch arm B', button B⁶, stationary contact a'', movable contact b'', wire 6 to magnets M, wire 7 to button c⁵, switch lever c', wire 8 to moving contact b'; thence stationary contact a', wire 9 to fields; thence wire 10 to ground.

[31] *Ibid.*, p. 698.

[32] Foucault, or eddy, currents were well known to dynamo and motor designers, for these currents, caused by an alternating magnetic flux in the core of the armature, field coils, and pole pieces, overheated the dynamos. To reduce the currents, the designers used laminated sheets in the metal core of the armature and in the pole pieces. The Foucault currents were produced in the moving mass of Sperry's brake from its cutting of the lines of force in the stationary electromagnet. See Sperry's patent, filed January 30, 1894 (498,511), "Electric Brake," issued February 26, 1895 (Patent No. 534,974).

[33] Sperry, "Electric Brake in Practice," p. 689.

[34] The patent illustrated was filed August 21, 1893 (483,611), "Electrical Controller," issued November 28, 1893 (Patent No. 509,776). His other controller patents, in order of application, were U.S. Patent Nos. 515,374; 535,511; 525,394; 525,395; 555,291; 562,501; 566,426; 569,305; 560,658; 562,100; 565,939; 571,410; and 577,081.

[35] The controller was an open-loop device; most of Sperry's automatic devices were closed loop.

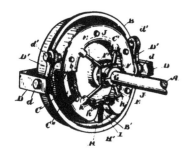

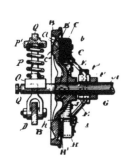

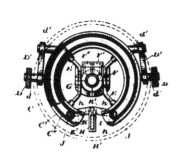

on his electric automobile.

Sperry not only invented, developed, and advised on the manufacture of his streetcar, brake, and controller, but he also helped to sell them. Coffin, president of General Electric, provided him with an engineering assistant and had them "work" the large cities.[36] Coffin also had a contract prepared for brake installations that provided "the services of E. A. Sperry to see that the installation is correct in all its details, instruct your men as to use, etc."[37] Sperry's reaction to these assignments in sales engineering is unknown, but in his free time, supported by his consulting fees, he concentrated on inventing.

After 1895 Sperry spent less time on streetcars. In that year, the Sperry Street Railway Company stopped manufacture, and the Sperry syndicate assigned to General Electric the brake and controller patents.[38] Before this, however, Sperry cars won a reputation for superior traction and were purchased by traction companies in Chicago, Youngstown, Cleveland, Pittsburgh, and other cities. The most publicized Sperry sale was to the Peoples Traction Company of Philadelphia which, early in 1894, bought 125 cars from Sperry and 125 from General Electric.[39]

Under the 1895 agreement with General Electric, Sperry continued to act as a consultant for several years while the company brought the brake and the controller into production. Early in 1896, General Electric completed 100 controllers and soon started on the manufacture of 200 more embodying modifications in the basic Sperry design. By March 10, 1897, the company shipped a total of 315 brake outfits.[40] The record shows a range of customer reaction, extending from enthusiastic acceptance to rejection. The most serious complaint was that the brake, after stopping the car, held only a few seconds on a steep grade (the motorman then had to resort to a hand-brake). Because the electric brake could not be used for all stops, the railway company had the considerable expense of duplicate equipment. Despite royalties from the brake in the first two or three years,[41] the members of the Sperry syndicate decided to dissolve in 1905 because the association was no longer profitable.[42] The success of Westinghouse in introducing the air brake on streetcars was probably

[36] B. F. Miles to EAS, January 20, 1896 (SP).
[37] Copy of a contract between General Electric and Cleveland Electric Railway Co., Cleveland, Ohio (SP).
[38] Agreement between Sperry Syndicate and General Electric, May 28, 1895 (SP).
[39] *Electrical World*, XXIII (1894), p. 167. In January 1895, during a very heavy snow storm in Philadelphia, the Sperry representative there reported that the line manager asked if there was a Sperry "in the house as that was the thing to pull the plow, much better than those things [alluding to the General Electric]. Unfortunately there was not a Sperry in the house so they took the other. The majority of the motormen at Germantown ask for a Sperry in this kind of weather." G. Patterson to EAS, January 29, 1895 (SP).
[40] Wm. J. Clark to James Parmelee, March 10, 1897 (SP).
[41] General Electric reported royalties of $3,186.45 during the three months from August 1 to November 1, 1897. Parmelee to EAS December 11, 1897 (SP).
[42] Instrument of settlement dated May 29, 1905 and signed by Sperry, Parmelee, Herrick, Miles, and Hayes.

a major reason for the declining sales of the Sperry brake.

As his consulting work tapered off, Elmer Sperry again enjoyed the freedom of action and endured the insecurity usually a part of the life of an independent inventor. Undoubtedly he could have joined the engineering staff of General Electric in a well-paid and responsible position, for the company continued to value him highly as a consultant. Other inventor-engineers of lesser achievement and promise had entered the engineering departments of the large electrical manufacturers. Sperry, however, preferred the role of independent inventor-entrepreneur. He was willing to do the promotion necessary to raise funds to finance inventions, and he wanted to invest his own income—insofar as his family responsibilities allowed—into new invention, development, and innovation.

iv

For years inventors had seen the obvious possibility of building a locomotive to run on roads rather than on rails. Steam road vehicles had been built by the Frenchman Nicholas Joseph Cugnot at the end of the eighteenth century and by English inventors early in the nineteenth. Experiments were so numerous that the British Parliament enacted the so-called Red-Flag Law, which required that a footman walk in front of a steam road vehicle to warn pedestrians and horsedrawn carriages. Because the roads were poor and the railroads were eminently successful in providing transportation, however, the road vehicle did not flourish.

Interest in the horseless carriage took on new life with the introduction of the practical internal-combustion engine. Nicolaus A. Otto, Etienne Lenoir, and other European pioneers improved the gas engine until it was a marketable product in the 1870's; a group of ingenious Germans—Gottlieb Daimler, Karl Benz, and Wilhelm Maybach—then adapted the stationary internal-combustion engine to transportable liquid petroleum fuel and built an appropriate road vehicle chassis and power transmission. By 1895, when Sperry decided to turn to the horseless carriage, the Germans and the French were already making automobiles; Sperry was interested by their activities.

Such an unlikely event as an automobile race in France seems to have brought Sperry—and many other American inventors—into the automobile field. Whereas there had been one inventor before reports of the race reached America, there were hundreds afterwards. *Horseless Age* estimated that, in the six months after the Paris race, there were 300 vehicles under construction in the United States.[43] Sperry obtained the report that described the June 1895 race, which ran from Paris to Bordeaux and included gasoline, electric, and steam propelled carriages.[44] The report predicted that "the time seems approaching when automatic

[43] William Greenleaf, *Monopoly on Wheels: Henry Ford and the Selden Automobile Patent* (Detroit: Wayne State University Press, 1961), pp. 35-36.
[44] C. W. Chancellor in *Consular Reports*, No. 180 (September 1895), pp. 25-27. Sperry made a note of this in his notebook (SP).

road carriages, propelled by steam, electricity, or petroleum, will come into general use and take away from the patient horse the worst part of his daily toil." After reading the report, Sperry wrote to the Thomson-Houston electrical company's Paris representative, asking for technical information from manufacturers of road carriages in France.[45]

Like many other Americans, Sperry decided to build an automobile and to enter it in the automobile race between Chicago and Milwaukee, sponsored by the *Chicago Times-Herald* and scheduled for November 1895.[46] Only eleven cars, however, were ready for the race—Sperry failed to make the deadline—and of these only three reached the finish line.[47] But Sperry, unlike many of the others, did not give up on automobiles. As an experienced inventor, he sensed that the time had arrived for the inventor-entrepreneur seriously intending to raise money and engage in patient development. He knew that the idea of a horseless carriage was widespread, and he also knew that the demands of development, refinement, and innovation would be the crucial test. Because other pioneers had demonstrated the promise and attracted considerable attention, Sperry believed that capitalists could be interested in development and manufacture. He also believed that with his experience and character he could solve the seemingly small problems of mechanical and electrical refinement that might frustrate the enthusiast.

For finances, Sperry turned to his Cleveland syndicate, the members of which anticipated the election of William McKinley in 1896 and consequently looked forward to profitable industrial investments. Furthermore, Washington Lawrence, Myron T. Herrick, James Parmalee, and Webb C. Hayes were all friends of powerful Marcus Alonso ("Mark") Hanna, adviser to McKinley. Hanna, who financed and managed McKinley's campaign,[48] favored mutually beneficial cooperation between politicians and industrial leaders. These circumstances were not lost on Elmer Sperry, for he predicted that the election of McKinley would favor his and his "Cleveland friends'" entry into the automobile field.[49]

Sperry's approach to the invention-development-innovation process was tried and proved. He wanted to obtain automobile patents that would encourage financiers to risk development funds in expectation of returns from a protected position on the market. The automobile field was still so young that inventions solving technological problems could be dramatic enough to insure a substantial market advantage in dealing with consumers. Sperry intended to patent his own improvements in the construction of the vehicle and especially the transmission of power, but he had also learned the advantage of supplementing one's own patents with those of

45 A. S. Garfield to EAS, September 5, 1895 (SP).
46 *Chicago Times-Herald*, July 16, 1895; and Sperry letters between EAS and the newspaper, July 27, August 2, and August 29, 1895 (SP).
47 EAS to A. S. Garfield, November 7, 1895 (Sperry letter book, September 21, 1895 to September 19, 1896; SP); Greenleaf, *Monopoly*, p. 36.
48 EAS to M. A. Hanna, December 31, 1903 (SP).
49 EAS to George Selden, October 28, 1896 (SP).

other inventors with less entrepreneurial ability and resources who would, therefore, consider assigning or licensing their patents. Before turning to others, however, Sperry, in 1896, applied for an auto patent on air-cooling the motor, a system of control for the vehicle, a special power-transmission gearing, a clutch, brakes, and other structural matters. These were critical matters going beyond the basic engine design and body form that the public associated with the invention of the automobile.

It is not clear how Elmer Sperry first learned of George Baldwin Selden and his remarkable application for a patent on an internal-combustion vehicle. Sperry realized the advantage of basic patent protection and was probably surprised to learn that the possibility existed of gaining rights to Selden's basic patent—one that promised a monopoly. In 1896 Selden and Sperry were corresponding about Selden's patent. After 1900 the patent would become famous—or infamous—but prior to that date, "only an alert and well-informed inventor or engineer could have been conscious of its existence."[50] A patent lawyer of Rochester, New York, who tried his hand at invention (a sideline not uncommon among patent lawyers). Selden began designing an internal-combustion engine and horseless carriage around 1873. In May 1879, he made a broad application covering a vehicle propelled by a liquid hydrocarbon engine. In an apparent effort to delay issuance of his patent until the automobile was marketable and to extend the patent's effective life, Selden, by repeatedly filing amendments to his application, kept his patent pending for sixteen years.[51] Selden has been called the Prince of Procrastinators.

Selden was keeping a number of irons in the fire in 1896, including a mechanical voting machine, and did not seem to consider his motor vehicle patent of overwhelming importance. He and Sperry first corresponded about a rotary engine of Selden's in 1895, and Sperry expressed interest in helping Selden protect it abroad. In June 1896, after referring to Selden's interest in his rotary engine, Sperry then asked: "By the way, have you formed any permanent relations yet on the motor-carriage matter as a whole?"[52] If not, then Sperry thought it "barely possible" he could get some of "our people" to look into the matter.

After this initial move, the correspondence about the "motor carriage" took on life. By June, Sperry was well along in perfecting his plans for a strong horseless-vehicle syndicate with factories and facilities for quantity production, beginning at 50 vehicles a day.[53] Probably encouraged by his Cleveland "people," Sperry brought General Electric into the exploratory negotiations.[54] Selden wanted some settlement and attempted to spur Sperry by predicting that the combination of a patent on Selden's new "rotary engine" and his basic patent of 1895 would control the " 'horseless

[50] Greenleaf, *Monopoly*, p. 55.
[51] John Rae does not consider Selden's tactics Machiavellian; see his *The American Automobile: A Brief History* (Chicago: University of Chicago Press, 1965), p. 34.
[52] EAS to Selden, June 13, 1896 (SP).
[53] EAS to Selden, June 18, 1898 (SP).
[54] C. A. Coffin, president of General Electric, to EAS, August 1, 1896 (SP).

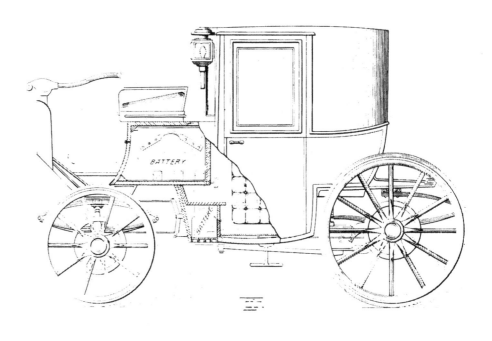

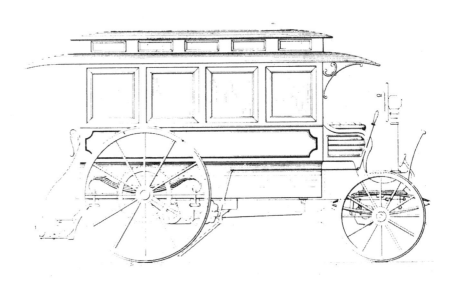

Figures 3.9 Sperry designs for his automobile.

carriage business' of the United States . . . for the next 17 years."[55] In Sperry's judgment, expressed confidentially to E. W. Rice, Jr., the General Electric vice-president, the broad Selden patent on the horseless carriage did not have much value in itself, but with backing in patents on special features, Sperry believed that "it could be made to add quite materially to an exclusive protection of the art."[56] Sperry had in mind patents on inventions, probably his own, such as steering, speed control, and safety features. He thought the combination of patents would "scare off" the smaller people proposing to manufacture horseless carriages and bring the larger ones to some profitable compromise. Sperry believed he could secure, for a small royalty, an exclusive license to work under the broad Selden patent.[57]

Sperry wrote Selden that Cleveland friends interested in the horseless-carriage syndicate would act after the election of McKinley, but the Rochester patent lawyer became impatient when no immediate offer of "something handsome by way of an option" was forthcoming.[58] Sperry was moving cautiously; he felt it important to involve General Electric and to acquire other patents supporting Selden's and placing it in a perspective that would bring Selden to lower the price.[59] Selden began to doubt that Sperry would be able to carry through the "scheme" to form a syndicate because he had not, as Selden suggested, built a vehicle embodying the patent. He knew, as did Sperry, that nontechnical capitalists wanted to see the "thing." Discouraged by the failure of another of his inventions and feeling the strain of having carried the "burden" of the vehicle patent "so long," Selden informed Sperry of an offer for his patent from another party.[60] "While personally having the most friendly feeling toward yourself," Selden wrote to Sperry, "I do not understand that I am under any definite obligations . . . which may not be discharged on notice." In the last letter to Sperry, written on February 1, 1897, a tired and harassed Selden asked for something even indefinitely deter-minate" from Sperry ("What can you say on the subject, or will you give up the struggle?").[61]

Sperry was not one to give up a struggle, but in this case, considering the reward not worth the effort, he dropped the negotiations. In 1899, Selden sold the rights to his patent on a royalty basis to the Electric Vehicle Company, later the Columbia Motor Car Co. The broad coverage of the patent was then exploited, and by 1906 Selden was receiving royalties from most automobile manufacturers in the United States. This situation obtained until the Ford Motor Company successfully challenged the scope

[55] Selden to EAS, September 17, 1896 (SP).
[56] EAS to Rice, January 29, 1897 (SP).
[57] EAS to G. R. Blodgett, January 29, 1897 (SP).
[58] Selden to EAS, February 1, 1897 (SP).
[59] A letter from C. A. Coffin to EAS, December 5, 1896, suggests that this was Sperry's strategy (SP).
[60] Selden to EAS, December 16, 1896 (SP).
[61] Selden to EAS, February 1, 1897 (SP).

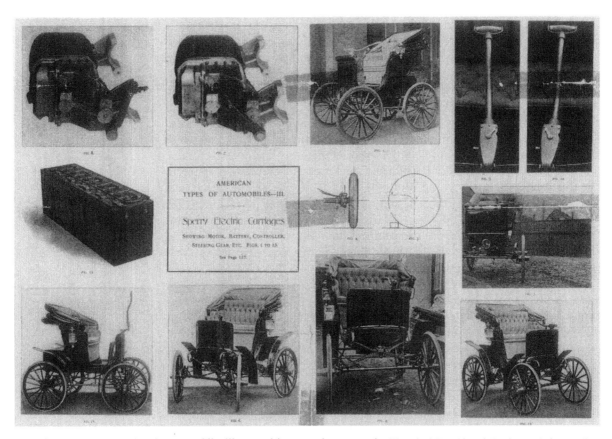

Figure 3.10 Sperry electric automobile, illustrated in a supplement to the Electrical World and Engineer, *July 22, 1899.*

of the patent in the courts in 1911 and the other companies stopped royalty payment. The patent expired at about the same time.

The negotiations with Selden may have been dropped because Sperry decided that the electric vehicle had a much brighter future than the internal-combustion vehicle. Generally, the electric car held public favor from 1895 to 1900, when the market was for the wealthy who objected to the vibration and fumes of the gasoline vehicle.[62] Sperry may also have been swayed by the loss of the only gasoline car he built. He had actually got the handsome machine on the Cleveland streets before a fire in his workshop, in February or March 1896, destroyed it. Sperry described the vehicle as "a thing of beauty with two cylinders (Otto cycle), steered by a tiller, upholstered in English Wilton and trimmed with aluminum." He thought it not only ahead of its field mechanically, but so beautiful "it would sell on looks alone." After the fire, Sperry looked at the twisted steel and said to himself, as he recalled to one of his sons, "that's what you get for deserting electricity."[63]

[62] Greenleaf, *Monopoly*, p. 56.
[63] The quotations are from a compilation of Sperry correspondence on this period prepared under the direction of Elmer Sperry, Jr. (SP).

Sperry also believed the state of technology favored the electric automobile. Use of the electric motor in streetcars had brought such improvement, he judged, that the motor could be used and maintained in a privately operated automobile. He also knew that the storage battery, while not perfect, had been improved remarkably. Furthermore, the spread of electric light and power generating stations throughout the nation meant that a battery charge was available in the smallest hamlet.[64] Sperry also thought that his experience with streetcars would help him introduce the refinements needed by the electric automobile.

Backed by the Cleveland syndicate, he rented machine-shop space in the Brush plant in 1897, and with the help of a few skilled mechanics and Franklin Schneider, formerly a superintendent of manufacturing for the Brush company, he built six electrics.[65] Sperry later remembered that Schneider helped greatly in overcoming the "multitudinous difficulties that we were working under."[66] Years later, Sperry nostalgically recalled the difficult period as "good old times."

In 1898 the six electrics were complete, and five patent applications were made. Sperry's design incorporated a number of advanced constructional features and a noteworthy control system. His emphasis on controls was well placed, for these were a major cause of failure in the early electrics. He placed the essentials of control in a single handle: the motorist steered by lateral movement of the lever, increased speed by depressing the lever, and applied the four-wheel brakes by pulling up the handle, as one would draw up on the reins of a horse.[67] Sperry described the control system as a means for interlocking various operations connected with the control and braking to provide what may be termed an 'automatic' speed regulation. Sperry's steering mechanism was praiseworthy because it absorbed sharp road irregularities and was self-centering—refinements suitable for the roads of the day.[68]

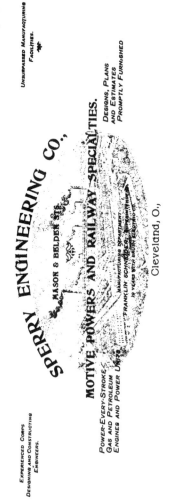

[64] "Sperry Electric Carriages; American Types of Automobiles," *Electric World and Engineer*, (July 22, 1899).

[65] A. Dana to W. B. Perkins, January 6, 1898 (SP). Six electric automobiles, however, did not absorb all of Sperry's energy. When the Brush factory decided to rent more space, he used it to establish the Sperry Engineering Company. ("Contract for renting space, etc." between EAS and the Brush Electric Company, and EAS to Brush Co., September 16, 1896; SP.) The company, founded in 1896, advertised gas and petroleum engines, an experienced corps of designing and construction engineers, and unsurpassed manufacturing facilities. Sperry was the manager and Franklin Schneider, who had had many years of experience working with the Brush company, was superintendent of the manufacturing department. For several years, Sperry and Schneider apparently had considerable success in selling an air compressor powered by a petroleum engine, but as was usually the case with Sperry, his heart was in the development of new products. An advertisement for the compressor ran weekly from November 1896, for six months in the *American Machinist*.

[66] EAS to F. Schneider, President of Van Dorn Electric Tool Co., March 21, 1917 (SP).

[67] Four-wheel brakes did not become common on American autos until Vincent Bendix manufactured them twenty-five years later.

[68] William W. Hudson, "The Modern Pleasure Electric Vehicle: Some Interesting Problems of Design and Construction and their Solution," *Scientific American*,

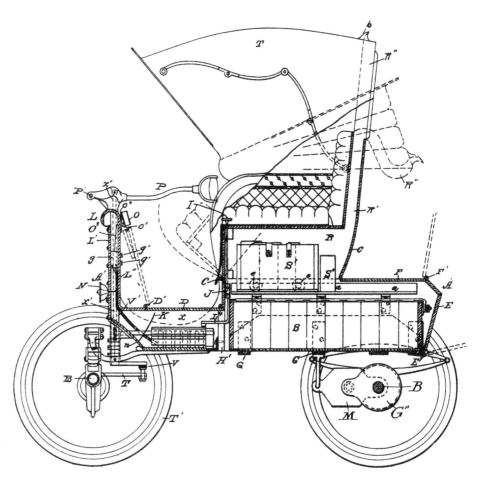

Figure 3.11
Drawing from
Sperry automobile patent
(Patent No. 640,968).

Sperry took John D. Rockefeller for what may have been his first automobile ride in one of his electrics. Rockefeller and the Sperrys attended the Euclid Avenue Baptist Church in Cleveland, and, while visiting Cleveland, Zula's father, Edward Goodman, introduced them to the famous Rockefeller, whom Goodman knew as a trustee of the University of Chicago. The next day Sperry and Goodman called on Rockefeller to take him for a ride through Forest Hills. Although Sperry tried to persuade the millionaire to take the controls, he refused, preferring that Sperry have the responsibility. After the excursion, the two men discussed Sperry and the future of the electric, and Rockefeller advised him to cash in on it and go on to invent other things.[69]

CIV (1911), pp. 31-32. On the design of the Sperry electric automobile, see his "Electric Automobiles," *American Institute of Electrical Engineers: Transactions,* XVI (1899), pp. 509-27 (a paper read at the 1899 annual meeting in Boston); for a list of his patents see Appendix.

[69] This account based on a letter from EAS to John D. Rockefeller, November 23, 1918 (SP) and some supplementary details in the Leonard manuscript. Leonard also tells of a futile effort of Sperry's to interest President McKinley (when the president was visiting Cleveland) in taking a drive.

After building the electrics, Sperry interested the Cleveland Machine Screw Company, a large manufacturer of automatic machine tools, ball bearings, and steel stampings, in tooling for quantity production in 1898. Sperry assigned his patents on the electric to the company, and he received stock in addition to becoming its electrical engineer. In the same year, control of the company passed to French capital; as a result, Sperry was sent to Paris (his first trip to Europe) to demonstrate the electric and to observe French automobile technology. The Cleveland Machine Screw Company's lawyer and all six electrics accompanied him.[70] Sperry later said that he was the first American to drive an American car on the streets of Paris; undoubtedly, he was the first American driving an American car who was arrested for reckless driving in Paris; he upset a vegetable cart. The cars were beautiful and ran well when Sperry was on hand, but they "did not turn out to be very seaworthy in the absence of professional leadership."[71] Overall, the trip and the demonstration were successful, for Sperry came back with an order for one hundred cars and the sale of the Sperry patent rights in France.[72]

The Sperry autombile made a very good impression in Paris because of the performance of its storage battery. Sperry had made substantial improvements in the mechanical design, and he later decided that his battery probably gave his electric its chief advantage over others. The Paris electrics of other inventors and manufacturers were giving about thirty miles per charge, while Sperry claimed his gave 87 miles and had a decided weight advantage. The excellence of his battery and the attention it aroused would soon bring him to concentrate his energies on it, but in the meantime plans for mass producing the automobile were foremost.

The Cleveland Machine Screw Company announced, in May 1899, that the first of the "Sperry Electric Carriages" was nearly complete and that jigs and tools to turn out Sperrys in lots of 100 were being readied. The subsequent history of the Sperry became somewhat complicated. The car manufactured by the Cleveland Machine Screw Company was exhibited in New York in 1900, and the company issued a catalogue of "Cleveland Automobiles—the Sperry System"; but negotiations were also under way to sell the patents on the Sperry to the American Bicycle Company, a large company with a Toledo factory where steam and petroleum automobiles were made. The American Bicycle Company in 1900 acquired Sperry's patents relating to storage batteries and to the constructional features for the electrically propelled vehicles.[73] The vehicle patents probably then became a part of the general patent structure American Bicycle used to protect its automobiles, including the line of electrics called "Waverleys." For this reason, the contribution of Sperry inventiveness to the development of the automobile after 1900 is difficult to isolate and evaluate.

[70] A compilation of Sperry letters done by Elmer Sperry, Jr., is the basis for this account (SP).

[71] H. H. Johnson to Elmer Sperry, Jr., 1935, quoted in compilation noted above.

[72] EAS to D. Beecroft, August 12, 1924 (SP).

[73] See Appendix.

The case with the storage battery patents, however, is different. Four months after the sale of the patents, the National Battery Company was formed—probably by the American Bicycle Company—to manufacture the Sperry battery. Between May and September 1901, Sperry was the manager of the National Battery Company and was "exceedingly busy" establishing the works and then presiding over a great increase in capacity. The company sold both a National battery and a Sperry battery, the latter receiving considerable publicity in the trade journals. Although some of these journals attributed the success of the Sperry to its chemicals, Sperry had, in fact, improved upon the construction of the battery. Dr. A. W. Burwell, a consulting chemist in Cleveland, who helped Sperry develop his battery, believed the diaphragms, or separators, the essence of the Sperry battery patents. These sheets of fiber composition placed between the plates carrying the peroxide kept the plates wet with acid and at the same time kept them from losing their peroxide, warping, or coming into contact with each other. A. S. Atwater, who later worked on the Sperry battery for the National Battery Company, wrote in 1908 that the Sperry grid and the Sperry mix method were superior to materials and processes used by other battery manufacturers. The 1902 catalogue of the company also stressed the value of the Sperry grid.[74]

When the National Battery Company moved to greatly enlarged works in Buffalo, New York, Sperry remained in Cleveland, where he was still retained as the electrical engineer for the company. The company not only sold batteries to the American Bicycle Company for its Waverley Electrics, but also made batteries for streetcars, and electric light and power stations. Sperry may have decided to remain in Cleveland despite the success of the company because he was reluctant to move his family, or because he felt that battery development had reached a plateau and his inventive genius could best be applied to industries whose development curves were more sharply upward.

In order to work in the battery field, Sperry had taught himself chemistry; both Zula and his daughter, Helen, often read to him from textbooks and technical journals. Although they found the reading impossibly dull, they read to him because Sperry believed he absorbed more by ear. His *forte*, however, remained electricity and mechanics. Soon, he found a dynamic field of technology, electrochemistry, in which all three of these facets of his expertise could be constructively applied.

v

Sperry's success with the battery encouraged him in 1901 to begin working in electrochemistry. His sense of timing was acute, for electrochemistry in America was developing rapidly and was not yet so well organized that an independent engineer-entrepreneur found access barred. The opening of the giant hydroelectric station at Niagara Falls in 1895 had accelerated the development of a field in which cheap energy was a

[74] Compilation of letters by Elmer Sperry, Jr. (SP).

prime requirement. Small enterprises trying to develop electrochemical processes gravitated to the Niagara-Buffalo area to use the power.

Sperry saw the development of the industry and foresaw the contribution he might make. He understood, as did others, that cheap electricity provided a new form of energy to the industrial chemist who had previously relied on heat for his reactions. Scientists, such as Michael Faraday in England, had shown how compounds could be broken down by electricity, but it was left to the Americans to realize the large-scale industrial implications of the science. Names such as H. Y. Castner, E. A. LeSueur, H. Burgess, H. H. Dow, C. P. Townsend, A. H. Hooker, and C. M. Hall, all testify to this.[75] In the first decade of the twentieth century, when the United States continued to import many chemicals from Europe, it produced almost all of its electrochemical products.[76]

Sperry saw as his contribution to this burgeoning industry his experience and excellence as an electrical and mechanical engineer. Most of those working in the field were chemists, and Sperry believed that the chemistry of their processes needed to be reinforced by refined electrical and mechanical design. His thinking in this respect was in line with the trend of the times, for after 1900 a new branch of engineering, chemical engineering, was evolving in response to problems whose solution required the combined resources of the chemist and the engineer. The rise of the new profession was indicated by the establishment of professional societies and by new courses in the engineering schools.[77]

In the first decade of the century, when Sperry concentrated upon electrochemistry, the electrolysis of salt was one of its most promising problems.[78] By brine electrolysis, the pioneers hoped to produce sodium hydroxide, or caustic soda, more economically than by existing processes. Because sodium hydroxide was an alkali, there was a large market for it in the manufacture of soap, wood alcohol, and other products. Electrolysis also yielded chlorine, which was used for bleaching. The first American electrolysis plant had begun operation in Rumford, Maine, in 1892. Not long afterwards a plant using Hamilton Y. Castner's process and another plant built by Herbert H. Dow were in operation, but these did not use cheap electricity and encountered early operational problems. Other pioneers were also balked by practical difficulties; so despite Sperry's tardy entry, there was a need for practical development.

Sperry learned, in 1901, of the electrolytic process of Clinton P. Townsend, a patent examiner in the electrochemistry section of the Patent

[75] Haynes, *American Chemical Industry*, pp. 269, 387. Haynes and other historians of the American chemical industry have noted the dependence of the complex, interwoven growth of electrochemistry on the availability of Niagara Falls power.
[76] Elmer Sperry, "Address to Chemist Club of New York," c. 1905 (SP).
[77] Sperry was a charter member of the American Electrochemical Society, founded April 3, 1902. He read papers before it in 1903, 1906, and 1908. *Transactions of the American Electrochemical Society*, I (1902), p. 27. See Appendix for a list of Sperry papers.
[78] William H. Walker, "How Electricity is Aiding the Chemist," *Scientific American*, CV (1911), pp. 243ff.

Office, who had probably come to know Sperry through his battery patents. His process was fundamentally simple, involving the passage of electric current through brine to liberate sodium hydroxide and hydrogen at one electrode and chlorine at the other electrode of an electrolytic cell. But he and Sperry saw in the Townsend process potential refinements that promised commercial success. Townsend, however, needed money so that he could give full time to the development of the project, and he also needed the assistance of an engineer to help design the apparatus for the chemical process. Because Sperry had had considerable experience in raising development funds and as an electrical and mechanical engineer, he and Townsend decided to pool their talents.

Townsend was willing to assign a share of his rights to Sperry and to a financial backer in return for funds and future profits,[79] so Sperry worked out an agreement and found a capitalist. The agreement specified an annual salary of $2800 for Townsend for three years, with the right to renew, and provided a small laboratory for him in a rented loft in Washington, D.C. Townsend was to work full-time in the laboratory on the caustic soda process and on developing an electrochemical process for producing white lead. The financier located by Sperry, Alden S. Swan, once had been the treasurer of the Brooklyn Bridge, but now he was especially interested in developing the white lead process for the paint makers. Swan apparently contributed a substantial sum, for he was promised 50 percent interest in both the white lead and the Townsend processes.[80] Townsend, Sperry, and Swan intended to sell the processes, after they were developed, to manufacturers.

At the start of the cooperative venture, the white lead process received most attention, but when seemingly insurmountable obstacles were encountered, attention was shifted to the Townsend process. In fact, the white lead process was not perfected until more than a decade later by Sperry, with the help of a young chemist, Preston Bassett; it was then sold to the Anaconda Company.[81] Progress with the Townsend process justified Townsend's filing the first patent application in April 1902.[82] Sperry then sent propositions to several companies that he thought might be interested in manufacturing caustic soda. He approached the Solvay Process Company, a producer of caustic; the Pennsylvania Salt Company, which was planning an electrolytic plant for caustic at Wyandotte, Michigan; and the Grasselli Chemical Company. For the rights to the Townsend process, Sperry asked $125,000 in cash at the outset and $75,000 annually for five years. This royalty would cover a plant producing ten tons of

[79] Townsend to EAS, August 17, 1901; an agreement between EAS and Townsend, September 2, 1901 (SP).
[80] Agreement between EAS and Alden S. Swan, March 1, 1902 (SP).
[81] See below, p. 208.
[82] An original application was filed April 12, 1902 (102,582) and then divided. One of the new applications was filed May 10, 1904 (207,227). "Electrolytic Apparatus," and issued January 3, 1905 (Patent No. 779,383). Also out of the division came the application filed May 11, 1904 (207,490); "Electrolytic Process," issued January 3, 1905 (Patent No. 779,384).

caustic daily; if the capacity were increased, payments would increase. In addition, the proposal asked that both Sperry and Townsend be employed as consulting engineers for five years with annual salaries of $4200 each. The price was thought too high for a process not yet proved in production.[83] After failing in their early efforts to sell the rights, Townsend and Sperry further refined the process.[84] Sperry concentrated on improving the design of the apparatus to increase efficiency and lower costs.

The essence of their electrolytic process was the cell. Townsend had initially conceived of a cell with two compartments separated by a diaphragm which allowed slow percolation of the brine solution toward the cathode. The products in the cathode compartment were hydrogen and sodium hydroxide. The caustic formed under oil and settled to the base of the compartment, from which it was tapped at intervals. Chlorine gas formed at the anode, in the other compartment; the oil and the diaphragm prevented the chlorine's recombining with the caustic and hydrogen. In 1903, during development and experimentation in Washington, Townsend and Sperry relocated the cathodes in side compartments and improved upon the design and the material of the diaphragm (see figure 3.12). They also looked for noncontaminating materials for the cell walls. Because the expense of this work was apparently too great for Swan, Sperry and William F. Dutton, another capitalist, assumed the burden, which reduced proportionately Swan's percentage of eventual profits.[85]

By the turn of the year, 1903–04, circumstances were more favorable. Sperry invited interested persons to Washington to inspect the improved caustic cell and found a financier and manufacturer bold and resourceful enough to provide needed support. Elon Huntington Hooker, just past thirty, charming, tall, handsome, and self-confident,[86] had attended the University of Rochester and the College of Engineering at Cornell University where he received the Ph.D. in 1896. On fellowships from Cornell, he studied in France and Switzerland, and as deputy superintendent of public works for New York State, he established a lasting friendship with Governor Theodore Roosevelt and other persons of influence and wealth. Sperry and Townsend needed both.

Hooker's rich resources and imagination led him to found, in 1903, the Development and Funding Company of 40 Wall Street, New York City. The purpose was to take up and develop meritorious enterprises.[87] Although the idea of the company was not original—the Elmer A. Sperry Company founded in 1888 was similar—Hooker's development company

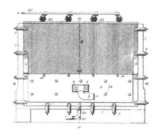

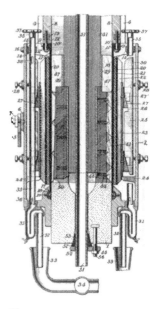

Figures 3.12
Two figures from Townsend and Sperry patent, Electrolytic Cell (Patent No. 1,097,826).

[83] Proposition from EAS to the Solvay Process Co., April 28, 1902; Solvay to EAS, May 29, 1902 (SP).

[84] Sperry filed a patent jointly with Townsend on December 5, 1903 (183,967), "Electrolytic Cell," issued May 26, 1914 (Patent No. 1,097,826).

[85] Agreement between EAS and Swan, November 24, 1903 (SP).

[86] Hooker and his companies are described in Robert E. Thomas, *Salt & Water, Power & People: A Short History of Hooker Electrochemical Company* (Niagara Falls, N.Y., 1955).

[87] R. G. Dun & Co. report on Development & Funding Co., December 27, 1904 (SP).

has been singled out as a forerunner of a type of investment-management group that became popular following World War II.[88] Before deciding to fund and develop the Townsend and Sperry process, the new company had examined over two hundred and fifty projects.[89]

Hooker sent his brother, Albert Huntington Hooker, a chemist, to Washington to evaluate the Townsend-Sperry process; on the basis of his favorable report, Elon obtained an option in February 1904.[90] The terms specified by Sperry called for the cash payment of $20,000 for his time and Townsend's and an additional sum of $30,000 to cover prior development expenses. Thus the development of the process had cost $50,000 and had not yet gone beyond the laboratory stage. In addition, Sperry and Townsend asked for a one-fifth interest in the company to be formed to use the electrolytic process.[91]

After reaching a preliminary agreement of the terms of the sale, Hooker employed (as an outside consultant) Leo H. Baekeland, the chemist later famous for Bakelite and Velox, to supervise the construction and operation of a pilot plant in a former Brooklyn boiler house of the Edison Electric Illuminating Company.[92] By December 1904, Hooker was convinced of the practicality of the process, so Sperry and Townsend signed over the patents to the Development and Funding Company for 350 shares of stock in the company and other considerations not recorded.[93] Hooker then disposed of enough Development and Funding Company stock to erect a plant costing approximately $236,000.[94]

Sperry and Townsend continued to advise Hooker, but the transition to full-scale production is a part of the history of the Development and Funding Company and the Hooker Electrochemical Company of Niagara Falls. The latter company, founded in 1909 to use the Townsend-Sperry electrolytic process, was an offspring of Development and Funding. After several years of operations difficulties, the enterprise proved very successful. Hooker was destined to become a major American chemical manufacturer. During World War I, like many other American chemical plants, his began producing new products, although chlorine and caustic soda remained the basic ones. A half-century later the company, still using a refined Townsend cell, produced about 100 chemicals and employed almost 7000 persons.

Sperry had moved on to a new problem, detinning, by the time the Townsend-Sperry process had been successfully introduced on the market. Before turning to Sperry's effort to develop a detinning process, one of

88 Haynes, *American Chemical Industry,* I, 278n.
89 Thomas, *Salt & Water,* p. 7.
90 *Ibid.,* p. 14.
91 EAS to E. H. Hooker, February 6, 1904 (SP).
92 The author was privileged to read a biography of Baekeland by Carl Kaufmann, completed in 1968 (manuscript).
93 Agreement between Development & Funding Co., Clinton P. Townsend, and EAS, February 16, 1905 (SP). This refers to the earlier agreement of December 31, 1904.
94 R. G. Dun report on Development & Funding Co., December 1904.

the most complex episodes in his career, note must be taken of a brief interim with palm oil and paint pigments—an interlude that nonetheless led into detinning. Sperry's interest in palm oil and pigments began with one of his friends, William F. Dutton of the American Tin Plate Company. Dutton told Sperry, in 1900, that the company was interested in reclaiming the millions of pounds of waste palm oil used in tin-plate melting pots.[95] The American Tin Plate Company, formed in 1899, controlled about 95 percent of all the tin plate production in the United States, and this concentration and scale bred an interest in economies that had seemed insignificant to small-scale enterprise. The tin-plate manufacturer used palm oil as a flux for the iron to be plated and as a protective cover for the coating. The oil, which covered the bath of tin into which the iron was immersed, lost, in time, some of its volatile fractions and was fouled by tin particles and other impurities. Dutton and Sperry thought the lost substances could be replaced and the impurities removed.

Sperry turned for advice to Dr. F. J. Burwell, an experienced chemist at the National Carbon Company.[96] Sperry agreed to furnish him with the materials he needed to develop the process in his small private laboratory; both agreed to turn over all patent rights to a prospective company that would use the process, each of them to receive ten percent of the stock. The remaining 80 percent of the stock would go to Dutton and others financing the company. That the company was never capitalized, however, suggests that the work of Burwell and Sperry did not culminate in a practical process.

Also in cooperation with Dutton, and in partnership with another chemist, Alexander S. Ramage, Sperry attempted to promote another project at the same time. Ramage thought it possible to obtain ferrous paint pigments by electrolysis from ferrous sulfate, another waste product of the tin-plate industry.[97] Again a company was formed, and both Ramage and Sperry sought patents on the process to provide assets for the company.[98] The outcome of this endeavor is not clear from the surviving records, but there was little or no profit.[99] In retrospect, the episode assumes significance primarily because Sperry came into closer contact with Dutton and became more familiar with the tin-plate industry; both figured prominently in the detinning process, a major project upon which Sperry embarked in 1904.

By then, Dutton was a director of the American Can Company, which was formed in 1901. After learning that the company had a detinning prob-

[95] EAS to F. J. Burwell, September 29, 1900 (SP).
[96] Burwell later discovered a method for processing Texas crude into higher, more-valuable, fractions of petroleum and received substantial payment and a position from the Standard Oil Company in return for the rights to his process.
[97] W. F. Dutton to EAS, June 13, 1901 (SP).
[98] Alexander Ramage application filed September 10, 1900 (29,477), issued January 14, 1902 (Patent No. 691,324); Sperry application filed March 14, 1902 (98,783); no patent was issued.
[99] Sperry and Ramage also worked on a process for making lead paint pigments and on one for substituting fish oil for linseed oil.

lem, he approached Sperry and Townsend, believing that they might be able to find a solution.[100] The company had a considerable amount of scrap from cutting round can tops from square sheets and had been using an electrolytic process for detinning scrap tin plate which left tin powder and iron. The iron recovered by electrolysis was used again, but the tin powder was marred by the presence of impurities, especially minute quantities of iron. The problem that Dutton initially passed on to Sperry and Townsend was to remove the iron from the tin powder.

Knowing that chlorine might be used to remove iron from impure tin powder, Dutton probably thought that Sperry and Townsend would be interested in using the chlorine produced by their electrolytic cell.[101] In 1904 Townsend began experiments in his Washington laboratory,[102] and his reports to Dutton apparently convinced the American Can Company that development beyond the laboratory scale was feasible, for Sperry and the can company signed an agreement in December which specified that the company pay for further development. In return for their contribution, Sperry and Townsend were to receive 20 to 40 percent of the stock issued by a new company formed to use the process. All Sperry and Townsend patents on the process were to become the property of the company.[103]

The future of the relations between Sperry and Townsend and the American Can Company were beclouded, however, by an infringement suit brought against the can company. The Vulcan Detinning Company claimed that the electrolytic process used by American Can infringed Vulcan patents.[104] If American Can were denied the electrolytic process, the action would undermine Sperry and Townsend because the process they were developing used tin powder, a product of the electrolytic process. The outlook was improved, however, when American Can sought to acquire the electrolytic detinning process of Th. Goldschmidt and Co., Essen-on-Ruhr, Germany, a concern that had pioneered in electrolytic detinning. American Can bought the rights to use the Goldschmidt process, a good move in the defense against Vulcan.[105] In addition, however, Goldschmidt and American Can discussed Goldschmidt's new detinning method using

[100] Townsend to EAS, October 17, 1904; and Dutton to EAS, November 9, 1904 (SP).
[101] Karl Goldschmidt, "Die Entzinnung der weissblech Abfälle und ihre wirtschaftliche Bedeutung," *Stahl und Eisen*; XXVIII (1908), pp. 1919-26, describes the history of chlorine detinning from earliest times through his own process.
[102] Townsend to EAS, October 17, 1904 (SP).
[103] Agreement between EAS and F. S. Wheeler, December 6, 1904 (SP).
[104] The complex state of detinning technology is discussed in F. Winteler, "Some Notes on the Uses of Electrolytic Chlorine," *Electrochemical Industry*, II (1904). 335–42; and *Scientific American Supplement*, "The Electrolytic Recovery of Tin from Scrap and Cuttings," (a digest of an article by Menicke in the *Zeitschrift für Electrochemie*), April 4, 1903, p. 22,788.
[105] Roy A. Duffus, Jr., *The Story of M & T Chemicals, Inc.* (New York: Codella Duffus Baker, Inc., 1965), pp. 6, 8. This is a history of a company that is a descendant of the Goldschmidt Detinning Company.

chlorine and tin-plate scrap as a raw material.[106] If American Can adopted this process, then the electrolytic process, tin powder, and Sperry and Townsend would be eliminated.

Sperry went ahead, nonetheless—with the support of American Can—to build a pilot plant for the process using tin powder. Apparently the company was encouraging both Sperry and Goldschmidt in an effort to have a strong position whatever the outcome of the litigation with Vulcan. After costs at the pilot plant turned out to be higher than anticipated, American Can, in May or June 1906, asked to renegotiate the agreement with Sperry and reduced Sperry's interest in the stock of the projected company.[107]

Earlier, Sperry had had difficulties with Townsend. In November 1904, Sperry wanted to renew the contract with the chemist, but Townsend thought his salary and share in eventual profits should be increased.[108] Sperry was willing to promise an increased share of profits, but he was reluctant, or unable because of a shortage of cash, to meet Townsend's salary demands. The two men negotiated, but in 1905 activity in the Washington laboratory declined, and Sperry concentrated on the pilot detinning plant at Niagara Falls, where he was represented by a young chemist, Ernest B. LeMare, a recent graduate of an English university. With a patent-law office in Washington, Townsend was probably unwilling to move to Niagara, but he and Sperry could have continued to promote and develop the white-lead project and develop new projects of promise—if they could have agreed on salary.[109]

In order to obtain chlorine, the pilot detinning plant at Niagara was located next to a caustic soda-chlorine plant of one of Hooker's rivals, probably because Hooker could not promise chlorine in time to meet the development schedule anticipated by the can company. Sperry commuted to Niagara as a consultant. Among other problems, he solved the troublesome difficulty of moisture in the air reaching the chlorine in the reaction vessel, by placing the vessel under positive pressure, thus preventing any leaks into it. Sperry also found a means of removing stannic chloride—a valuable by-product—by running the process at a temperature below its boiling point. When the detinning was complete, the stannic chloride was drained off and sold to the silk mills for use in weighting silk cloth.

[106] The first of these was filed January 30, 1905 (243,403) by Hans Goldschmidt, Essen-on-Ruhr, and issued November 14, 1905 (Patent No. 804,530). Other American patents on the process were issued September 18, 1906 (Patent No. 831,223), May 29, 1906 (Patent No. 822,115), and November 20, 1906 (Patent No. 836,496).
[107] EAS to F. S. Wheeler, October 18, 1906, mentions this contract (SP).
[108] Townsend to EAS, November 1, 1904 (SP).
[109] The negotiations between Sperry and Townsend over renewal of the contract and further work in the Washington laboratory can be found in Eugene Byrnes and Townsend (firm of Byrnes and Townsend, patent lawyers, Washington, D.C.) to EAS, January 4, 1905; Townsend to EAS, January 14, 1905; and agreement between EAS, Townsend, and E. A. Byrnes, December 1, 1905 (SP). Townsend later became head of the patent department and research director of a forerunner of the Union Carbide and Carbon Corporation.

Sperry also improved the design of the reaction vessel to ensure thorough agitation and contact between the tin powder and the chlorine, and to feed tin powder and chlorine in a continuous process. Concentrating on mechanical design, Sperry applied for nearly 20 patents from 1905 through 1908.[110]

Sperry realized that his bargaining position with the American Can Company worsened as the development at Niagara was prolonged. To provide for the eventuality that American Can might adopt the Goldschmidt detinning process using tin-plate scrap, Sperry devised material-handling apparatus to be used with it.[111] Also, probably warned by Dutton that a decision of American Can to use the Goldschmidt process would dry up development funds for his process, in October 1906, Sperry renegotiated his agreement with the can company: instead of taking stock, he agreed to sell his patents outright, giving American Can an exclusive option.[112] Sperry's sale price was $60,000; the option price was $10,000 and considered the first installment towards purchase. Sperry's major objective then was to persuade American Can to exercise the option. This became more difficult when American Can lost the suit to Vulcan in 1907 and abandoned electrolytic detinning. Sperry's approach to American Can was then to suggest the value of the company buying his patent to prevent it from falling in the hands of a competitor (Vulcan) and to use it in interference or infringement suits against the Castner Electrolytic Alkali Company, another competitor.[113]

When, in 1907 and 1908, a series of patent interferences were declared involving Sperry, American Can, Castner, Goldschmidt, and others, American Can came to value the Sperry patents.[114] Having bought the Goldschmidt patents and formed the Goldschmidt Detinning Company in December 1908, American Can turned over its option on the Sperry patents[115] to the new Goldschmidt company; Sperry then allowed his interference with Goldschmidt to be lost by default and another inter-

[110] See Appendix.
[111] See especially patent filed April 11, 1908 (426,613), "Process of Preparing Merchantable Iron from Tin-Plate Scrap," issued December 8, 1908 (Patent No. 906,321); and one filed March 13, 1908 (420,946), "Process of Preparing Merchantable Iron from Tin-Plate Scrap," issued October 13, 1908 (Patent. No. 901,266).
[112] Agreement dated October 29, 1906, between EAS and the Chemical Reduction Co., the American Can subsidiary organized in 1905 to develop the Sperry process at Niagara (SP).
[113] Dutton to EAS, October 19, 1905; Max Mauran to Dutton, November 9, 1905; Dutton to EAS, December 18, 1905; EAS to Townsend, November 11, 1905 (SP). Patent interference 27,321, *Sperry* v. *Townsend* v. *Baillio;* patent interference 28,111, *Sperry* v. *Baillio;* interferences declared March 26, 1907.
[114] For example, patent interference 28,131, *Karl Goldschmidt and Joseph Weber* v. *Elmer Sperry,* declared October 17, 1907.
[115] Poor's Railroad Manual Company, *Poor's Manual of Industrials,* (New York: Poor's, 1912), pp. 2047–48, shows that the company incorporated on December 16, 1908, obtained most of its officers from the American Can Company. In 1916 the information on the Goldschmidt company was given with that of the American Can Company.

ference to be dissolved.[116] Goldschmidt Detinning then exercised its option to buy Sperry patents.[117]

With the American Goldschmidt Detinning Company in possession of the Sperry patents, Sperry and this company shared an interest in selling rights abroad. The value of his patents is suggested by his success in selling his detinning rights in Europe to the German Goldschmidt company.[118] When selling the rights abroad in the summer of 1909, he also investigated techniques for producing liquified chlorine, which would make possible shipment of chlorine from Hooker at Niagara Falls to a new detinning plant at Cartaret, New Jersey.[119] Sperry also took the occasion to investigate the state of European gyroscope technology. His interest was shifting from chemistry to this new and esoteric field of mechanics. Sperry sensed that industrial chemistry was becoming so structured by the large companies, especially by their patent departments and their research laboratories, that it was becoming increasingly difficult for an independent to survive. The patent controversy and the business competition in which he found himself involved must have convinced him that his prime talent, his inventiveness, might flourish better in new or less cultivated fields. Before following Sperry into the new field, however, some attention will be given to the Sperry family. They had not only read aloud to him from chemistry textbooks so that he could enter, and survive in, the world of industrial chemistry, but had reminded him that there was another world.

vi

In 1893, the Sperrys had moved to Cleveland and settled in a Victorian house at 855 Case Avenue, between Cedar and Central Streets; in 1905, Sperry's chemical work carried them to Brooklyn, where they settled on comfortable Marborough Road. The physical environment of these houses and their neighborhoods say much about the Sperry style of life. It was undeniably upper middle class, stable, and child-oriented. Case Avenue was a neighborhood of roomy frame houses with substantial yards, large enough for even the four Sperry children. Helen Sperry, the eldest child, recalled the atmosphere established by her mother:

She believed in letting the children work out their desires, and the back yard was full of messy building operations and pits, the grounds riddled with tunnels and deep boy-made gullies with precarious bridges over

[116] The Patent Office Examiner decided that Baillio, Townsend, and Sperry were anticipated by the work of Kugelgen and Seward. Patent Office Examiner to EAS, April 28, 1910 (SP). In January 1904, possibly unknown to Sperry and Townsend, Franz Von Kugelgen and George O. Seward had filed a patent on a process for detinning scrap tin-plate using chlorine. They not only claimed they could produce pure tin by chlorine detinning but also stannic chloride, a compound used by silk mills to weight the cloth, as a by-product.
[117] Agreement between the American Can Company and EAS, February 20, 1909 (SP).
[118] EAS to W. B. Horne, August 1909 (?) (SP).
[119] Funding and Development paid one-quarter of his travel expenses.

which a small cart rumbled with a hollow sound which delighted us. Above ground rose fearfully flimsy structures housing goats, guinea pigs, rabbits, pigeons. There were fires and gardens, grubby children and happy times in the old back yard in Cleveland.[120]

The newer four-bedroom house in Brooklyn lacked the spaciousness of the Cleveland home, but the neighborhood offered compensations. Prospect Park South, in Flatbush, was an attractive development and "the bedroom of New York" for many bankers, brokers, businessmen, and their families.[121] As in Cleveland, the new Sperry home was a vital setting for teenage children. Helen again recalled the scene:

Indoors in the Flatbush house, I wonder now that Mother stood the bicycle shop in the basement and Lawrence's venture, the aeroplane built in the attic (the stairs had to be taken down to get it out, and the furnace was burnt out when he made a summer fire to steam the ribs for bending). The alarm clock went off at unearthly hours on some Saturday mornings while the clan gathered to try the new glider at Sheepshead Bay race track in the still air of dawn. And the motorcycle stage, when two boys reeked with oil from homemade engines on bicycles—the motorcycles spit, stank, coughed, gasped, hesitated, trembled with a mighty ague, and never went far. The oil under two boys' fingernails was ineradicable; oil trickled on the cellar stairs as the cycles were tugged up and down. The servants of that period were young and tolerantly Irish. Nowadays such an understanding of the need of children to work out their schemes and experiments and giving of the materials and environment is called progressive education.[122]

Zula Sperry was the vitalizing influence in the home. She found vicarious fulfillment in the activities of her children and her husband and was committed to the principle that the wife and mother's primary role was to create an environment in which life flourished. To another age she might appear self-sacrificing; Zula seemed to think her life self-fulfilling. To her, a happy creative family was "about the best thing there is."[123]

Zula shielded her husband from the problems of raising four zestful teenaged children and managing a home.[124] Her daughter could not remember her ever having carried household and family problems to her husband. She had a clear head and excellent judgment about such matters, and by taking responsibility for them, she believed she could sustain the creativity of her husband, about whose work she was intelligently informed and in whose successes she shared. Sperry's letters show that he discussed his work problems with her, especially those arising from misunderstandings with his business associates. When she felt that Elmer had been stymied in his work by financial and managerial problems, she sometimes turned for

[120] Mrs. Helen Sperry Lea wrote several pages of reminiscences about her mother not long after Zula's death in 1930 (SP).
[121] Ralph Foster Weld, *Brooklyn is America* (New York: Columbia University Press, 1950), p. 79.
[122] Helen Sperry Lea reminiscences (SP).
[123] Zula to EAS, July 31, 1899 (SP).
[124] Interview with Helen Sperry Lea, January 28, 1966.

Figure 3.13 The Sperry home in Cleveland, Ohio, on Case Avenue.

Figure 3.14 Sperry home in Brooklyn after 1915, on Albermarle and Marlborough Roads.

Figure 3.15 The Sperrys' first home in Brooklyn, on Marlborough Road.

advice to her brother Herbert, head of the Goodman mining machine company. Zula was also more tolerant than her husband of human foibles, and he recognized these virtues:

Serenity and patience are the great crown jewels in your coronet. . . . Thru all the strife of battle we all have you to lean upon, so be stout of heart. . . . You are the most necessary person in the world to a lot of rustlers and they resemble nothing short of a mob at times, yet you will notice it all centers about you. . . .[125]

Zula also reminded her charges that there were worlds other than the technological. Living in an era when inventors were American heroes, they might well have forgotten that. Zula read aloud to them from novels and played the piano for them. She also encouraged Helen's interest in art and occasionally persuaded the busy family to attend concerts. There were signs that the members of the family were not entirely unresponsive. In fact, her husband's letters from Europe when he was abroad selling the Sperry electric auto revealed more than superficial interest in European culture. He was enthusiastic about London, writing to Zula after a day traveling about the city, "I am in love with London."[126] Although Sperry found the morals of the Parisians "simply outrageous, disgusting," that

[125] EAS to Zula from aboard the *Lusitania*, June 1911 (SP).
[126] EAS to Zula, October 22, 1898 (SP).

101

did not close his eyes to the magnificence of the city, and he looked forward to showing Paris to Zula "on our great trip, which will surely come off sooner or later."[127]

Zula and Elmer would travel to Europe, but in the meantime the children had to be placed in prep schools and colleges. In Cleveland, Helen attended Miss Middleburger's private school and then transferred to Packer Collegiate Institute in Brooklyn, from which she graduated in 1909. Packer was noted for a solid, demanding curriculum and for its "refined and elevated influences."[128] Both Edward and Elmer, Jr., matriculated at Philips Exeter Academy shortly before World War I; Lawrence was sent away in 1911 to Professor Evans' School in Mesa, Arizona, because it was hoped that once he was separated from his home-built airplane his consuming passion for airplanes might cool. All three of the boys shared their father's interest in invention and engineering, but Lawrence's commitment was the most intense.[129] When one of the boys was asked why they emulated their father, the answer was, "What more exciting was there to do?"[130] The answer, of course, flattered Elmer Sperry and led him to encourage their mechanical interests by providing them tools, shop space, bicycles, and so on. Lawrence did not allow college to delay his return to aviation, but Edward and Elmer, Jr., entered Cornell University, where both decided to study engineering. Edward graduated in 1915, whereas Elmer, Jr., withdrew after two years to join the engineering staff of the Sperry Gyroscope Company, then meeting an unprecedented wartime demand for invention and development. Years later a building at Cornell would be named for the father and his sons.[131]

Elmer Sperry kept informed, although not as closely as Zula, of his sons' college life. He was pleased when Edward wrote that he found mechanics especially interesting because of its practicality, but many years earlier Sperry had sensed that Edward was as interested in business as in technology. From the time they were small, the father noted with an intuitive eye the distinguishing characteristics of his children; in 1901 he wrote a poem for them, as he often did, that captured the essence of the boys' personalities as he saw them. He cast Lawrence in the role of the inventor and engineer ("a dynamo builder by nature was he"); Edward he saw "for a merchant . . . much better fit"; and Elmer, the baby of the family, he saw working in overalls in a "wee shining" shop. The careers of the boys with the Sperry Gyroscope Company, the next Sperry venture, proved him remarkably prescient.

[127] EAS to Zula, London, November (?), 1898 (SP).

[128] Henry R. Stiles, *History of the City of Brooklyn* (Brooklyn: Published by subscription, 1870).

[129] See below, pp. 223–30.

[130] Interview with Mrs. Helen Lea, January 28, 1966.

[131] Sperry Hall was dedicated on November 1, 1966, with the inscription: "Sperry Hall is named in memory of Elmer A. Sperry, inventor, especially remembered for his useful applications of the gyroscope, and his sons, Edward Goodman Sperry, and Lawrence Burst Sperry." Elmer Sperry, Jr., died on December 21, 1968.

CHAPTER **IV** The Gyrostabilizer:
A Very Powerful Fulcrum in Space

With the gyroscope it is possible to create and maintain a very powerful fulcrum in space effective for the heaviest kind of mechanical duty. ELMER SPERRY, 1910

CHEMICAL MATTERS WERE SPERRY'S MAJOR CONCERN AFTER THE MOVE TO Brooklyn; but in 1907, in his late forties, he embarked upon a phase of his career that brought him national, even international, fame as an inventor and engineer in an entirely new field. He became, within a decade, the most famous of the inventors applying the gyro to useful purposes. Although the public associated him—and most famous inventors—with the romantic myths of invention, the account of Sperry's invention and development of gyroscopic instruments is more one of drive, resourcefulness, persistence, and courage than of intuitive flashes, dramatic acts of creation, and overnight development.

What follows is an account of the invention and development of the Sperry marine stabilizer, an account extending over half a decade. It is concerned with the origin of Sperry's interest in gyro stabilization; his original conception of the active stabilizer; and the development of his invention from a mathematical model and physical scale-model to full-scale shipboard equipment. The development of the gyrostabilizer was a new experience for Sperry because he obtained governmental support for the first time. He had previously depended on private enterprise, as had most inventors, but in the first decade of the twentieth century, the United States navy found its technological needs increasing along with its funds for development. Sperry's interests corresponded with the navy's needs; together they helped to open a new era in the history of research and development—an era characterized by the increasing influence of the government in determining the shape of technology.

Figure 4.1 Sperry testing the stabilizing effect of a small gyro, similar to Beach stabilizer of 1910.

i

The origin of Sperry's sustained interest in the gyro is difficult to establish. The problem is challenging because his transition to this field seems on the surface one of the most illogical decisions of his career. In the past, he had shifted from arc lights, to mining machinery, to streetcars, but the thread had been electric light and power. He had moved from streetcars to motor cars, but by way of an electric-motor automobile. From the motor car to electrochemistry, the stepping stone was the storage battery, which introduced him to the general field of electrochemistry.

After Sperry became famous for his many useful inventions applying the principles of the gyro, his teachers and acquaintances recalled incidents which they believed explained the origins of his interest. A normal-school teacher recalled that "when we came to the gyroscope, you were enraptured." Professor Dayton C. Miller, of the Physics Department of Case Institute of Technology, remembered that when Sperry lived in Cleveland he had borrowed a small electrically driven laboratory gyroscope to test its stabilizing effect on a row boat.[1] Sperry's notebooks, diaries,

[1] Report of a conversation with Dr. Miller; John R. Weske of the Department of Mechanical Engineering, Case School of Applied Science, to Elmer Sperry, Jr., May 9, 1939 (SP).

and letters, however, do not indicate that he followed up this test. Sperry also recalled that in Cleveland he bought toy gyros for his children and enjoyed demonstrating their puzzling characteristics. All of these incidents show that Sperry was fascincated by the device early in his career, but they do not provide the basis of a decision to develop gyroscope applications commercially.

His notebook entries give the best evidence of the beginnings of Sperry's professional interest. He made his first entry relating to the gyro early in 1907, when he cited an 1874 essay entitled, "Problems of Rotary Motion Presented by the Gyroscope, the Precession of the Equinoxes, and the Pendulum."[2] Since the author's object was to deduce the precession of the equinoxes from the theory of the gyro, it seems unlikely that Sperry was stimulated or helped by it. A few pages later Sperry cites three more articles on the gyroscope. This suggests that in the spring of 1907 he made a search for published material on the subject, which dates reasonably well the onset of his serious interest. The additional references were to articles in *Electrical Engineering, Engineering,* and the *Scientific American,* all leading technical journals to which Sperry referred often during his career.[3] The article by W. R. Kelsey in *Electrical Engineering* interested him in particular, for he placed a typewritten copy in his files along with several pages of his own computations based on the problems posed by the author. The article was a closely reasoned explanation of "the nature and magnitude of the forces called into action when the plane of rotation of a revolving mass is turned through any angle." Using the formula provided, Sperry could calculate the magnitude of the inertial forces of a gyro and therefore its stabilizing effect.

Sperry's search for earlier articles was probably stimulated by news items in the spring of 1907 on the applications of the gyro, especially on its stabilizing effects. The most influential immediate cause of his shift to the new field was probably this emphasis in the technical journals and the conversation it undoubtedly stimulated among engineers at their professional meetings and in private conversations.[4] Sperry would have responded quickly to the current of excitement about the gyroscope's potential application to practical problems. In the spring of 1907, the *Scientific American*

---

2 Sperry Notebook No. 47 (May 6, 1906 to December 5, 1907). By the location in the book, the entry appears to be spring 1907. The article cited was in *Smithsonian Contributions to Knowledge* (Washington, 1874), XIX, pp. 1–52. The author was J. G. Barnard.

3 W. R. Kelsey, "Gyrostatic Action and Its Bearing on Certain Points of Engineering Design," *Electrical Engineering,* London (July 18, 1902), p. 86; "Admiral Fleuriais' Gyroscopic Horizons," *Engineering,* London, LXXIX (1905), pp. 361–62; and H. Dailey, "Electrical Gyroscopes," [driven by static electricity], *Scientific American,* LXXXIV (1901), pp. 245–46.

4 Several years later, Sperry wrote to Louis Brennan, a fellow inventor in the gyrostabilizer field, that he had come across his work during discussions with engineering friends at the New York Engineer's Club, the Railroad Club of New York, and during meetings of the Society of Mechanical Engineers and the Institute of Electrical Engineers. EAS to Brennan, June 5, 1909 (SP).

ran four articles on the practical application of the gyro.[5] This interest was reflected in other journals and suggested, as the editor of *Scientific American* noted in May 1907, that a new field of activity was opening. The *New York Times* observed in 1909 that "the evidences are multiplying rapidly—some experts believe conclusively—that the great wonder worker in the field of mechanics during the next quarter-century is to be the gyroscope."[6]

The editor of the *Scientific American* thought that use of the gyro for steadying ships promised its most valuable practical application and noted the strong endorsement that the device had received from the well-known and authoritative former Chief Constructor of the British Navy, Sir William H. White, whom Sperry soon encountered. White thought that the gyro would be particularly useful on yachts and coastal passenger steamers because of its relatively small size, but he did not preclude its use on ocean steamers. White envisaged the gyro's providing warships a steady gun platform, the advantages of which "are already well known."[7] He also remarked on the use of the gyro to stabilize a monorail. By 1907 there were at least three American patents in the field of gyrostabilization.[8]

ii

In Europe, Ernst Otto Schlick of Hamburg, Germany, and Louis Brennan of England had found means of using the gyro in the solution of stabilization problems. Schlick, a well-known naval engineer, had described in 1904 a model of his device for reducing the amplitude of a ship's roll.[9] He became interested in the possibility of using the gyro as a stabilizer while investigating the effect of large rotating masses, such as paddle wheels, on the behavior of a ship. (This was also the subject of the W. R. Kelsey article that had interested Sperry.) In 1906 Schlick had installed a full-scale gyrostabilizer on the *See-Baer*, an inactive torpedo boat of the Germany navy. The ship, known as a heavy roller, underwent a series of tests, including sea trials in rough weather, and the report was that "the boat behaved very well, much better, indeed, than when the gyrostat was not acting." During one phase of the test the gyrostabilizer

5 Frederick Collins, "The Gyroscope as a Compass," XCVI (April 6, 1907), p. 294; "New Uses for an Old Device—The Gyroscope Railroad Train," (May 18, 1907), p. 406; "The Brennan Gyroscope Monorail," (June 1, 1907), pp. 449–50; and "Practical Tests of the Schlick Gyrostat for Ships" (June 15, 1907), p. 494.
6 *New York Times,* December 12, 1909, editorial.
7 *Scientific American,* XCVI (1907), p. 406.
8 T. C. Forbes, issued September 13, 1904 (Patent No. 769,693); E. O. Schlick, issued September 6, 1904 (Patent No. 769,493); Louis Brennan, issued August 8, 1905 (Patent No. 769,893).
9 A description of the model was presented to the Royal Institution of Naval Architects. Alfred Gradenwitz, "An Apparatus for Preventing Seasickness," *Scientific American,* XCI (1904), p. 40; E. O. Schlick, "Gyroskopischen Einfluss Rotierender Schwungräder an Bord von Schiffen," *Zeitschrift des Vereines Deutscher Ingenieure,* L (1906), pp. 1466–68. On Schlick, see Conrad Matschoss, *Männer der Technik* (Berlin: VDI Verlag, 1925), p. 240.

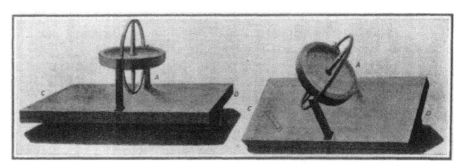

Figure 4.2
A simple illustration
of precession.
If the platform is tilted,
the gyro wheel will tilt
according to the direction
of its rotation
and resist the movement
of the platform.
(From the Scientific
American, *January 22, 1910.)*

reduced roll from thirty degrees of amplitude to one degree.[10] Sir William White, who had been appointed by the British Admiralty to study the performance of the *See-Baer,* reported so favorably that the British firm of Swan, Hunter & Wigham Richardson Ltd. of Newcastle-on-Tyne, builders of the *Mauretania* and *Lusitania,* acquired the Schlick patents and bought the 52.2 ton *See-Baer* from the German government for display and test purposes. Subsequently the firm equipped two coastal vessels with gyrostabilizers.

Louis Brennan's application of the gyrostabilizer also attracted considerable attention in technical circles. He demonstrated before The Royal Society in the spring of 1907 a monorail car stabilized by a gyro which carried a boy along a tight wire strung around the society's meeting room at Burlington House. The next month Brennan demonstrated the car on the grounds of his estate, using both a wire rope and a single track. Brennan said that the invention had evolved from thirty years of continuous experimentation and that the model railway had been completed nearly two years before the demonstration. Publicity was delayed "in deference to the requests of the British and Indian governments," which helped finance the development.[11] The sponsors anticipated that the monorail would prove economical on the light railways of India and in new construction projects on the difficult terrain of the country. The *Scientific American* tended to belittle the achievement, believing that what had been done with a small car could not be achieved with a full-sized car because the weight and necessary power of the gyro would prove impractical.[12] Less than two years later, however, Brennan successfully demonstrated a full-sized gyroscopically stabilized railroad car.[13] Brennan's work so interested Sperry that in 1909 he visited him in England and later acquired his patents.

iii

The flurry of activity and publicity surrounding gyro stabilization

[10] "Practical Tests of the Schlick Gyrostat for Ships," *Scientific American,* XCVI (1907), p. 494. Before gyroscope became the general term in the literature, the expression gyrostat was often used for stabilizing devices.
[11] "The Brennan Gyroscope Monorail," *Scientific American,* XCVI (1907), pp. 449–50.
[12] *Ibid.,* p. 449.
[13] *New York Times,* December 12, 1909.

in 1907 and Sperry's interest in the fascinating device brought him into the field, but his first efforts did not suggest the importance that this new venture would assume for him within several years. Sperry was not the first inventor, and by no means the last, to see an invention culminate in developments not at first anticipated. When he began working with the gyro, he may have had in mind nothing more than a peripheral effort to develop an ingenious invention quickly and to sell the rights for manufacture and marketing by others.

His first attempt to commercialize the gyro was related to a circus stunt. In August 1907, he wrote to Thompson & Dundy of Luna Park, Coney Island, inviting discussion of an entirely new device that "some engineering friends and myself have been engaged in for some time."[14] He believed that it touched a "high mark" as a "thriller" and also as an engineering feat. In a style far more vernacular than customary in his letters, he observed that he was "casting about to see whom we had better hitch up with."[15] As the basis for discussion, he prepared a three-page description of the stunt described as "the wonderful trained wheel-barrow." It involved the use of what appeared to be an ordinary wheelbarrow, although it actually concealed a gyro, with which an acrobat would perform a number of seemingly miraculous feats of balance on a tightrope. Sperry attempted for several years to interest someone in this ingenious device, including Barnum and Bailey Circus and Sam & Lee Shubert Co. By 1912, however, when a performer requested special gyros for a vaudeville stunt, Sperry replied that he regretted that his company was so busy manufacturing large gyro apparatus for the government that there was no time to design special small instruments.

Within a few months, he applied himself to another device and made his first successful patent application in the field. He no longer attempted to exploit the superficial mystery of the gyroscope; he attempted to introduce a gyrostabilizer into an expanding industry he knew well—the automobile. From having designed his own cars and having ridden in others, he knew that the auto was unstable—an instability that assumed dangerous proportions on the uneven roads of the period. (On one occasion Sperry's son, Elmer, Jr., narrowly escaped injury when he was thrown from a careening auto.)

He listed as the objectives of his vehicle stabilizer, for which he made a patent application on December 2, 1907,[16] greater safety, increased comfort, longer tire life, and greater design freedom (resulting from the possibility of a higher center of gravity). Even though he had the auto foremost in his mind, he drew up the patent to relate "to apparatus for

[14] Sperry applied for a patent on an "amusement apparatus" on August 24, 1907 (389,952). No patent was issued.
[15] EAS to Thompson & Dundy, August 6, 1907 (SP).
[16] The patent (Patent No. 907,907) was issued on December 29, 1908. The idea for land-vehicle stabilization was not new. In 1888 Carl Benz had secured a U.S. patent for a "self-propelling" vehicle stabilized by a flywheel on a vertical axis; *Scientific American*, CV (1911), p. 232. There were at least four U.S. patents before Sperry's in which the gyroscopic principle was used to stabilize autos.

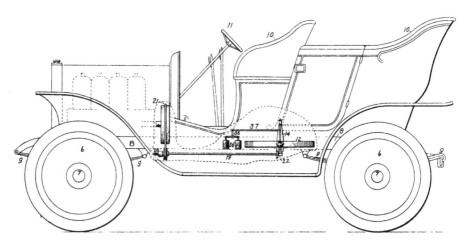

Figure 4.3 Patent drawing of Sperry's design for an automobile stabilizer (Patent No. 907,907).

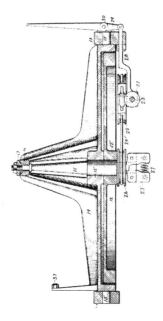

*Figure 4.4
Detail of
Sperry automobile stabilizer.
"Steadying Device for
Vehicles"
(Patent No. 907,907). (Key:
gyro wheel, 12; pivot, 13';
frame of auto, 13; swinging
frame, 14; vertical axle, 15;
hand-controlled lever to
engage and disengage gyro
from power source, 29).*

steadying bodies susceptible of movement and especially susceptible to angular movement."[17] He explained that the essence of his invention was a gyro that created a force opposing the undesirable one causing, for example, lateral tipping. He illustrated his "steadying device for vehicles" in terms of an automobile (see figures 4.3 and 4.4).

Sperry acknowledged that the gyrostatic influence of a rotating flywheel was well known, but he stressed that his "flywheel" was capable of precessing—or capable of angular motion in a plane at right angles to the upsetting or tilting force. In the example given in his patent, the wheel was mounted with a vertical spin axis and pivoted in a housing that could rotate about a horizontal axis running transversely to the fore and aft line of the automobile. In operation the gyro wheel and housing were intended to precess, or swing on the pivots, when the vehicle tilted to right or left. Sperry calculated that a 200-pound wheel properly precessing could develop more than four tons of resistance to the tilting force. This was enough, he believed, to prevent an automobile from turning over.

Precession was basic to Sperry's invention. He knew that a rigidly mounted flywheel, paddlewheel, or turbine wheel on a ship did not stabilize the ship because without freedom to precess, or do substantial work, the wheels could not absorb the disturbing energy. For the same reason, the engine flywheel or wheels on the automobile did not stabilize it. The tendency of such rigidly mounted revolving masses to precess was prevented by the mountings or bearings. In later applications of the gyro, the precession phenomenon was also critical and Sperrry often needed to define it in explaining his devices. One of his best definitions was:

If I impress a force on one end of the axis of a gyroscope it will resist this impressed force but will turn in a direction at right angles to the force impressed. This motion at right angles to the impressed forces is called "pre-

17 Sperry Patent No. 907,907.

109

cession." It will be observed that the gyroscope does not resist any forces impressed by linear motion; nothing but angular motion causes precession, and nothing but the forces impressed by angular motion are resisted. This angular motion may be about a point within the gyro or a point at any distance from the gyro, but must be angular motion [see figure 4.5].[18]

The Gyroscope.

Sperry tried to convince the automobile manufacturers of the value of his precessing stabilizer. He wrote a persuasive letter to Henry D. Joy, president of the Packard Automobile Company in August 1908, showing, by calculation, that a skid often generated enough force to upset a car.[19] Although some manufacturers had lowered the center of gravity to increase resistance to the upsetting force, this expedient, in Sperry's opinion, adversely affected appearance and comfort. His invention, he argued, would solve the problem without lowering the center of gravity. He noted that the "Gyroscope Automobile Company" advertised a so-called gyroscopic car, but because its gyro wheel was rigidly mounted—unable to precess—the device was "worse than useless."[20] The sales manager of Packard replied that "no gyroscope, nor anything but a block and tackle or a preserving angel" could prevent the average accident from occurring.[21] Sperry wrote one more letter pleading for his device in the name of safety; he even prepared a four-page analysis of "energies operating in the skidding of vehicles,"[22] but the automobile manufacturer turned a deaf ear.

The wheel rotates rapidly.

Suppose the rim is split into segments.

Although unsuccessful, Sperry's approach to the automobile industry was in accord with his style of invention; he saw the possibility of using a new device (the gyro) to solve a particular problem arising from the introduction of a new technology (the automobile). His activity in this case, and in others, was an effort to adapt an invention to an environmental condition (bad roads) that the original inventors and developers had **not** foreseen.

Attend to two segments.

iv

Sperry next turned to ship stabilization. Owners and operators of ships were interested in stabilization because the introduction of the iron ship had increased the stability problem. A ship under way with a broad exposure of canvas enjoyed a natural equilibrium, for not only did the force of the wind across the canvas reduce the amplitude of roll, but a broad beam and a large metacentric height did so as well. With the coming of the iron ship, these stabilizing forces were lost. Narrow-hulled and with

The segments are connected to an axle.

As they turn about the axle, one moves up, the other down.

---

18 Sperry in a lecture at the U.S. Naval Academy in 1912 (mimeograph; SP). The spin bearings of a gyroscope prevent the transfer of external torque to accelerate or decelerate the rotor. The torque impulse, therefore, must be effective about an axis orthogonal to the spin axis to cause a deflection of the spin axis toward alignment with the torque axis. This deflection is precession. (I am indebted to F. D. Braddon for these observations.)

19 EAS to Joy, August 21, 1908 (SP).

20 The "Gyroscope" Automobile Company, Inc., of 231 West 54th Street, New York, was the general sales agent for "The Gyroscope Car," described as the simplest and safest car on earth. The invention was based on the patents of J. P. Lavigne of Detroit, Michigan.

21 Waldon to EAS, August 27, 1908 (SP).

22 Typescript in Sperry Papers (unsigned).

*The axle is given
a horizontal twist.*

*The segments then are pushed
to the right and left.*

*They thus have horizontal
and vertical motion.*

They therefore move diagonally.

The axle must therefore tilt.

*Thus, when the gyroscope is
pushed, it tilts at right angles to
the direction of the push.*

Figures 4.5
*Precession of the gyroscope.
(From G. P. Meredith,
"Visual Education in the
Air Age," Aeronautics, XII,
April 1945.)*

a small metacentric height, this type of ship had a large amplitude roll. If the metacentric height were increased, the amplitude of the roll was reduced, but the ship then became "stiff" with short, fast-period rolls that were extremely unpleasant for the passengers.[23]

Sperry knew first-hand the discomforts of an iron steamer at sea. On his voyage to Europe in 1898, he was unable to eat for two days, and most of the other passengers were in worse condition. He found some diversion in analyzing the roll of the ship, probably thinking then how he might invent a device that would answer this pressing—at least to him—social need. When working on the stabilizer years later, he still recalled the discomfort of that voyage.

Aware of the general problem of ship stabilization, encouraged by Schlick's successes with the gyrostabilizer, and perhaps stimulated by his intimate acquaintance with a "need," Sperry ventured into the realm of marine technology. His notebook for the period from December 5, 1907 to December 1, 1908—begun about the time he filed the automobile stablizer patent—opened with a statement of the formula for calculating the angular momentum of a gyroscope couple. With this formula he calculated the weight, diameter, and rpm of a gyro wheel necessary to counter the force of waves. Within a few pages he analyzed "Dr. Schlick's plan" and began to search for the invention that would improve upon it.

The basic principle of the Schlick stabilizer was simple and well known. If a torque is imposed over time on a gyro free to precess, the gyro wheel will respond by precessing. Because the spinning wheel has mass and angular velocity, its angular momentum is substantial. The angular momentum is proportional to the angular velocity, the mass, and the radius of the wheel (the concentration of the mass at the circumference of the wheel also increases the momentum). This angular momentum gives the wheel an inertia, and in precessing, or changing its plane of rotation, the wheel counteracts and absorbs the momentum of a wave at sea. If the ship has no gyrostabilizer, the wave momentum is absorbed by the roll of the ship; when equipped with a gyro which has less inertia than the ship, the gyro will respond and absorb much of the disturbance.

By spring 1908, Sperry had improved upon Schlick's stabilizer by developing an active type stabilizer. He inserted, among data on chemical problems, a two-page notebook entry giving his most original thoughts on his stabilizer.[24] These pages show, for the first time, that he knew he

23 In response to the stability problem, a number of inventors before Schlick and Sperry designed gyro and nongyro ship stabilizers. Among these were Henry Bessemer and Sir John Thornycroft, whose devices, though ingenious, were considered impractical. Sperry discusses the work of several of his predecessors in "The Gyroscope for Marine Purposes," *Transactions of the Society of Naval Architects and Marine Engineers,* XVIII (1910), 143–54.

24 Unfortunately for the historian, Elmer Sperry did not date the entries in his notebooks during this important period of his career, but fortunately, in the case of the invention of the active-type ship stabilizer, Sperry had William White, the naval architect, date the entries for the stabilizer when he saw them on April 21, 1908. Notebook No. 48 (December 5, 1907, to December 1, 1908; SP); therefore, the invention came between December 5, 1907, and this date.

had improved upon Schlick and that he could articulate his improvement. Not only did he note "active instead of passive," which summarized his distinction between his invention and Schlick's, but he also wrote, "gy. used to *transmit* energy round a corner," a metaphor that he often used to express the essence of his stabilizer's performance. He also observed that he had eliminated the lag from stabilizer performance. Sperry calculated the anticipated performance of his active stabilizer and compared it with the performance of the "passive" Schlick; he concluded that his active-type gyro would be at least 100 times more powerful than a Schlick stabilizer of the same weight. In private, Sperry called the heavier Schlick machine an "English blood ugly . . . a brute of a machine."

He filed an application for a "ship's gyroscope" on May 21, 1908; the patent, although his first for the ship stabilizer, proved basic. The complexity and sophistication of the patent, compared with the earlier auto stabilizer patent, was remarkable.[25] The patent was more than a description of Sperry's invention; it was a lecture on gyrostabilizers, emphasizing how he had advanced the art. "It has been known," he wrote, "that gyroscopes mounted in swinging frames pivoted in a plane transverse of a ship would reduce the rolling to some extent" (an obvious reference to Schlick); but, he continued, "such apparatus was sluggish, did not act effectively until the ship had acquired considerable momentum and hence was only partially successful in dampening the rolling. . . . My invention becomes *active* promptly on incipient rolling [italics added]" and counters the force without waiting for the roll of the ship to develop appreciably. The "active" characteristic was the primary invention embodied in this patent. Sperry should be remembered as the inventor of the active-type gyrostabilizer for ships—not as the inventor of the gyrostabilizer. Usually inventions should be so qualified, and this is the case with Sperry as with Edison and other major inventors. Edison can no more be thought of as the inventor of the incandescent bulb than Samuel Morse of the telegraph, or James Watt of the steam engine—in each case, the invention must be qualified (the high-resistance, carbon-filament incandescent bulb; the high-inertia telegraph with relay; and the steam engine with separate condenser).

*Figures 4.6
Pages from Sperry notebooks.*

On many occasions, Sperry explained his active principle to laymen, and his clear and bold concepts captured the imagination of his audiences. His stabilizer, he explained, "headed off" a roll—it reacted to each incipient disturbance of a ship's stability with a counter-effort of exactly "the same dimensions created by the waves or swells." He reasoned that since "all rolling of ships is due to an accumulation of individual wave increments . . . it only becomes necessary to check each increment as it arrives to prevent the ship ever starting to roll." Although this explanation was an idealization of the performance of his stabilizer, it captured the essence of its operation.

In his patent, Sperry described the active principle as the use of fully controllable precession (figs. 4.9, 4.11). Schlick's gyrostabilizer, free and

25 Filed May 21, 1908 (434,048), "Ship's Gyroscope," issued August 17, 1915 (Patent No. 1, 150,311). He filed an earlier application for a ship stabilizer on January 15, 1908 (411,017), but no patent was issued.

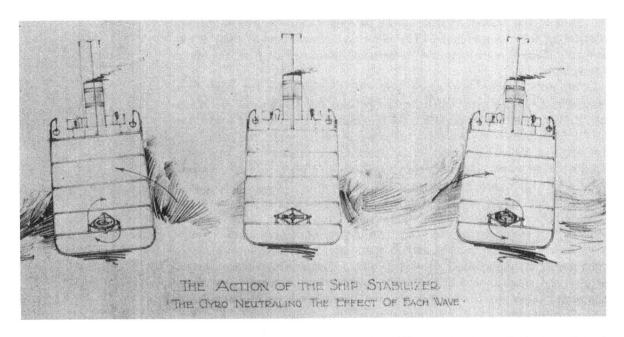

THE ACTION OF THE SHIP STABILIZER.
THE GYRO NEUTRALING THE EFFECT OF EACH WAVE.

Figure 4.7 From The Sperry Gyroscope Co., "The Sperry Ship Stabilizer" (undated typescript in Sperry Papers).

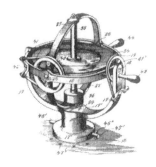

Figure 4.8
Figure from
Ship's Gyroscope patent
(Patent No. 1,150,311).

uncontrolled, precessed only in reaction to the torque put upon it by the roll of the ship. Sperry used a motor to precess the gyro artificially, by which he could overcome the gyro's inertia and begin precession before the unaided "passive" gyro could feel the torque from the roll. He controlled the precession motor, and the precession of the stabilizing gyro wheel, with what he then called an actuator—what would later be called a sensor. The actuator was a pendulum, sensitive to extremely small rolls and yet powerful enough to operate switches, valves, or other means for controlling the precession motor. The sensor felt the roll before the large gyro and initiated precession. He mounted this sensor to respond to the roll of the ship, but not to the pitch or yaw. As was his custom in drawing up patent applications, he described a particular device—here the controlling pendulum—as one of the means that could be used to carry out an invention. Later he used a small pilot gyro as a sensor (figure 4.12).[26]

He included what he called a "phantom" in his control system (see "traveling nut" (58) in figure 4.10), a moving part that "shadowed" the pendulum. When the pendulum moved (when, in fact, the ship rolled about the pendulum), it sent a signal to a small motor that moved the phantom. Because Sperry wanted the signal to cut off when the command had been followed, the phantom opened the signal circuit when its position corresponded exactly to the position of the pendulum and closed the signal circuit again when the pendulum moved. This cutting off of the signal was a "feedback." He used the phantom to obtain the feedback and also

[26] Filed January 6, 1915 (716), "Ship Stabilizing and Rolling Apparatus," issued July 10, 1917 (Patent No, 1,232,619).

to avoid taking strong control signals directly from the pendulum and interfering with its movement. The phantom, powered by a small motor, sent the control signals to the precession motor by way of the drum controller.

Located between the sensor and the precession motor, the drum controller (a device with which he had become familiar during his street-car days in Cleveland) operated and programed a sequence of operations of the large precession motor (starts, stops, reversals, speed variation, etc.). In adapting the drum controller to automatic control of the precession motor, he gave one more demonstration of his ability to apply general solutions to particular electromechanical problems.

Although Sperry did not emphasize the importance of the automatic control system in describing his active stabilizer, it was central to the device. The essentials involved the combination of a pendulum to sense a variable "error" (ship roll) and to send signals via a follow-up device (phantom) to a programed controller (drum controller), which operated a powerful electric motor (precession motor) that corrected the error (stabilized the ship). Sperry did not use the terminology of automation, but he incorporated in his stabilization system a sensor, feedback, a programed controller, and a servomotor. Drawing upon his Chicago experience with automatic regulators, he was writing an important chapter in the history of complex automation—and he would write more later.

Sperry often acknowledged his debt to Schlick for the general idea of gyroscopic ship stabilization, but he never explained the inventive act that brought him to incorporate the active principle in his design. Although the notebooks state the active principle, they do not throw light on its origins. Louis Brennan's work may, however, provide a clue. It is probable that Sperry studied Brennan's monorail-stabilizer patent when he first became interested in stabilizers.[27] To maintain his monorail car in a position that would counter the upsetting forces of its own loading, the accelerations of curves, and gusts of wind, Brennan accelerated the precession of the gyro.[28] It seems likely that Sperry's inventive leap was the combination of Schlick's basic idea with Brennan's particular device to arrive at a new combination.[29]

The idea for using a pendulum as a sensor to activate his stabilizer may have come to Sperry after he read about Sir John I. Thornycroft's efforts to stabilize a ship. Thornycroft, a noted marine architect, had introduced the use of a pendulum to sense incipient wave motion, after which a weight, weighing about five percent of ship's displacement, was shifted by heavy hydraulic machinery in the hold of the ship to counter

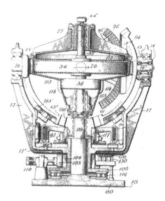

Figure 4.9
The active-type
marine gyrostabilizer
(Patent No. 1,150,311).

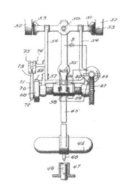

Figure 4.10
Controlling pendulum
for active-type marine
stabilizer
(Patent No. 1,150,311).

27 Brennan, American patent, issued August 8, 1905 (Patent No. 796,893).
28 John Perry, *Spinning Tops* (London: Society for Promoting Christian Knowledge, 1908), p. 48.
29 Later the Sperry Gyroscope Company purchased the basic Brennan patents, and the company's patent attorney observed that they covered "broadly the idea of accelerating precession of the gyro at certain periods"; an undated company memorandum, "Summary of Patent Situation on Gyroscopic Stabilizers for Ships" (internal evidence dates the item as the early 1920's) (SP).

114

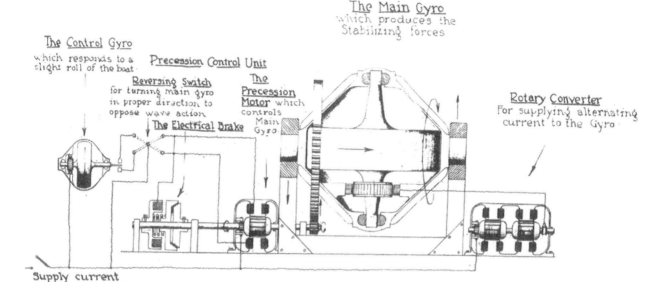

The Control Gyro
which responds to a
slight roll of the boat

Precession Control Unit

Reversing Switch
for turning main gyro
in proper direction to
oppose wave action

The
Precession
Motor which
controls
Main
Gyro

The Electrical Brake

The Main Gyro
which produces the
stabilizing forces

Rotary Converter
For supplying alternating
current to the Gyro

Supply current

Figure 4.11 From The Sperry Gyroscope Co., "The Sperry Ship Stabilizer" (undated typescript in Sperry Papers).

the roll. This, according to Sperry, was the first effort to steady a ship by controlled reactionary force.[30] Thornycroft did not use a gyro.

v

In April 1908, a month before applying for his patent, Sperry arranged an interview at the Belmont Hotel in New York with one of the world's leading authorities on naval architecture, Sir William H. White. White was in America representing the firm of Swan, Hunter, and Wigham Richardson Ltd., for whom he had reported on the trials of the Schlick stabilizer. With Sperry was Calvin W. Rice, Secretary of the American Institute of Mechanical Engineers, no doubt present to witness Sperry's disclosure of his invention—for which he then had no patent protection.[31] White initialed the Sperry drawings and the notebooks disclosing the details of the active stabilizer.[32] Sperry was so far along in his preparation for patenting that the sketches shown to White were almost the same as those submitted with his patent application. Later correspondence shows that White's reaction encouraged Sperry, but there is no evidence that White recommended that Swan, Hunter, and Wigham Richardson acquire the patent, the invention, or the right to manufacture or market it in Britain.

[30] Elmer A. Sperry, "The Gyroscope for Marine Purposes," *Transactions of the Society of Naval Architects and Marine Engineers*, XVIII (1910), pp. 144, 146.
[31] Rice recorded the substance of the meeting in two typewritten pages signed by Calvin Rice, May 6, 1908 (SP).
[32] Notebook No. 48 (December 5, 1907, to December 1, 1908; SP).

Although still lacking a backer, Sperry began work on the model needed to convince others of the feasibility of his invention. In 1907 he learned that Charles E. Dressler & Co. of New York City, a firm that made instruments and apparatus for inventors, had constructed an electrically driven, laboratory-demonstration gyro designed by Professor A. G. Compton, of the College of the City of New York.[33] Sperry realized that Dressler's shop was capable of the fine machine work now required, and in the fall of 1908, he used Dressler's as a development shop. On November 18, he demonstrated there a model based on, and exhibiting, the active characteristic. The model had two tripod stands supporting an electrically driven gyro mounted on a pendulum made from a length of gas pipe and weighted with a heavy bench vise. The pendulum simulated a ship. Using strings and gears, Sperry did the work of a precession motor and artificially precessed the gyro. He showed for the first time that artificial precession would oscillate or dampen the pendulum oscillations; the gyrostabilizer could therefore counter a roll and artificially stimulate a roll. He recorded that with active precession he dampened the oscillation of the heavy pendulum in fifteen to twenty seconds. "This, of course, is a positive demonstration and proof of my principle of precessional reaction. . . ."[34] In the previous spring, when he explained his stabilizer's performance to William White and filed his patent, Sperry had relied on mathematical analysis and projections; he now had his first physical model. After the experiment of November 18, he wrote: "Model active gyro works wonderfully, must digest all that I have seen."[35] (See figure 4.15.)

More money was needed, however, before Sperry could build the additional models and carry out the experiments that would enable him to design a full-scale gyrostabilizer. Fortunately, Sperry's needs coincided with those of the U.S. Navy. In fiscal year 1908–09, following Sperry's application for the stabilizer patent, United States Navy expenditures reached an all-time high of $156,401,161; in 1909–10 a new peak would be reached ($181,936,341). The navy was involved in a naval armaments race with Great Britain, Germany, and other nations, and had budgeted increased funds for technological development. Furthermore, during his term in office (1901–09), Theodore Roosevelt initiated an ambitious program of naval development. Prior to 1906, Congress authorized ten battleships, four armored cruisers, and seven vessels of other classes;[36] after Britain introduced the H.M.S. *Dreadnought* in December 1906, Roosevelt obtained authorization for six ships of the dreadnought class: the *Dela-*

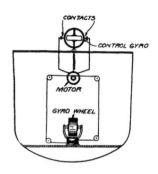

Figure 4.12
Sperry described a pendulum control in his basic patent on the gyrostabilizer; he later used a small gyro control, shown in this schematic diagram. "This sketch illustrates diagrammatically the arrangement of a stabilizing installation. Should the ship roll, one or the other of the contacts would be closed by the control gyro. A relay switch, omitted for sake of simplicity, would energize the motor in the proper direction, precessing the gyro by means of the steel wire cables. The control gyro is extremely sensitive and responds to the slightest movement of the ship. As soon as the ship begins to roll the stabilizing gyro is called upon to oppose the wave force and the ship is therefore never allowed to roll." From Albert W. Stringham, Transferring Forces "Around the Corner," 1916.

---

[33] Dressler advertised the electrically operated gyroscope at $125. In the advertising literature, Dressler predicted that "the person who discovers the formula governing the rotation of gyroscopes . . . may reveal the long sought goal: one fundamental physical law governing the universe, and may receive the Alfred Nobel $40,000 cash prize."

[34] "Gyroscope Experiments of Elmer Sperry." A two-page type-written report of experiments of November 18, 1908, signed by EAS and witnessed (SP).

[35] Sperry Appointment Diary, November 18, 1908 (SP).

[36] Harold and Margaret Sprout, "*The Rise of American Naval Power, 1776–1918*" (Princeton: Princeton University Press, 1939), p. 260.

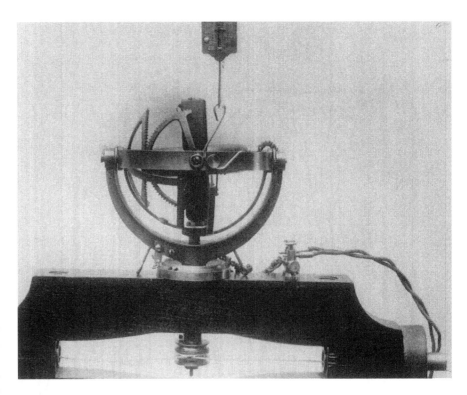

Figure 4.13
Test gyro mounted on a pendulum simulating a rolling ship. With this, Sperry demonstrated the active principle. In an informal talk, he explained as follows: "Now, then, here is another model of a ship. Why do we use a pendulum as a model of a ship? There doesn't seem to be any connection with a pendulum. They are usually with clocks. Now we use a pendulum to illustrate a ship because the moment a ship is launched she takes up a period of roll from which she never departs. The period of a modern battleship is 16 seconds over and back, no matter whether it is at Madeira or here; it is always 16 seconds, perfectly dependable. Therefore, a pendulum, which is another thing that has a given period and never departs from it, illustrates a ship just about as good as anything can. Now then, I have no motor on this ship model to work this precession, but if I pull on the frame with these strings right over the center of the gudgeon, and at right angles to the swing of the pendulum, you see I have no influence whatever on the pendulum. If I pull in other directions, then I can influence the pendulum, but I am not doing that; I am pulling right parallel with the gudgeon of that pendulum and not affecting the pendulum at all. Now see how different it is when the wheel is spinning. When I pull I can roll the ship or I can stabilize the ship, and you will notice that this is real stabilization. That puts the ship absolutely under control of the gyro, which in an ordinary ship is so small you can't find it. It doesn't look as if it had any right to even speak to the ship, let alone telling that ship it shall never roll."

ware, South Dakota, Utah, Florida, Arkansas, and *Wyoming.* Elmer Sperry would come to know the first four of these well.

The *Dreadnought* incorporated technological advances that made naval officers throughout the world keenly aware of the effects that inventive skills would have on the armaments race. The status of inventors and engineers appreciated accordingly. Roosevelt tried to end a long-standing feud between the engineering bureaus and the line officers,[37] which had frustrated improvements in naval technology. The line officers could not make their needs felt, and the engineering staff could not make the line aware of what was available for filling those needs. Roosevelt also supported the work of a number of technologically informed officers such as William Sims.[38] During the administration of President Taft (1909–13), when Sperry began invention and development for the navy, emphasis was upon efficiency.[39] The navy, therefore, was interested in technological innovation, but it was not yet ready to sponsor new devices from invention through development. It preferred that the inventor demonstrate

[37] *Ibid.,* p. 271.
[38] On Sims, especially his reform of gunnery practices between 1901 and 1905, and on the difficulties in bringing about such reforms in the navy then, see Elting Morison, *Admiral Sims and the Modern American Navy* (Boston: Houghton Mifflin, 1942), chaps. 8 and 9.
[39] "The Decline of the United States Navy," *Scientific American,* CIX (December 20, 1913), p. 466. During the last two years of the Taft Administration, Congress made appropriations for only one battleship each year rather than the two recommended by the Naval Board. See also Sprout, *American Naval Power,* p. 289.

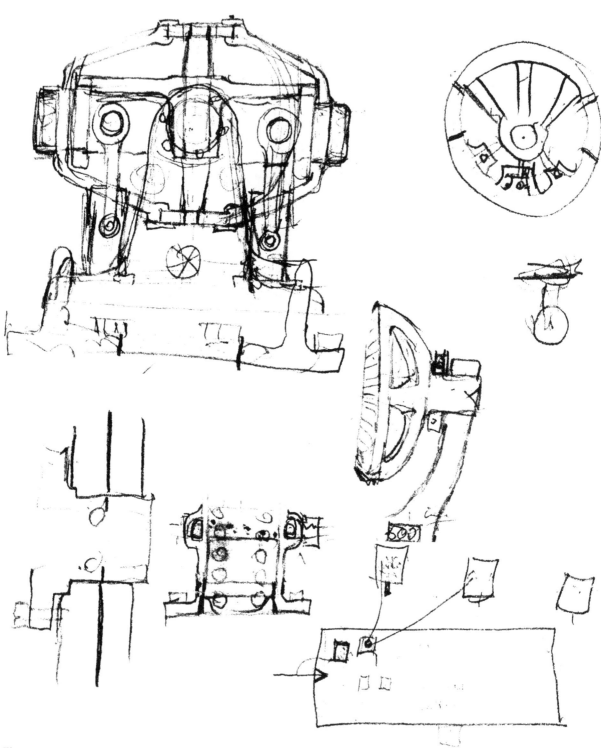

Figure 4.14 Sperry sketches of the ship stabilizer.

Figure 4.15
Dressler made
Sperry's experimental model
of an active stabilizer
and ship simulation.

Figure 4.16
Dressler urged experimenters
to use his gyros,
discover new scientific
principles, and win a prize
—the Nobel Prize.

a developed working model. In 1911, however, Rear Admiral **Bradley A. Fiske**, a successful inventor himself, made an eloquent plea for a more liberal attitude toward inventors. In the *United States Naval Institute Proceedings,* he urged the Navy to take the risk of developing inventions.[40] "Does anyone deny," he asked, "that our electric lights, torpedoes, guns and engines were invented before they were developed; that they were conceived before they grew and waxed into maturity?" By the fall of 1912, the *Scientific American* noted an improvement in the navy's attitude toward invention,[41] a change that worked in Sperry's favor.

Of greatest help to Sperry during these years was Captain (later Admiral) David W. Taylor.[42] Sperry later recalled Taylor's "helpful counsel" and "patience" when he was working to develop his gyrostabilizer.[43] Sperry first encountered Taylor when he was in charge of the navy's experimental model basin, a position Taylor had been given because of his known advocacy of applied science. Taylor was a remarkable officer who had attained the highest scholastic record in the history of the U.S. Naval Academy and had won similar distinction for advanced study in naval construction and marine engineering at the Royal Naval College at Greenwich. As a young officer, he had made scientific studies of hull design and propulsion units; in 1914 he became Chief Naval Constructor and the Chief of the Bureau of Construction.

Before Sperry met him, Taylor had conducted extensive experiments with antirolling tanks for ship stabilization. These large elongated sea-water–filled tanks had been incorporated into ships' hulls by several ship designers, including the German, Hermann Frahm. The water was expected to assume oscillations from the ship's roll and create opposing movements. Taylor concluded, after tests in the model basin and at sea, that the antirolling tanks were prohibitively large, filled strategic athwartships space, and that neither the United States Navy, nor in fact any navy or marine, possessed the personnel requisite to keep the period of the tanks in phase relation with the natural period of the ship and also with the period of the sea. Taylor, "probably the greatest living authority on these matters," therefore turned with interest to Elmer Sperry's invention.[44]

The gyrostabilizer interested Taylor and the navy for several reasons. If the stabilizer could counteract the roll of a man-of-war, then the ac-

40 Bradley A. Fiske, "Naval Power," *United States Naval Institute Proceedings,* XXXVII (1911), pp. 683ff.

41 William Atherton Du Puy, "Inventors and the Army and Navy," *Scientific American,* CVII (September 14, 1912), p. 227.

42 On Taylor, see William Hovgaard, "Biographical Memoir of David Watson Taylor, 1864–1940," *National Academy of Sciences: Biographical Memoirs,* XXII (1943), pp. 135–53; and Edward L. Cochrane, "Taylor, David W.," *Dictionary of National Biography,* Supplement II (1958), pp. 652–53.

43 EAS to Taylor, June 29, 1922 (SP). He also wrote of the inspiration he had always received from Taylor.

44 Elmer Sperry, "Non-Rolling Passenger Liners—Observations on a Large Stabilized Ship in Service, Including the Plant and Economies Effected by Stabilization." Paper read before the general meeting of the Society of Naval Architects and Marine Engineers, New York, November 13–14, 1919 (SP).

curacy of gunfire might be improved—especially on heavily rolling ships. If the stablizer quenched the roll, the entire ship would become a stable platform. In very heavy seas, the gyrostabilized fleet might have a decided advantage. The stabilizer could also reduce seasickness and improve morale on such heavy rollers as destroyers and submarines. There was the additional possibility that a gunnery officer on a surface ship could induce roll to increase the range of fire by increasing the gun elevation or to establish a predictable pattern with which firing could be synchronized.

Sperry first described his plans for a stabilizer to Captain Taylor in May 1908, about the time of his interview with Sir William White. In December, Taylor met Sperry in Washington and the inventor showed him the model built at Dressler's. After long discussion and extensive calculations, the two decided that new models were needed for testing. Sperry agreed to build the gyro model and Taylor undertook to construct the ship model at his model basin.[45] At Taylor's suggestion, Sperry wrote to the head of the Bureau of Construction and Repair, Taylor's superior, requesting the navy's authorization for tests along the lines that he and Taylor had agreed upon, and in February approval was given. Thus began Sperry's association with the navy, an extended association that brought Secretary of the Navy Charles Francis Adams to say "that no American has contributed so much to our naval technical progress."[46]

vi

Taylor's pendulum simulation of a ship and Sperry's model stabilizer were not ready for testing until the fall of 1909. In the meantime, Sperry had an opportunity to investigate the state of gyroscope technology in Europe and to compare his stabilizer with those of Schlick and Brennan. The trip was originally planned to promote his detinning process in Europe. Sperry was also to read two papers at the Seventh International Congress of Applied Chemistry, in London.

After his arrival in London in May, he met Sir William Crookes, the chemist and physicist noted for his work with cathode rays. Shortly thereafter, he wrote to Crookes, who was on the Council of The Royal Society, offering to demonstrate the gyrostabilizer with models brought from America.[47] Sperry also called on Professor John Perry, author of a book on the gyroscope, *Spinning Tops*, which Sperry admired. After Perry heard of the inventor's work, he suggested that Sperry present his models at the *conversazione* of The Royal Society on June 24. Sir William Crookes made the same suggestion to the Secretary.[48]

[45] Notebook No. 48; Sperry Diary, December 22, 1908; EAS to Chief Constructor, Bureau of Construction and Repair, January 27, 1909; and Taylor to EAS, January 28, 1909 (SP).
[46] *New York Times*, June 17, 1930.
[47] EAS to Crookes, June 2, 1909 (SP).
[48] The yearbook of The Royal Society does not record a *conversazione* on June 24, 1909. *Yearbook of the Royal Society* (London: Harrison and Sons, 1910), Sperry refers to the suggestion of Crookes and Perry to the Secretary in EAS to Secretary of The Royal Society, June 16, 1909 (SP).

While in Britain, Sperry also visited the ship-building firm of Swann, Hunter, and Wigham Richardson. Under the Schlick patents, the firm had installed a stabilizer on a Scotch coastal vessel. The installation seemed particularly suitable because the vessel had to lie off the coast in rough seas to take on sheep from scows. While the firm cordially received him and showed him the *See-Baer* with its Schlick stabilizer, he was not permitted to see the Scotch installation. Some years later Sperry was told that in a heavy storm the passive-type Schlick gyro, not provided with stops, had "flopped completely over" and tipped the vessel bottom side up. He believed that the disaster had set back the cause of the gyro-stabilizer in England "tremendously."[49]

Sperry also planned a visit to Louis Brennan. His intent was to acquire the right to use Brennan's car at an amusement park. Sperry did not anticipate that his work for the navy would soon leave little time for amusement devices. In his initial letter, Sperry presented himself more as a promoter than as an inventor and engineer; perhaps he thought this stance would be more persuasive. Sperry wrote that "from our view point in the States, we feel that these matters [gyroscope application] outside of their scientific interest are only, or at least mainly, useful in proportion as they can be made profitable or made to become 'money earners.' "[50] Brennan replied cordially, but the meeting was delayed. When Sperry did go to see Brennan in Kent, Brennan showed interest. He, in turn, journeyed to London to continue their discussions. The two agreed to correspond, and Sperry promised to have his "organization" in Berlin send Brennan "full literature on what is being done and contemplated in the German metropolis."[51] They did correspond during the next year, although Sperry's pursuit of the plan became somewhat desultory.[52] As development of the gyrocompass and gyrostabilizer for the navy progressed, Sperry lost interest in this kind of peripheral commercial exploitation of the gyro.

In Germany, Sperry spent more than a week with Schlick and others developing and testing the Schlick stabilizer; he also visited the Krupp marine engineers at the Germania Works at Kiel, as well as the faculty at the Charlottenburg *Technische Hochschule* in Berlin, to discuss gyroscopic matters. The North German Lloyd Line allowed him to experiment with the Schlick 18-ton stabilizer installed on their *Silvania*. For twelve hours in rough weather, he observed the stabilizer—the largest in the world—and made notes that he thought would be helpful when he conducted his tests with Taylor.

Sperry's friend, Leo Baekeland, arranged for Sperry to meet Krupp engineers, "and before he knew it, his gyrostabilizer was examined,

49 EAS to Charles Doran, January 23, 1928 (SP).
50 EAS to Brennan, June 5, 1909 (SP).
51 EAS to Brennan, August 12, 1909 (SP).
52 EAS to George Tilyou at Steeplechase Park, Coney Island, New York, June 22, 1910 (SP); and a series of letters from EAS to a group in Los Angeles that proposed to develop a residential section and a pleasure resort using a Brennan monorail car for access (SP). The intent was to acquire the Brennan rights through Sperry.

dissected, discussed, and its business possibilities were arranged all within 24 hours."[53] Baekeland was an admirer of German firms because scientists and engineers helped manage them, not just "bankers or so-called business men." He predicted that the Sperry gyrostabilizer would be adopted by the German navy. The Krupp engineers wanted Sperry to stay in Germany to install a large plant on two torpedo-boat destroyers and to discuss the possibility of an installation on a submarine of 300 tons.[54] Krupp proposed to acquire exclusive right to make and install the stabilizer in Germany (for naval ships) if the test installations proved the efficacy of his design. Sperry would be remunerated for each installation on the basis of the displacement of the vessel on which the gyrostabilizer was installed.[55] Krupp also asked that Sperry guarantee that he would not grant licenses to others outside of Germany on more favorable terms; Krupp thought this stipulation merited, considering the expensive risks the firm would undertake in the trial installations. Sperry did not find the proposition acceptable, but he was encouraged that the Krupp engineers, whom he found the keenest he had met on the continent, thought well of his design.[56]

While in Germany, Sperry also met Hermann Anschütz-Kaempfe, designer and builder of the gyrocompass.[57] From his talks with the German inventor and with others who knew him, Sperry concluded that Anschütz-Kaempfe had little mechanical ability and depended greatly on his engineering staff. Sperry had already begun on the design and patent application for his gyrocompass and thought his own design would result in a "far cheaper instrument . . . than Anschutes [sic] is turning out."[58] He returned to the United States unimpressed by the general state of gyro technology in Europe and convinced that his work was more original, more practical, and more promising. That an American inventor could go to Europe, with its reputation for science-based engineering, and not only hold his own with the inventors and engineers but even make them take note, attests to the vitality of American technology and of Sperry.

vii

Upon his return to America in September, Sperry found that Taylor had a model of a dreadnought-class battleship ready for testing Sperry's stabilizer. Taylor's model was a heavy pendulum with weights that could be adjusted horizontally and vertically to represent ships with various metacentric heights. Arrangements were made so that the amplitude and frequency of roll could be varied. Initially, the pendulum was weighted and the metacentric height adjusted to represent the 26,000-ton dreadnought.

[53] Baekeland to E. H. Hooker, June 30, 1909 (SP).
[54] Friedr. Krupp-Germaniawerft to EAS, July 5, 1909 (SP).
[55] Friedr. Krupp-Germaniawerft to EAS, July 8, 1909 (SP).
[56] EAS to Henry Howard, July 29, 1909 (SP).
[57] See below, p. 130.
[58] EAS to Henry Howard, July 29, 1909 (SP).

Sperry supplied a model of his active stabilizer and a comparable model of a Schlick passive type. Air pressure precessed the active gyro, and electricity powered the gyro wheel, the angular velocity of which was measured by Taylor's stroboscope. The experimenters found that with practice they could determine revolutions up to 7000 rpm at which the Sperry model was intended to run. Taylor devised a means of recording the angle of precession on the same strip of recording paper with the angle of roll of the pendulum "ship," thus showing graphically the relationship between ship roll and precession.

Together, Sperry and Taylor were a remarkable combination of ingenuity and experimental finesse. The navy provided men and facilities to help them with the experiments. Having no company and no staff of technicians, Sperry found this assistance of considerable value. Later it was estimated that the gyrostabilizer tests at the Washington model-basin could not have been carried out by Sperry for less than $50,000.[59]

Sperry later recalled a story about one of the bluejackets who helped them. After the experiments, the sailor asked Sperry what they were all about. When told that the experiments proved that the rolling of ships could be prevented, the sailor said that if the inventor had been to sea he would know it was foolish to try such a thing: "Why when you get out there in the middle of the ocean what have you got to hang on to to hold her?" Sperry, perhaps without success, explained that the gyro created a "fulcrum in space."

Taylor performed an invaluable service for Sperry, who was no mathematician, by preparing a mathematical study of gyro behavior and its application to ships.[60] Taylor appended this study to his report on the tests, after he was unable to locate a straightforward analysis of gyro action for his superiors. Sperry, too, had been bothered by the abstruse character of earlier analysis and said of Taylor's:

In this most unique and valuable work, Captain Taylor has given an original mathematical treatise on practically all the phases and bearings of this question [gyrostabilization of ships], including an original investigation of the underlying phenomena of the gyroscope itself. It is of the greatest value to this important art that its problems should have come under the observation and been reviewed by so able a mathematician, experienced in all branches of experimental research.[61]

Taylor's 40-page analysis, among other things, verified a formula for the reaction couple of a practical gyro (the couple is the product of the weight of the gyroscope wheel, the square of its radius of gyration, its velocity of rotation, and the angular velocity of tilt, or precession, of the gyroscopic axis). He also derived a formula for determining the "roll

[59] Memorandum prepared by Sperry Gyroscope Company, "The Sperry Gyroscope, Its Commercial Uses and the Results Already Obtained in the Various Fields," c. 1911 (SP).
[60] "Theory of the Gyroscope and its Application to Ships to Prevent Rolling." (Mimeograph, SP.)
[61] Sperry, "The Gyroscope for Marine Purposes," pp. 150–51.

quenching power" of a gyrostabilizer, when the characteristics of the ship and gyroscope are known. By roll-quenching power, he meant the roll that the stabilizer would eliminate by a single impulse under the most favorable conditions. Sperry used these formulae in subsequent developmental work.

Taylor's report to the head of the Bureau of Construction and Repair[62] summarized his major findings and recommended further tests. He found the agreement between the experimental results and his theory "in general as close as to be expected." Furthermore, the report confirmed Sperry's assertion that the active gyrostabilizer had advantages over the passive, for Taylor was convinced that the force of the gyro couple should be applied when the ship was nearly upright and that only the active type could be controlled to do this. Taylor also concluded, as Sperry had, that the most efficient gyrostabilizer should respond sensitively to incipient rolls—and the active type had the advantage here, too.

In conclusion, Taylor suggested that one of the older destroyers be used for sea trials of the gyrostabilizer. He presumed that "Mr. Sperry would undertake to supply the gyroscopic portion of such an experimental installation at a reasonable price and within a reasonable time." For the gyroscopic portion, Sperry would receive $25,000. Far more helpful than the money, however, would be the information acquired as a result of the full-scale tests, information to be used later to solve problems of general design involved in the application of the stabilizer to ships.

viii

Early in 1911, the Bureau of Construction and Repair designated the USS *Worden*, a 433-ton torpedo-boat destroyer, for the sea trials. This would be the first installation of a gyrostabilizer on a man-of-war in active service and the largest investment made by any navy in gyro development. (The importance of naval support is perhaps clearest here. It is inconceivable that Sperry could have provided a ship with crew from his own funds. He later estimated that the cost to equip and operate a vessel for the contemplated tests would have been $150,000.) From years of experience, Sperry—and the navy—knew that complex devices intended to perform in complex environments could not be produced on the basis of scale model tests in artificial or simulated environments. Tests of full-scale devices in real conditions were necessary.

Constructing a gyrostabilizer of unprecedented size presented formidable problems. These included the design and manufacture of such subcomponents as a gyro wheel able to withstand severe internal stresses and bearings able to withstand extremely high pressures. The solutions to these problems depended on the existing state of technology. Sperry's major problems were no longer related to design, but centered on the search for suitable construction materials, machines, and processes to make the components. Sperry had to subcontract the work. He negotiated the contracts,

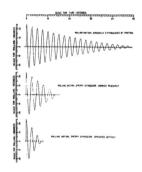

Figure 4.17
Comparative tests of active and passive ship stabilization. From E. A. Sperry, "The Gyroscope for Marine Purposes," Transactions of the Society of Naval Architects and Marine Engineers, *XVIII, 1910.*

Figure 4.18
Captain David W. Taylor, USN, made these calculations in analyzing the Sperry marine stabilizer. From Appendix to a report from Taylor to Chief Constructor, Bureau Construction and Repair, July 15, 1910.

[62] Report dated July 15, 1910 (SP).

scheduled the construction, provided engineering drawings, and monitored acceptance tests. Finally, he supervised, in close cooperation with the navy, installation on the *Worden*.

Sperry employed Carl Lukas Norden, an engineer twenty years his junior, to help him. (Norden became known in World War II for the Norden bombsight.) Norden, a round-faced Dutchman who smoked cigars (to Sperry's discomfort), had apprenticed for three years in a machine shop before entering the world-famous Zurich Federal Polytechnic School. After graduation in 1904, the twenty-four-year-old sailed for America, as did many young engineers at the turn of the century. They knew that European technical education was well regarded in America, and they also had heard that the expanding economy of the United States allowed the engineer great opportunity.[63] After arriving in America, young Norden worked two years for the Worthington Pump and Machine Company in Brooklyn and then became a design engineer for the J. H. Lidgerwood Manufacturing Company. Sperry probably met him in this connection, for Sperry negotiated for a time with Lidgerwood for construction of the precession motor.

The Hyde Windlass Company of Bath, Maine, was selected to build the precession motor; the New Jersey Steel Company was engaged to cast the gyro wheels (the stabilizer would have two wheels); and the Midvale Steel Company of Philadelphia contracted to forge the shafts and machine the wheels.[64] General Electric was selected to design and construct the main bearings because in manufacturing high-speed and heavy wheels for turbines, the company had gained valuable experience in designing and constructing bearings able to withstand high pressure and temperature. General Electric would also supply a small steam-turbine electric generator set to supply alternating current to drive the gyro wheels. By June 15, 1911, Sperry could report to the Bureau of Construction and Repair that practically all of the subcontracts had been let and the work was going forward.

The difficulties encountered with the gyro wheel caused much anxiety for Sperry and Norden. The wheels were monsters, weighing 4000 pounds and measuring 50 inches in diameter. Besides drawing on the experience of the companies, Sperry brought in as consultant Aurel Stodola, one of the world's experts on turbine design.[65] The navy carried out a series of tests on the wheels during casting and forging, and Taylor was interested enough to travel from Washington to Philadelphia to witness one of the tests. They involved testing the strength of small sections removed from

[63] Despite American romanticizing of the inventor or engineer who had studied at the "great university of hard knocks," well-educated graduates of European polytechnic schools found excellent positions in the U.S. This was especially true in electrical engineering, chemistry, and advanced machine design.

[64] Frederick W. Taylor, the scientific management pioneer, first made a name as chief engineer at Midvale. Before leaving Midvale in 1890, he had designed and constructed for the company the world's largest steam hammer for forging.

[65] See A. Stodola, *Steam and Gas Turbines* (translated from the 6th German ed.; New York: McGraw-Hill, 1927).

Figure 4.19
The USS Worden,
showing the deck mounting
of the two gyros in housings
astern of the lifeboats.
From the Sperry Gyroscope
Co., "The Sperry Ship
Stabilizer" (undated
typescript in Sperry Papers).

Figure 4.20 (Right)
Testing the ability of the
active stabilizer to roll the
USS Worden, which is tied
to the pier. From The
Sperry Gyroscope Co., "The
Sperry Ship Stabilizer."

Figure 4.21
The stabilizer
for the USS Worden
undergoing shore tests
at the Brooklyn Navy Yard,
May 1912. From E. A. Sperry,
"The Active Type
of Stabilizing Gyro."

the shaft forging and the wheel casting. Tests of wheel balance in the presence of naval inspectors were frequent. Although dimensions were held to very close tolerances,[66] the naval inspector found fault with the wheels machined at Midvale, rejected both, and New Jersey Steel had to cast new ones. On September 13, a month after the rejection, new castings were ready. Sperry reported to the navy: "We have the pleasure of stating we have finally secured two steel wheels which should be first class, inasmuch as they are poured without a core and each forms the bottom of an 8-ton casting."[67] The weight in excess of the two tons specified for the wheel consisted of huge "risers" subsequently removed from the wheel.

Despite many other problems, the gyrostabilizer was delivered to the navy in April 1912 for shore tests before installation on the *Worden*. These tests involved operating the gyros at gradually increased speeds and rate of precession. Sperry noted with great satisfaction that the main shaft bearings performed well, that the precession engine moved the heavy gyro to and fro at a rate exceeding specifications, that the turbo generators gave every indication of success in service, and that the balance of wheel and shaft was good. "We now have had not only considerable experience with the gyros," he wrote, "but find that their operation is simple and much more powerful than originally planned."[68] On shore, the stabilizer favorably impressed everyone who witnessed its performance.

The shore test brought some newspaper publicity, which Sperry probably welcomed because he hoped to develop a market in the merchant marine and especially on passenger vessels. Headlines on the *Evening Sun* proclaimed, "New Physical Law Revealed; Strange Force Shown by Gyroscope at Navy Yard; Stunts that Startled; 'Unique Reactions in Pure Mechanics,' Says Inventor."[69] The imagination of the reporter had been stimulated when he learned that the gyro, when precessing, had risen with its steel base from the floor. To him this was as remarkable as if a man "sitting in the centre of a heavy bed should raise his hand in the air and lift the bed beneath him." Undoubtedly Sperry talked to the reporter, for the latter used the metaphors of the inventor, such as "mysterious beast," and "tremendous muscles." Other favorable newspaper accounts, some stressing the gyro as a boon for the seasick, followed, but Sperry's immediate problem was his navy contract.

After the gyrostabilizer was installed on the *Worden*, Sperry wrote to Taylor: "That Monday, July 15, marks an epoch in history inasmuch as a ship was first rolled by an active type gyro."[70] Sperry was excited because he had witnessed a convincing demonstration of the validity of his active principle. While tied at the dock, the *Worden* had been rolled in dead

[66] Sperry Gyroscope Company to Chief of Bureau of Construction and Repair, June 19, 1911 (SP).

[67] EAS to Chief of Bureau of Construction and Repair, September 13, 1911 (SP).

[68] EAS to Chief of Bureau of Construction and Repair, undated (probably June 1912; SP).

[69] The New York *Evening Sun*, Wednesday, May 22, 1912.

[70] EAS to Taylor, July 16, 1912 (SP).

water by the artificially precessed stabilizer. Sperry assumed, then, that the stabilizer could counter a wave-produced roll comparable to the artificial roll. The *Worden* stabilizer tests were successful, but not conclusive. Sperry predicted that the installation would quench a five-to-seven degree roll, and it performed as well as, or better than, that.[71] (Unofficially, some years later, a naval officer familiar with the installation said, "The stabilizer did pretty good work; cutting a total roll of thirty degrees down to about six degrees.") The chief complaint seems to have been the heavy stresses the stabilizer imposed upon the structure of the *Worden*. The crew found the resulting creaks and groans extremely disturbing.[72] In order to obtain more conclusive results, the navy later considered designing a compartment of a battleship to receive the stabilizer and actually installed another stabilizer on a 10,000-ton transport, the *Henderson,* during World War I (see below, pp. 223-30).

The Sperry Gyroscope Company began offering the stabilizer on the commercial market in 1915, a date that may be taken as the beginning of innovation. Development, however, did not end then, for gyrostabilizers of different characteristics performing in different environments disclosed the need for further adaptation.

[71] Lt. R. E. Gillmor, USN, "Active Type Gyro Applied to USS *Worden,*" *American Machinist* (August 7, 1913), p. 231.

[72] The official report on the *Worden* trial has not been located in the navy records at the national archives. The quotation is from a letter to Sperry by an S. Myers (probably one of his employees), who talked in 1923 to a naval officer who had been, it seems, on the *Worden*. Sperry's informant reported that generally the officers familiar with the stabilizer, though they acknowledged its roll-quenching powers, did not think a stabilizer "would do much good" on a destroyer, even for ordnance work. S. Myers to EAS, December 21, 1923 (SP).

CHAPTER **V** Harnessing the Rotation of the Earth:
The Gyrocompass

*You see it is very necessary to leave magnetism entirely and to reach out and lay
hold of some other equally prevalent field of force.*
ELMER A. SPERRY TO JOHN D. ROCKEFELLER,
November 23, 1918

*One said that 'true as the needle to the pole' was a bad figure, since the needle
seldom pointed to the pole. He said a ship's compass was not faithful to any particular point, but was the most fickle and treacherous of the servants of man. It
was forever changing. It changed every day in the year; consequently the amount
of the daily variation had to be ciphered out and allowance made for it, else the
mariner would go utterly astray. Another said there was a vast fortune waiting for
the genius who should invent a compass that would not be affected by the local
influences of an iron ship. He said there was only one creature more fickle than a
wooden ship's compass, and that was the compass of an iron ship.*
MARK TWAIN, "Some Rambling Notes of an Idle Excursion,"
Atlantic Monthly, XL (1877), pp. 444–45

WORK WITH THE ACTIVE STABILIZER FAMILIARIZED SPERRY WITH THE ESSENTIAL
characteristics and applicability of the gyro and also brought him into close
relationship with the navy. These circumstances greatly facilitated his entry
into the gyrocompass field. He began development of his gyrocompass
while still at work on the stabilizer models, and when he built the *Worden*
stabilizer in 1911, he also installed his first gyrocompass. Sperry's development of the gyrocompass paralleled that of the stabilizer not only in time
but also—as will be seen—in the way he analyzed prior work, identified inadequacies, and improved upon these in designing a practical device
fulfilling an imperative need.

i

The magnetic compass functioned satisfactorily on wooden ships. It
depended on the earth's lines of magnetic force, and although variations
in the direction of these lines frequently caused the needle to point away
from true north (declination), correction tables allowed the navigator to
compensate for the error. Although the lines of magnetic force were
weak, the wooden hull did not interfere with the operation of the compass.
The iron and steel ship changed that situation. The metal hull and super-

structure did shield the compass from the earth's lines of force and created a false magnetic field that the compass could not distinguish from the earth's. To counter this error (deviation) the compass was compensated by a complex procedure; however, changes in cargo and the swing of large steel turrets could nullify compensation procedures. Thus, the navigators found that steering a course by a magnetic compass often led to substantial error, which they sometimes attributed—with tongue in cheek—to unknown ocean currents. These "unknown currents" multiplied in the final decade of the nineteenth century when electricity was introduced aboard naval ships. The electric wires, generators, and motors created an electromagnetic field that also disturbed the magnetic compass. Considering the warship as a system, one could clearly perceive that the magnetic compass was a weak component by 1900.

The problem of magnetic compass errors became increasingly critical as the navies of the leading powers improved their gunnery. The introduction of all big-gun ships carrying ten- and twelve-inch cannon firing at great ranges, continuous-aim firing, and other gunnery advances increased the need for complementary improvements in the compass."[1] "Ordnance has been developed at great expense," the Chief of the Bureau of Navigation wrote, "ballistics have been improved, but these are to a certain extent nullified if the arrangements for fire control, of which the compass is essential, are not improved along with them."[2]

The need for a gyrocompass in submarines was especially crucial. When submerged, the submarine's course was virtually guesswork, for the steel hull shielded the magnetic compass from the earth's magnetism and the electric motors propelling the submerged craft created a magnetic field that hopelessly confused the compass. Without an improved compass the submarine could not make extended underwater runs in restricted waters or through charted mine fields.

The need to equip the submarine with a serviceable guidance device resulted circuitously and unpredictably in the first practical gyrocompass. After taking part in an arctic expedition, Dr. Hermann Franz Joseph Hubertus Maria Anschütz-Kaempfe (1872–1931) planned, in 1902, to use a submarine to reach the North Pole. He published his plans in a brochure entitled, "The U-Boat in Service of Polar Exploration." Some thought that his scheme was absurd; however, when the Krupp shipbuilding firm, *Germaniawerft,* stated that it could build a submarine to navigate beneath the ice, the only remaining obstacle seen by Anschütz-Kaempfe was the lack of an effective compass or guidance system. Although his degree was in art history and his graduate studies in medicine, Anschütz-Kaempfe tried to construct a guidance device which could establish a fixed reference line by utilizing the characteristic of a gyroscope, freely

[1] To improve the performance of the magnetic compass, the hull and fittings near it were made of brass or other nonmagnetic materials, but this solution was expensive and "at best was only a palliative and not a remedy."
[2] Chief of the Bureau of Navigation (BuNav) to the Secretary of the Navy (NavSec), December 15, 1910, National Archives Record Group 24, #5778–84. Hereafter cited BuNav and NavSec.

Figure 5.1
Leon Foucault's gyroscope
by which he demonstrated
the rotation of the earth.
So that minute relative
motion could be observed,
the microscope was focused
on a scale mounted
on the gyroscope.
Foucault's use of a filament
to suspend the gyro
influenced Sperry's design
of the gyrocompass.
Later, he visited
the Conservatoire National
des Arts et Metiers in Paris
to see the historic instrument.

suspended at its center of gravity, to maintain the orientation of its axis of rotation. In 1903 tests on the Starnberger See, a lake near Munich, revealed a serious weakness: friction at the gyro's bearings introduced disruptive forces, especially when the vessel accelerated.[3] Furthermore, constructing a gyro with a center of rotation coincident with its center of gravity was difficult. Because the device could be used, however, for short periods to establish a reference from which a course change was observed, the German navy became interested. By 1904 Anschütz-Kaempfe and the navy had tested his guidance device on the cruiser *Undine*.[4]

Anschütz-Kaempfe turned to the development of a gyrocompass—in contrast to a guidance device—in 1906. Theoretically, the gyrocompass would, at any place on the earth's surface (other than the two poles), align its axis of rotation parallel to the axis of the earth (along the meridian). Such a device would not serve for the inventor's polar expedition, but now he was developing his invention with the help of the navy, which did not place high priority on polar expeditions by submarine.

Encouraged by the German navy, Anschütz-Kaempfe formed the Anschütz Company at Kiel in 1906. A group of scientists and engineers, led by the expert mathematician Max Schuler, assisted in the development of the compass, which had its sea trials on the *Deutschland*, the fleet flagship, in 1908.[5] After the tests, the captain reported that during the 28-day trial the compass ran without interruption and that errors in bearing were infrequent and easily correctable.[6] This performance attracted the attention of the naval personnel, engineers, and scientists throughout the world—including that of Elmer Sperry.

ii

Anschütz-Kaempfe's inventions were based on prior science, and Sperry would also apply these principles in inventing his compass. Léon Foucault, the French scientist famous for his demonstration of the earth's rotation, had published an account of his experiments with the gyro in 1852. He called the device a gyroscope (literally, "to view the turning"), for he used it—as he had used the pendulum earlier—to show the rotation of the earth. He also predicted that the gyro could be used as a compass.[7]

3 Anschütz-Kaempfe to the Chief of the Nautical Section of the German navy, December 22, 1903 (copy in Deutsches Museum, Munich).

4 Letter of Anschütz-Kaempfe, March 17, 1904 (copy in Deutsches Museum, Munich).

5 For more on the development of the Anschütz compass, see *The Anschütz Compass* (London: Elliott Brothers, 1910).

6 Anschütz-Kaempfe to State Secretary von Tirpitz, May 20, 1908 (copy in Deutsches Museum, Munich).

7 See Léon Foucault, contributions in *Comptes Rendus des Seances de l'Academie des Sciences*, XXXV (1852), pp. 421, 424, and 496. See also Foucault communication to the editor of *Journal des Débats*, September 22, 1852, and a communication in the same journal, October 1, 1852. All of these are reprinted in *Recueil des Travaux Scientifiques de Léon Foucault* (Paris, 1878), pp. 401–5, 406–9, and 528–33. Sperry had a translation of the communications made in 1911. In 1836 a certain "Lang of Edinburg" suggested, but did not carry out, experiments similar to Foucault's.

To understand the work of Foucault—and of Sperry and Anschütz-Kaempfe—it is necessary to contrast two basic gyros according to their mounting. A gyro with three degrees of freedom—such as the Anschütz-Kaempfe guidance device—is free to turn about three axes including its own axis of rotation. Foucault achieved this by mounting his gyro in two rings, one rotating on a horizontal axis within the other, which in turn rotated on a vertical axis. The gyro wheel's center of gravity was at the intersection of all three axes of rotation. Foucault rotated the wheel with a string, and the rapidly rotating gyro wheel maintained its orientation in space for a brief period. In his experiment, Foucault aligned an optical sighting device on the gyro wheel. The relative displacement of the sighting device and the gyro demonstrated the motion of the earth, for, as Foucault assumed, the sighting device turned with the earth and the gyro maintained its orientation in space during the few minutes when the wheel rotated at high speed. The gyroscope, Foucault wrote, is

supported by a sort of Cardan's suspension, analogous to that which supports chronometers and ships' compasses, with this difference only that the concentric circles, instead of being in the same plane, are normally, or in their average position, perpendicular one to the other. One of them plays round a horizontal . . . while the other is movable about a vertical axis represented by a suspension and consisting of a torsionless filament. If, at the moment it is put in rotation, this body by its axis points at a star in the sky, during the whole time that the movement (spin) lasts this axis will remain pointing toward the same point of the firmament . . . if we aim at one of the points of the heavens which appear to be moving quickly, the axis of rotation . . . will be found to share the same apparent movement [as the heavens].[8]

Foucault also experimented with a gyro having two degrees of freedom and predicted its use as a compass, but was unable to demonstrate this because he could not maintain a high-speed rotation of the wheel.[9] To achieve two degrees of freedom, the interior ring of the gyroscope was weighted so that the freedom of rotation about the axis of this ring was restrained, leaving only two degrees of (complete) freedom. The gyro then became pendulous; that is, it felt the force of gravity. Foucault predicted that a gyroscope designed along these lines would feel the force of gravity and that this force would cause it to precess until the axis of rotation of the spinning wheel was parallel to the axis of rotation of the earth. With its axis so oriented the gyroscope would be pointing to true north. Although Foucault could not maintain the spin of the gyroscope, several scientists, including G. Trouve in France (1865) and G. M. Hopkins in America (1878), solved this problem by making the gyro wheel the armature of a direct-current motor.[10]

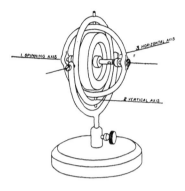

Figure 5.2
Demonstration gyro mounted to allow free movement about three axes, or three degrees of freedom. From Elmer Sperry, Jr., "The Gyro-Compass," The Sperryscope, IV, 1925.

[8] Léon Foucault, quoted in EAS Patent No. 1,279,471.
[9] This discussion of Foucault's work is based on O. Martienssen, "Die Entwicklung des Kreiselkompasses," Zeitschrift des Vereines deutscher Ingenieure, LXVII (February 24, 1923), pp. 182–87. A. Schein, an expert assistant of Sperry's, cited the Martienssen article for its excellence.
[10] G. Trouve, "Gyroscope electrique pour la demonstration du movement de la Terre," Comptes Rendus des Seances de l'Academie des Sciences, CXI (1890),

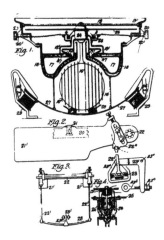

Figure 5.3
Elmer Sperry's first patent
on a gyrocompass.
On the first compass he built,
he abandoned mercury
suspension (shown here in
tank [18]) and used the
torsion wire. (Patent No.
1,242,065.)

Spurred by the possibility of using a gyro with an electric drive as a compass, a number of inventors in the late nineteenth century attempted to build practical devices. William Thomson (Lord Kelvin) described his before the British Association for the Advancement of Science in 1884 and although the British navy tested the device, the navy did not adopt it.[11] About the same time, a Frenchman, Dubois, tested his compass on board a French frigate, but his efforts did not culminate in a practical device. Among the other nineteenth-century pioneers were Marinus Geradus Vanden Bos of Leyden and Werner von Siemens of the German electrical manufacturing firm of Siemens and Halske. Elmer Sperry later credited Vanden Bos with the first patent disclosing the fundamentals of a true gyroscopic compass.[12] All of these men contributed to the development of the gyrocompass, but none of them persisted until a practical shipboard device was produced.[13]

iii

The experience Sperry had acquired in developing the gyrostabilizer gave him an advantage that increased his chances for successfully producing a gyrocompass. The immediate cause of his decision to invent a gyrocompass seems to have been the publicity given Anschütz-Kaempfe's efforts.[14] In November 1908, at a meeting of the Society of Naval Architects and Marine Engineers, Sperry commented on the gyrocompass. After hearing a lengthy paper on the errors of the magnetic compass, he asked for a few minutes to inform the meeting of the success of "a German instrument-maker" during "the past two years" in developing a compass that did not have the weaknesses of the magnetic compass. Sperry said that a friend of his had examined the machine during the summer of 1907 (probably at the Anschütz factory) and that improvements had been made in the interim. Sperry knew of the sea trials on the *Deutschland* and believed that the instrument operated with extreme accuracy. Then he prophetically commented upon the future of the gyroscope:

The Marine interests may certainly expect great and material aid from this wonderful instrument. Not only, I predict, will it guide our ships but it will be found to have other important and far-reaching bearing upon the operation of ships at sea.

Our knowledge of its powers, which are no less than wonderful, and

pp. 357–61; and G. M. Hopkins, "An Electrical Gyroscope," *Scientific American*, XXXVIII (1878), p. 335. When Sperry credited Hopkins, the American physicist, with the electrically driven gyroscope, Elihu Thompson recalled, "I think it is much older as I remember there was such a gyroscope in the collection of apparatus of the Central High School in Philadelphia when I was teaching there . . . certainly in existence there in the early seventies"; Thomson to EAS, May 4, 1912 (SP).
11 Sir William Thomson, "On the Gyrostatic Working Model of the Magnetic Compass," *Report of the 54th Meeting of the British Association for the Advancement of Science* (London: John Murray, 1885), pp. 625–28. In this article, no mention is made of naval tests.
12 Vanden Bos, British patent issued 1884 (8394); German patent number 34,513.
13 Martienssen, "Entwicklung des Kreiselkompasses," p. 182.
14 See, for an example of the publicity, Frederick Collins, "The Gyroscope as a Compass," *Scientific American*, XCVI (April 6, 1907), p. 294.

acquaintance with the true laws of its action and functions, are gradually increasing and now that the Darius Greens have made good, we should not be surprised if this instrument which has for centuries been a scientific toy and a 'plaything of mathematicians' should become one of the most useful in the arts and industries, even outside shipping and naval interests, in which field I look upon it as having made great practical advances, if not having reached the point of complete demonstration.[15]

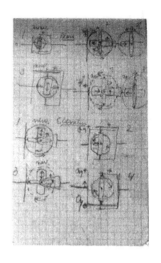

His remarks do not show how far his interest in the gyrocompass had carried him in the invention-development process, but only a month later, it is clear that he was planning experiments, for he wrote to a friend, Henry Howard of the Merrimac Chemical Company in North Woburn, Massachusetts, thanking him for suggesting that the Waltham Company do the fine precision work necessary for a test model of a gyrocompass.[16]

Sperry's notebooks indicate that he became interested in inventing and developing a gyrocompass late in 1907, or early in 1908, when his first gyrocompass entry describes a spring action to counter precessional error in a compass.[17] In 1908 work on the gyrostabilizer delayed the compass, but his interest assumed new life in 1909 when he learned from a friend "high in the Navy" (Taylor?) of the navy's need for a compass.[18] Compass entries in his notebook then proliferated. He devised a way of dampening unwanted compass oscillations with magnets and a means of driving repeater compasses. He also made calculations for the size of a compass gyro wheel.[19] In September 1909, he applied for a gyrocompass patent (see figure 5.3)[20] that did not prove basic, for Sperry fundamentally changed the design before filing his basic patent in 1911.

iv

Since he was working in 1909 on both a stabilizer and a compass and had other gyro applications in mind, Sperry rented space in a machine shop at 18-20 Rose Street, New York, in the shadow of the Brooklyn Bridge. and hired a small design staff to work in it. He probably used the money from the sale of his detinning patents for this purpose. The owner of the shop, the Fred K. Pearce Co., was one of the model-builders upon whom Sperry depended heavily.[21] An early Sperry employee, Carl F. Carlson, who

[15] EAS commenting on a paper read by Lt. Cmdr. L. H. Chandler, USN, "Deviation of the Compass Aboard Steel Ships—Avoidance and Correction." Paper read at the 16th General Meeting of the Society of Naval Architects and Marine Engineers in New York City, November 19–20, 1908; *Transactions of the Society of Naval Architects and Marine Engineers,* XVI (1908), pp. 55–104. Sperry's comments are on pp. 112–13.

[16] Howard was probably the "friend" mentioned at the SNAME meeting, for he had been abroad; he spoke glowingly of the precision work of the Germans and had an interest in the gyrocompass. EAS to Howard, December 14, 1908 (SP).

[17] The first reference to a gyrocompass comes early in Notebook No. 48, which covers the period from December 1907 to December 1908 (SP).

[18] EAS to Henry Howard, July 29, 1909 (SP).

[19] Notebook No. 50 (March 10, 1909, to December 31, 1909; SP).

[20] Filed September 25, 1909 (519,533), "Ship's Gyrocompass Set," issued October 2, 1917 (Patent No. 1,242,065).

[21] A letterhead described the company as a maker of gauges, jigs, and special tools:

Figures 5.4
Pages from Sperry notebooks.

later became supervisor of the electrical department of the Sperry Gyroscope Company, recalled the screen-enclosed room in which he worked with "Mr. Sperry" and the one other employee, Frederick C. Narveson, a draftsman who later became assistant to the chief engineer.[22] In the ten-by-fifteen-foot enclosure, the inventor, with the help of his two employees, prepared drawings for Pearce mechanics to use in making models and tested the models as they were built.

The inventor added substantially to his staff when he acquired the services of Hannibal Ford, an electrical engineer who had graduated from Cornell University. He helped develop the gyrocompass and other Sperry devices and was, until he left Sperry in 1914 to form his own company, Sperry's chief engineering assistant. Ford is now recognized as another pioneer in guidance and control devices.

Sperry first met Ford, who was also born in Cortland County, at an engineering society meeting and later began a correspondence with the young engineer and inventor.[23] Ford worked for the Smith Premier Typewriter Company in Syracuse, New York, inventing and designing components and attachments, but this job left him energy enough to invent on his own during free hours. Among other things, he had patented a "flexible block" railway system and had also made application for a patent on a gear-cutting machine. Invention excited him, and he wanted to be a professional. "Have had the *Patent Gazette* regularly this year," he wrote, "and am keeping in closer touch with inventions than ever before, also spending considerable time out of hours on some special problems."

Promotion at Smith Premier came too slowly and, in February 1909, he asked Sperry if he knew some challenging "outside" problems he might tackle. This presented a good opportunity for Sperry to test the young inventor, so he sent him a complex problem of designing a gear train incorporating elliptical gears. Ford sketched out a general solution the first evening. There were further ramifications and the technical exchange continued with Sperry finally writing, "it is certainly a great comfort to talk these matters with some one who knows mechanics as you do." That was in March; on September 27, 1909, Ford signed a contract to start work with Sperry on October 18.[24]

Sperry kept Ford and the draftsmen at Rose Street busy. As was his custom, he utilized fully the potential of the telephone. Often, while at his Wall Street office or on business trips, he telephoned lengthy instructions to his staff. Sometimes a sketch—one of Sperry's famous back-of-an-envelope or convention-program sketches—would arrive at the shop to be translated into an engineering drawing. One employee recalled that Sperry might appear a week or so after he had sent the sketch to examine the resulting drawings. His comments always showed that he remembered

---

a general, light, accurate manufacturer, and a developer of inventions—the last mentioned "a speciality."
22 Interview with Carl F. Carlson in *The Sperry Searchlight*, I, no. 10, p. 8.
23 The Sperry-Ford correspondence extends from May 18, 1907 to March 25, 1909. There are eight letters (SP).
24 Ford's pocket diary shows that he actually began on October 11. Preston Bassett, of the Sperry Company, saw this diary on May 10, 1949, at Ford's house.

every detail of his original instructions. "He had a wonderful memory and still has."[25] Sperry's ideas often came to him outside of the normal working day or under unusual circumstances. Relatives and employees recall that while he was developing the gyrocompass he often jotted notes of new ideas and made sketches in his notebook while commuting on the train or subway. Over the weekends, he continued work at home on his drawing board and over the dinner table in conversation with his sons, although Zula frowned on this.

v

During 1910 Sperry concentrated on the development of the gyrocompass. The navy was showing more interest in a compass than in a gyrostabilizer because the Germans and the British were installing Anschütz compasses on their vessels. While a stabilized ship would have some advantage during an engagement, the ship with a reliable compass would have more.

In June 1910, Sperry wrote the Naval Observatory that D. W. Taylor had informed him of the navy's pressing need and that "I have one of these compasses nearly completed."[26] Earlier in the year, however, the navy had given an Anschütz compass a satisfactory sea trial on the USS *Birmingham* and considered installing gyrocompasses on the *Utah*, *Arkansas*, and *Wyoming*, and on a 1908 class and 1909 class submarine. With funds available for these installations, the navy began negotiations with E. S. Ritchie and Sons, the American representative of the Anschütz company, in the summer of 1910.[27] Ritchie already was supplying the navy with magnetic compasses. Sperry, however, had an advantage, for the navy, though apparently willing to buy compasses made in Germany, preferred equipment manufactured in America. In the case of the new dreadnoughts *Delaware* and *North Dakota*, congressional authorization specified American-made equipment. Sperry wrote that he hoped his compass would make it unnecessary for the navy "to go to Germany."[28]

He was formally invited to submit an instrument for preliminary shore tests and subsequent sea trials; purchase of his instrument, or that of a competitor, would depend on the outcome of such tests. The expense of installation on board ship was to be borne "by the maker of the instrument."[29] Sperry had to risk the expense of accelerated development because Ritchie had acquired the rights to manufacture the Anschütz compass in America and was pressing for a contract. By December, when the

[25] Carlson interview in *The Sperry Searchlight*, I, no. 10, p. 8.
[26] EAS to Lt. Cmdr. D. W. Blamer, June 27, 1910, National Archives, Record Group 24, #5778-31-W.
[27] Lt. Cmdr. J. H. Sypher, Superintendent of Compasses, to Chief BuNav, July 14, 1910 (endorsed by Acting Bureau of Navigation Chief), National Archives, Record Group 24, #5778-23-12.
[28] EAS to D. W. Blamer, June 27, 1910, National Archives, Record Group 24, 5778-31-W.
[29] Acting BuNav Chief to EAS, August 10, 1910, National Archives, Record Group 24, #5778-31.

Bureau of Navigation requested from the Secretary of the Navy authorization to purchase five Anschütz compasses, Sperry's compass appeared to be out of the competition.[30]

How Sperry—a newcomer to the compass field—then bested the Ritchie-Anschütz combination is not made clear in the records. Yet, less than a month after the Bureau of Navigation asked the Secretary of the Navy for the Anschütz compass, the department had not authorized purchase and, instead, was testing Sperry's compass and negotiating with him. On January 19, 1911, Sperry wrote the Bureau of Navigation that he was ready for a factory inspection of his gyrocompass in operation.[31] Within two days, the bureau had established a board for that purpose.

The correspondence does not record the full story of Sperry's relations with the navy or the exact chronology of invention and development. Sperry did not have a compass "nearly ready" as he said he did in June, but it seems that the navy nonetheless encouraged him to continue his work. His appointment diary does suggest some events not noted in the official correspondence. On July 5, 1910, about the time negotiations began with Ritchie, Sperry spent a day at the Naval Observatory "on compass" matters. He made the trip after learning from a "very high" navy official (Taylor?) that the navy was ready to buy gyrocompasses from Germany and that he should come to Washington to present what he had.[32] Sperry's notebooks suggest that he then devoted extensive thought and experimentation to the compass.[33]

Some of Sperry's ideas were undoubtedly influenced by talks with the navy experts at the observatory, for they had access to all of the information on the sea trials of the Anschütz compass. In a manner that Sperry later described as his inventive technique, he looked for the weak points in earlier devices invented by others—in this case the Anschütz compass—and concentrated on surmounting those. By fall, he had recorded in his notebook—in a hand that often becomes illegible—a rough draft of the patent application forming in his mind and sketches for the design of a compass. In October or November, he completed the construction of his first compass,[34] but before turning it over to the navy for trials in January 1911, he tested it on a device that simulated a ship's pitching, rolling, yawing, changing course, and changing speed at different latitudes at sea. The tests revealed additional problems leading to further improvements in his design. Finally, in a paper before the Society of Naval Architects and Marine Engineers, given on November 17 or 18, 1910, he described the characteristic features of his design and argued that his solutions to basic compass problems were more practical than the Anschütz solutions.[35]

[30] Chief of BuNav to NavSec, December 15, 1910, National Archives, Record Group 24, #5778–84.
[31] EAS to BuNav, January 19, 1911, National Archives, Record Group 24, #5778–99.
[32] EAS to Hecht, July 30, 1910 (SP).
[33] Notebook No. 51. (January 1, 1910 to November 5, 1910; SP).
[34] EAS to Lewis Brennan, November 14, 1910 (SP).
[35] Sperry, "Gyroscope for Marine Purposes," pp. 143–54. Sperry's experimental Journal.

Figure 5.5 Sperry experimental gyrocompass, installed first on the Princess Anne, *and subsequently on the USS* Drayton *and the USS* Delaware.

Figure 5.6 Sperry experimental compass of 1911.

Figure 5.7 Sperry examining the repeater compass for his gyrocompass installation aboard the Princess Anne, *1911.*

The navy's board of officers examined the compass in operation at Sperry's shop January 24, 1911, the first step toward acceptance. The board consisted of Commander J. W. Oman, Lieutenant-Commander J. H. Sypher (Superintendent of Compasses), and Lieutenant-Commander B. B. McCormick. Their report of January 28 concluded: "The Board finds the operation of this compass satisfactory, and believes it can be adapted to service afloat; and it recommends that the apparatus be installed on board ship at the maker's expense for the purpose of further tests."[36] Sperry must have read the report with relief, followed closely by elation, for he had risked time, funds, and professional reputation in a new and complicated field, against formidable competition.

After this success, Sperry and the navy moved quickly to the first shipboard installation and sea test. The navy intended to use the USS *Birmingham,* but her itinerary delayed installation. Lieutenant-Commander Sypher, instead, asked Sperry in May 1911 to install his compass aboard a destroyer. Before this, however, Sperry and Ford, both impatient, had made a preliminary sea trial aboard the *Princess Anne,* one of the Old Dominion Line ships running from New York to Hampton Roads. The gyro was first started on April 21, and ran continuously for five days. Sperry and Ford reported the trial a reassuring success. (Some years later, Sperry confided that one of the high points of his career came when he first saw his compass aboard ship, running admirably, turn slowly toward and settle on the meridian.)

The ship chosen for the navy shipboard test was the *Drayton,* an 800-ton destroyer. Ordered to berth in New York for several weeks, the *Drayton* was to put to sea daily to test the gyroscope under various conditions. Commander Sypher predicted that "this is a severer test than it would be even on the *Birmingham,* but I have no fear of the result. When it is passed," he wrote, "we can get down to business in regard to putting them in the Navy."[37]

Installed in the powder magazine of the ship, surrounded by heavy steel that would have prevented a magnetic compass from functioning, the compass went to sea on May 22. The officers and Sperry found its performance "very successful." Tests running the gyro rotor at various speeds were also "successful." Generally Sperry reported the compass "running excellent on the meridian."[38]

The sea trials demonstrated the remarkable advantages of the gyrocompass. The officers and crew had grown accustomed to the vagaries of the magnetic compass—even to the *Drayton's* compass being thrown off twenty degress by the heat from the smokestack nearby—but the supposed conservatism of naval officers in response to new equipment was not dis-

gyrocompass, installed on the *Princess Anne,* the *Drayton,* and the *Delaware,* is at the Smithsonian Museum of History and Technology (accession #106664, catalogue #309636).

[36] Board to BuNav, January 28, 1911, National Archives, Record Group 24, #5778–101.

[37] Sypher to EAS, May 1911 (SP).

[38] Sperry appointment diary, May 21 and 22, 1911 (SP).

cernible. Sperry would find at first that the merchant marine, disinterested in precision gunfire, was less interested in the expensive compass.

The test compass was then tried on the USS *Delaware*, first of the Navy's dreadnoughts and the ship chosen to represent the navy at the celebration of the coronation of King George V. The *Delaware* displaced 20,300 tons, attained a speed of 21 knots, and carried a main battery of ten twelve-inch guns and an eleven-inch armor belt. She was a dramatic ship, and to equip her with a Sperry compass roused considerable excitement among Sperry and his colleagues—and also in the Sperry home, where Zula Sperry wrote enthusiastically to her brother of the installation.

Hannibal Ford took the test compass to the Boston Navy Yard, where the *Delaware* had put in after a European cruise. The officers had seen an Anschütz compass on the British cruiser *Indomitable* and looked forward eagerly to the Sperry installation. The *Delaware* was well suited for the compass, for she had an excellent electrical system to power the gyro. Two of the officers were well versed in electrical engineering, Ensign Reginald Gillmor, Annapolis, 1907, and a petty officer, electrician Thomas Morgan. These two would have chief responsibility for the compass during the trials. Each would one day become president of the company that made the compass.

Installation was complete July 15. The compass installed on the *Delaware* was a sophisticated instrument in 1911; years later, however, it was remembered as a crude affair. P. K. Westcott, who became an installation engineer for the Sperry Gyroscope Company, thought the "old outfit was a curio, the binnacle consisted of a square frame set up on pipe legs . . . the compass was supported by a spring from each Corner."[39] He recalled that a Fels Naphtha Soap box served as a cover by the time the compass left the ship, and that the slate power panel for the system had been modified so often that it would have made an excellent sieve.

The master compass, linked to two repeater compasses, underwent sea trials in August and early September. The repeaters were electrically driven devices that visually reproduced in various parts of the ship, including the bridge, the movements of the master gyro below decks. Ensign Gillmor wrote a detailed report of the trials on September 22; his superior, Commander W. R. Gherardi, the navigator, submitted a summary report. After noting that the compass was the "instrument used in the development process" and predicting that many errors of construction would be eliminated in subsequent compasses, Gherardi concluded that the compass as a whole had functioned well and had demonstrated its value as a battle and maneuvering compass. He thought, however, that the sea-going officers would accept it as a navigational compass—a compass for ordinary steering—only after the compass won the confidence which came "with long performance."[40]

During the 300 hours that the gyrocompass ran during the test, Gill-

[39] P. K. Westcott, "What I Have Seen and Heard," *The Sperry Searchlight*, I (1918), p. 17.
[40] Gherardi to Commanding Officer of the *Delaware*, September 25, 1911 (SP).

mor recorded a number of performance failures.[41] All of the "casualties" noted in his report were due, he firmly believed, to simple defects to be expected in the experimental compass. He hoped that his observations would help Sperry improve the design of the production model. The Gillmor report and a Sperry memorandum written in response reveal the improvements Sperry made.[42] Gillmor noted that on two occasions the gyro frame caught a suspension ring and threw the master compass off 30 to 40 degrees; Sperry replied that this resulted from the close tolerances and the differential expansion of the two metals when the temperature rose above a certain point—a malfunction for which he provided on the production model compass. Gillmor reported a vacuum drop within the gyro case from twenty-seven inches to sixteen inches in about five hours; Sperry said that this had been corrected in the new compass; "for instance, a vacuum this morning is higher than it was last night owing to the change in barometric pressure."[43] Gillmor observed that when the turrets were fired, the repeater compass went off several degrees; Sperry responded that he and his "engineers," finding this due to an unsymmetrical, unbalanced element in the repeater, had eliminated it. The exchange continued in this way as development progressed. Sperry formed a high opinion of the naval officers with whom he dealt in developing the compass and other devices: their standards were high and their judgments impartial.[44]

Prior to the test on the *Delaware*, the navy had contracted for six gyrocompass systems. On June 10, Commander Sypher had initiated the requisition for the compasses, four for battleships and two for submarines. The contract of June 28 called for the delivery of the first set in 120 days and the remaining sets at 30-day intervals. Because the contract would tie up considerable Sperry development money,[45] Sperry read with relief that the navy agreed to pay the $10,000 for each set as delivered rather than waiting for the fulfillment of the entire contract. When, within a month, the navy asked him to manufacture ten additional compasses, Sperry proudly wrote that "the United States Government has placed with me the largest orders ever placed by any government for gyroscopic apparatus."[46]

The compass installations under the initial contract were placed on the navy's newest and proudest ships—the four dreadnoughts *Utah*, *Florida*, *North Dakota*, and *Delaware* (the test compass was replaced). The *Utah* and the *Florida* were the latest dreadnoughts, larger (21,825 tons) than the *North Dakota* and *Delaware*. Sperry installed the two other compass systems of the first contract on the submarines E-1 and E-2. By June 1912,

41 Gillmor to Commanding Officer of the *Delaware,* September 22, 1911 (SP).
42 EAS to Superintendent of Compasses, November 9, 1911 (SP).
43 *Ibid.*
44 After contracting for the first production-model Sperry compasses, Lt. Cmdr. Sypher wrote Sperry that he had told the Anschütz people that he would be glad to purchase an Anschütz outfit in order to test it thoroughly in competition with the Sperry. Sypher to EAS, June 17, 1911 (SP).
45 Sypher to EAS, June 17, 1911 (SP).
46 EAS to Gerald Christy, Managing Director, Lecture Agency Ltd., London, England, July 28, 1911 (SP).

the first sets under the second contract were being given shore tests on newly designed naval test apparatus.[47]

Compass 101, the first serially produced, was installed on the USS *Utah* November 13, 1911. Prior to installation, the compass had had a preliminary acceptance trial at the Sperry plant on November 11. The trial was witnessed by a board of officers appointed by the commandant of the New York Navy Yard. This board had the authority to accept the compass for trial installation, but final acceptance depended upon the decision of the commanding officer of the *Utah,* made after three months of satisfactory shipboard operation. After satisfactory performance during the shore-based test, Sperry and Hannibal Ford supervised installation aboard the dreadnought.

Sperry concerned himself with the details of installation; the work needed his expert and experienced eye. Furthermore, the inventor could not miss the installation of his first production model. An early employee of Sperry's recalled that when the gyrocompass 101 was being installed, "Mr. Sperry" came down to the central station of the *Utah* with "his usual cheery smile, although I know from later development that he carried a load of care. . . . Mr. Sperry promptly made himself known to me when he learned that I was in charge of the wiring . . . [and] very kindly offered a number of suggestions very useful in making the installation."[48]

The ship's officers kept the system under close observation while the ship sailed to Hampton Roads, Pensacola, Galveston, and returned to Boston. A log was kept of the compass operation: "at some times it operated satisfactorily, and at other times it was several degrees off or else precessing continually."[49] In Boston, Sperry had the compass completely overhauled; after final trials, it was accepted.[50]

The year 1911 had been an eventful one for the inventor, now in his fifties. When the year began his invention had yet to be tried by the navy, and months of intensive and exhausting testing and development followed before he placed his first production-model gyrocompass aboard a ship. In less than twelve months, Sperry had carried the compass from development through innovation.

vii

In June 1911, after the tests on the *Princess Anne* and the *Drayton,* Sperry made two patent applications.[51] One of the patents (1,279,471)

[47] *Annual Reports of the Navy Department for Fiscal Year 1912* (Washington, 1913), p. 183.
[48] Westcott, "What I Have Seen and Heard," p. 17. Westcott, apparently a navy yard employee at the time, was shortly employed by Sperry.
[49] R. C. Welles, memorandum for the superintendent, October 29, 1936 (SP).
[50] Compass 101, modernized in 1913, served aboard the *Utah* until April 1925 (after 1913 it functioned as a duplex compass for another Sperry master compass). Subsequently moved to the New London Submarine Base for instruction purposes until 1936, it was then overhauled and placed on exhibition by the navy; compass serial number 109, Mark I, Model 2, is at the Smithsonian Museum of History and Technology (accession #66742, catalogue #313403).
[51] Filed June 21, 1911 (634,594), "Gyroscopic Compass," issued September 17, 1918 (Patent No. 1,279,471); filed June 21, 1911 (634,595), "Gyroscopic Navigational

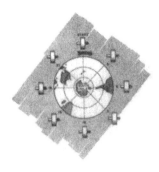

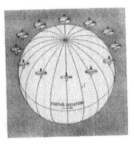

covered compass construction and operation. The other (1,255,480) covered the device automatically correcting errors that resulted from changes in the ship's course, speed, and latitude; it was subsequently considered as important as the first.[52] In these patents, Sperry describes the essence of his invention. The text of the more general patent (Patent No. 1,279,471) begins with Sperry's acknowledgement of his—and by implication other gyrocompass inventors'—indebtedness to Foucault. Quoting Foucault's paper of October 1852, Sperry explained how a gyroscope with three degrees of freedom could be made into a compass by limiting the degrees of freedom to two. To do this, Sperry suppressed the freedom of movement about the horizontal axis by hanging a weight on the gyro. So suppressed —and supported in compass gimbals—the device became a gyrocompass. Sperry described the weighting of the gyro as making it "pendulous" and subject to a "ballistic effect." It was pendulous because the gyro, when weighted, hung from its suspension like the bob of a pendulum feeling the force of gravity. Sperry, it seems, used the expression "ballistic effect" because the gyro when pendulous not only felt gravity but also felt the accelerations and decelerations of a ship in motion.

Once made pendulous, the gyro seeks to align its axis along the meridian, Sperry explained, because of the relative motion of the gyro axis and the axis of the earth. This can best be understood by assuming a gyroscope at the equator with its spin axis aligned in an east-west direction. The gyroscope will *apparently* tilt away from the earth's horizontal as the earth rotates (a gyroscope with three degrees of freedom, when situated on and viewed from the earth, makes one revolution in 24 hours; viewed from a fixed point in space, however, the axis would maintain its alignment—hence *apparent* tilt). The tilt of the gyro wheel about its horizontal axis lifts the weight contrary to the force of gravity; but the countereffect of the weight (acted upon by gravity) is to create a torque in a plane including the spin axis. Because of precession, the gyroscope will respond to the torque by turning toward the meridian. This tilting and precessing process will continue until the gyroscope spin axis is parallel to the axis of spin of the earth—at which point there is no relative motion of earth and gyroscope, no tilt, and no precession (see above, p. 109, for Sperry's definition of precession).

The performance of the pendulous gyro aboard a ship differed, Sperry knew, from this theoretical model. In accelerating or decelerating (speed changes, course changes, rolling, yawing, and pitching), the ship disturbs the pendulous gyro. Because the weight does not distinguish between gravity and these ship motions, the gyro precesses away from the meridian under the effect of these forces. It attempts to align its spin axis with an axis resulting from the combination of the earth's spin and the ship movements (the ship, a little world of its own, is rotating about a bewildering complex of axes). Sperry's resolution of this problem was a compromise:

Apparatus," issued February 5, 1918 (Patent No. 1,255,480).
52 Robert L. Wathen of the patent department, Sperry Gyroscope Company, "Sperry Gyro Historical Patents," memorandum of the Sperry Gyroscope Company, December 29, 1958 (SP).

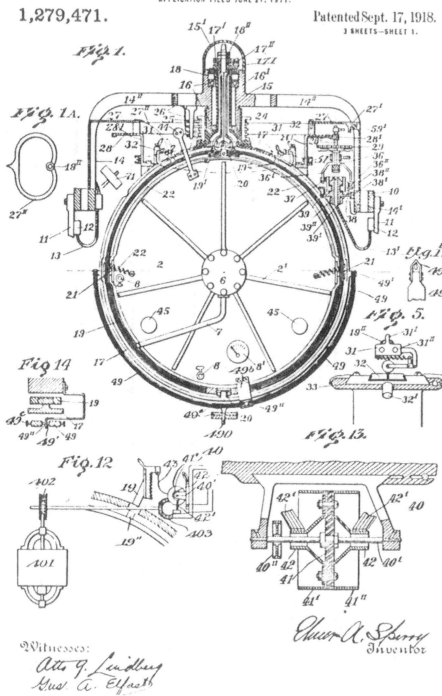

Figure 5.9
First sheet of
Sperry basic gyrocompass
patent. (Key: Gyro wheel
case, 2; cardan ring, 10;
phantom, 17; suspension or
torsion wire, 18; gimbal ring,
19; horizontal pivots, 21;
compass card, 27; gears
driven by servomotor (not
shown), 28, 29; trolley, 31;
contacts, 32; bail, 49;
yielding connection, 49b.)

144

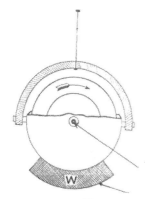

Figure 5.10
Gyro with one degree
of freedom restrained
(the pendulous effect).
From The Sperry Gyroscope
Company Ltd., The Sperry
Gyro-Compass, 1917.

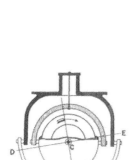

Figure 5.11
The phantom from which
the sensitive element
is suspended and on which
the weight (W.) is carried.
The phantom is connected
to the sensitive element
by the pin or eccentric pivot
(B). From The Sperry
Gyroscope Company Ltd.,
The Sperry Gyro-Compass,
1917.

he suspended the weight (the bail) from a horizontal axis and connected the gyro and bail by a loose or yielding connection.[53] Within the limits allowed by the loose connection, the gyro had three degrees of freedom; but when it reached the limits of play allowed by the connection, the gyro had only two degrees of freedom because the bail restrained its movement about the horizontal axis (see figure 5.9).

The compromise appealed to Sperry's imagination. In his patent, he wrote of his gyro as both a three-degree-of-freedom device (when on the meridian), and a two-degree-of-freedom device (when moving toward or off the meridian). He wrote in his patent:

According to the present invention the disadvantages inherent in the ordinary pendulous gyro, due largely to its being directly affected by acceleration pressures, are avoided: and by providing what may be termed latent restraining means for developing a positive orienting force, the erratic performance of the ordinary gyro having three or even only two uncontrolled degrees of freedom is transformed into perfectly dependable indicating action. At the same time all the important advantages of the pendulous gyro having but two degrees of freedom are retained in the present apparatus, and notably the tendency of the gyro always to seek the north.[54]

The Sperry compass, as he described it, had two major systems: "For convenience I term the gyro-wheel and the parts directly connected therewith so as to be moved thereby in azimuth, the sensitive element [the first system], while the other parts of the azimuth movable unit are termed the follow-up [the second system]." The bail was a component in the follow-up system, and the yielding connection linked it with the sensitive-element system. The follow-up system involved some of Sperry's sophisticated control techniques, similar to those he used in his gyrostabilizer (Figure 5.8).

The follow-up system took command signals from the sensitive element, amplified them with a servomotor, employed the output from the servomotor to do work, and then used closed-loop feedback to compare the work done with the work desired. As with the gyrostabilizer, the follow-up system (including a phantom) shadowed, or followed, the sensitive-element system. When the sensitive element moved to seek, or retain, a position along the meridian, the follow-up system did the same. As long as there was a relative displacement between the two systems, the sensitive element sent signals to the servomotor which activated the follow-up. The signals were generated when the trolley contact, which was attached to the sensitive element, moved off the neutral position on to the contacts attached to the follow-up system. If the trolley moved to one contact, it completed an electric circuit that drove the servomotor in one direction; if it moved to the other contact, the motor reversed (Figs. 5.8 and 5.11).

Driven by the servomotor, the follow-up system did an appreciable

[53] He mounted the gyroscope in gimbals, as was customary with a magnetic compass. In theory, the Cardan suspension (the gimbals) would have allowed the gyroscope to remain stationary despite ship movements, but in reality the friction of the mounting bearings communicated the movements (lessened considerably in force) to the gyroscope.

[54] Patent No. 1,279,471, p. 2.

amount of work, which the sensitive element with its weak movements could not perform. The follow-up system had several functions: it carried the compass card, controlled the repeater compasses, and aligned the attachment points for the yielding connection. The follow-up also carried the support for the vertical torsion wire from which the sensitive element hung. Because the follow-up copied instantly the movements of the sensitive element, appreciable torsion, which would have restrained the sensitive element, could not develop. Able to prevent this, Sperry had a practical alternative to the mercury-bath suspension used by Anschütz and earlier by Kelvin.

Sperry's other patent application of June 21, 1911, described the correction device for his gyrocompass. This device corrected compass errors arising from course, speed, and change of latitude. As the ship changed latitude, the spin axis of the gyroscope changed its orientation in space to maintain an alignment parallel to the horizontal surface of the earth.[55] As a result, Sperry introduced a correction for latitude. If the ship's course is other than east-west, another error is introduced because the ship has an angular motion on the surface of the earth about an axis other than the axis of rotation of the earth. The gyroscope axis aligns itself with an axis which is a resultant of the earth's axis of spin and the axis about which the ship is moving because of the course. The greater the speed in such a situation, the greater the error; therefore, an additional need for a speed correction arose.

Sperry invented a mechanical analogue computer to correct these errors automatically (see figure 5.12). The equation for the correction was

$$D = \frac{a\,K\cos H}{\cos L} \pm b\tan L$$

D = total correction necessary,
K = linear speed of ship in knots,
H = angle in degrees of ship's heading or course from the true geographical north,
L = latitude,
a and b = constants of the individual compass.

Sperry's computer was a mechanical embodiment of the equation. The variable inputs to the computer were: the speed, taken directly from a tachometer driven by the ship's engine; the ship's latitude, set by a hand-operated lever; and the ship's course, taken directly from the follow-up system of the master compass. The trigonometric functions were embodied in the computer by means of pivoted bent levers (see figure 5.11) with cam slots carrying rollers. The slotted levers were connected by links. Sperry wrote in his patent: "It is evident that all the various corrections corresponding to different factors . . . are combined into one movement which is impressed on the bar or correction device." The correction could be

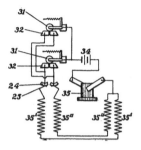

Figure 5.12
Wiring diagram for the servomotor, trolley, and contacts of the Sperry gyrocompass. From "Gyroscopic Navigation Apparatus" (Patent No. 1,255,480).

[55] By horizontal surface of the earth is meant a plane normal to a line passing through the center of the earth and the compass at the surface.

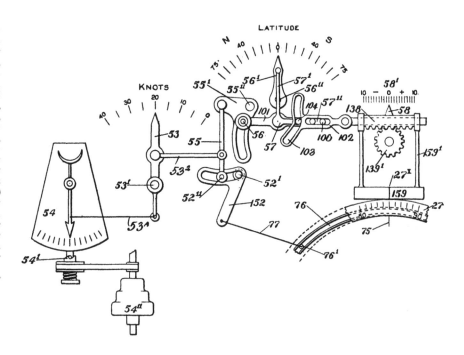

Figure 5.13
Patent diagram
of an analogue computer
providing course, speed,
and latitude correction
for Sperry gyrocompass.
The ship speed is entered
either by hand,
using the pointer (53),
or automatically
by the tachometer (54),
which is driven by
the ship's engine or turbine
(54″). The course is
introduced by moving a
dummy compass card (27)
and the latitude
by setting the pointer (56′).
The components of the
computer perform the
functions resulting in
the correction:

$$D = \frac{a\,K\cos H}{\cos L} \pm b\tan L.$$

read from a pointer on the bar, or it could be applied directly to the compass.[56]

viii

Development of the gyrocompass did not end after it was accepted and installed by the navy, for its use on ships that had various rolling, yawing, and pitching characteristics sometimes revealed unforeseen inadequacies. Sperry and his slowly growing staff of engineers adapted the design of the compass to meet these problems; the modifications required inventiveness in themselves, and some of them were patented. The additional patents created a family of interdependent patents, which gave Sperry wide protection for his compass. The sequence of invention followed by important modifications illustrates the fact that invention often breeds invention. Sperry described this phenomenon well when he wrote: "Necessity is the mother of invention, so the reverse would also seem true, each development in science and engineering necessitating the development of some device or devices to increase the efficiency of the prime development."[57]

Modification and refinement of an invention did not always result in a more complex design. Sperry liked to recall Thomas Edison's observation that when in the development of machines the inventor or engineer found himself proceeding toward greater and greater complexity, he could be almost certain that his efforts were along the wrong line. In the develop-

[56] Patent No. 1,255,480.
[57] Elmer A. Sperry, "Engineering Applications of the Gyroscope," *Journal of the Franklin Institute,* CLXXV (1913), pp. 447–82.

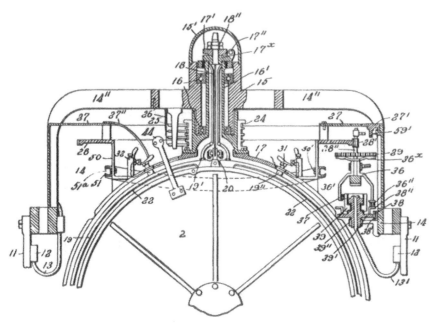

Figure 5.14 Detail of upper portion of Sperry gyrocompass. From Gyroscopic Navigation Apparatus (Patent No. 1,255,480). (Key: components of the follow-up system, 17, 17′, 27, 32; components of the sensitive element, 2, 19, 31; suspension or torsion wire, 18.)

ment of the gyrocompass, Sperry thought this had been especially applicable. As his familiarity with the invention increased, a process of analysis took place that brought the inventor a better understanding of the essence of his invention; as he redesigned with his focus upon that essence, he invariably removed superfluous accretions. Additional development, whether toward simplification or not, could be extremely expensive. Because the production process had already been set up, modifications in design probably meant changes in tools, dies, patterns, and the like. Realizing that changes might be interpreted by the uninformed as compromising the reputation of the inventor and manufacturer, Sperry had a memorandum prepared in 1916 that explained the post-innovational phase of development. "In the development of an apparatus such as the Sperry Gyro-Compass," it stated,

two courses are open to the inventor; first, to make private tests of the instrument extending over a long period and under conditions as near those of service conditions as possible, or, second, to place the instrument in actual service and eliminate defects as they become apparent.

The perfection of the instrument by the first method, that is by private tests, is inexpensive but cannot be depended upon to develop a perfect instrument; on the other hand, if the instruments are placed in actual service, the development will be more expensive, will somewhat compromise the reputation of the instrument, but can be depended upon to absolutely bring out all defects and make it possible to correct these defects, so that the result will be a perfect instrument, requiring little care or supervision while at the same time being perfectly accurate and dependable.

In developing the Sperry Gyro-Compass Mr. Sperry chose the latter method. Eight compasses of the original design were placed in service and no printed instructions were issued for their care. The Navy Department instructed the navigators using the Compass to try it and report very thoroughly on the results. The engineers of the Sperry Company kept in close touch with all ships using the instrument, so that they were instantly aware of any developed defect.[58]

The report continued by listing, in summary, defects that had been found in the gyrocompass since 1911; it then explained how each had been corrected. During these years the navy had the use of a compass which, despite minor defects, improved dramatically upon the magnetic compass.

Competition with Anschütz also stimulated further development of the Sperry compass. Before World War I, the Anschütz company had hoped to install its compasses on the ships of foreign navies, as well as on German vessels. Similarly, Sperry wanted the European market, including the German.[59] When Sperry decided to compete actively with Anschütz for the German naval market, the Nautical Department of the German Imperial Marine Bureau organized and conducted tests in 1914. The history of the comparative tests between the Sperry and Anschütz compasses is a good illustration of competition as a stimulus for post-innovational improvement.

The German navy elaborately staged the tests. A specially designed travelling and turning table simulated ship movements during the shore phase of the tests; the sea test involved temporary installation on a German torpedo boat. The sole Sperry representative, O. B. Whitaker, confronted Anschütz-Kaempfe and a sizeable staff of engineers when the tests were held at the Kiel navy yard in May 1914. The tests were not without incident—technical and personal. Whitaker thought that Anschütz-Kaempfe almost snarled with contempt every time their eyes met and that the naval examining board consulted unduly in low German tones with the Anschütz contingent when decisions were made.

On the first test, maladjusted test equipment damaged the Sperry compass. The test equipment was adjusted but, according to a subsequent report of the Sperry company, its lone representative could only "patch up" his damaged compass for subsequent tests. The Anschütz compass, also damaged, was replaced by another.[60] The "patched" Sperry compass withstood a spinning test without excessive deviation, but according to Whitaker, Anschütz would not allow his compass to be subjected to the same rate of turn. Anschütz pointed out that the test specifications did not call for the higher rate of turn; Whitaker then concluded that the test had been designed by Anschütz and his engineers. The examining board pro-

[58] "Addendum to Reports on the Sperry Gyro-Compass, Now in Use in the United States Navy," 1916? (typewritten copy; SP).

[59] It seems likely that the tightening of the international alliance system and the increased tension between the systems influenced Anschütz and Sperry sales to foreign navies.

[60] The Sperry version of the test is based upon a typewritten memo in the Sperry file entitled, "Comparative Test Made Between the Sperry and Anschütz Gyro-Compass, in May, 1914 at Kiel Germany" (no date), and on an article by O. B. Whitaker in *The Sperry Searchlight*, I (March 1919), p. 1.

vided an insubstantial mounting for the Sperry compass when the sea test began, and Whitaker attributed some of the compass errors during this test to the mounting. By this time the Sperry representative in London had arrived to request a rerun of the Sperry test and permission to witness the sea test of the Anschütz. Both requests were denied.

The Anschütz report of the tests claimed a brilliant victory for the Anschütz company, especially in the rolling tests.[61] The report emphasized that the Sperry compass was a new model specifically intended to reduce rolling error and that, as a result, it sacrificed simplicity of construction. (Sperry claimed that the simplicity of his compass eliminated operational and maintenance problems.) The Anschütz report asserted that the rolling error of the Sperry compass had been lessened as a result of the design changes but that the error remained greater than that of the Anschütz. The report also described Whitaker as a "particularly well experienced engineer," who was nonetheless unable to keep the compass well adjusted and on the meridian .

Sperry received the results of the comparative tests while traveling abroad. He reacted by instructing Ford and Tanner to conduct experiments to find a way to reduce the rolling error further; he instructed another of his engineers, Elemer Meitner, to do research in the foreign and domestic scientific and engineering periodicals to the same end. Having ordered experimentation and research, Sperry then had his patent lawyer spend two days in Washington searching in patents for pertinent information on damping.[62]

Working evenings and weekends, Ford and Tanner found a simple remedy by June. In a special report to Sperry, Tanner described how the problem was solved. They first analyzed the problem mathematically; a simulated compass was then constructed involving no more than a bar carrying weights and suspended on a string. The testing consisted of swinging the bar and moving the weights. The research engineers compared the test data with their mathematical analysis, found the analysis validated, and modified a real compass by adding weights at strategic locations.[63] The actual compass was then swung on a test stand to simulate

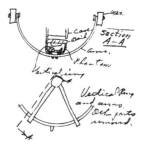

Figure 5.15
Sketch in a letter
from H. L. Tanner to EAS,
June 5, 1914,
explaining efforts to correct
the compass rolling error
(SP).

[61] The Anschütz version is based upon a letter in the Sperry file signed Anschütz & Co. and dated Kiel, May 13, 1914. Internal evidence suggests that the letter was addressed to the British market.

[62] These instructions are in letters Sperry sent to C. Horace Conner, whom he left in charge in Brooklyn (SP).

[63] The most important improvement made in the compass before World War I was the introduction of the "floating ballistic" to replace the yielding connection between the sensitive element and the bail. The floating ballistic was a small gyro that stabilized the connection between the sensitive element and the bail. As a result, less disturbance was transmitted from the bail to the sensitive element during rough weather when the compass tended to swing in its gimbals. The error had been especially large on intercardinal headings. The device was patented by H. C. Ford and H. L. Tanner (filed March 7, 1914, issued July 23, 1918 [Patent No. 1,273,799]). The floating ballistic was placed on the Mark I (battleship compass) and the Mark II (submarine compass.) Another major improvement in the Sperry compass resulted from the incorporation, shortly after World War I, of the "liquid ballistic"

the rolling. "We are at present running tests on the compass to confirm this theory and are getting excellent results," Tanner wrote early in June, and "we shall continue the tests throughout the night, using the same compass both with and without weights to find the exact effect and see how closely it checks with calculations."[64] The modification proved successful.

ix

Sperry would continue to develop his gyrocompass—and gyrostabilizer—in close cooperation with, and supported by, the navy. The pre-World War I naval armaments race had resulted in increased funds for technological development, and Sperry had been swept along. When he first entered the gyro field, however, he had sought the support of private capital. If development capital had been forthcoming—from the merchant marine and passenger ship lines, for example—Sperry might have continued to work in the private sector. His initial applications, however, were and, until after World War I, continued to be military. Instead of amusing circus crowds or increasing safety, the stabilizer created a stable gun platform; instead of bringing a merchant ship to port, the compass generated a reference for gunfire at sea. Shortly, a Sperry gyrostabilizer for airplanes, instead of simplifying flying, would become a guidance and control system for a flying bomb.

Other major inventors of the period also felt the influence of the armaments race and the war. Anschütz-Kaempfe had been turned from polar exploration to compass installations on battleships. Wilbur and Orville Wright were encouraged in the development of their aircraft by their anticipation of military contracts (their first such contract was with the United States Army Signal Corps in 1908). About the same time, early pioneers in radio such as Lee De Forest and Reginald Fessenden depended on military contracts for their development funds. In the early years of the century, then, the development of several areas of technology was closely intertwined with, and stimulated by, armament needs and war.[65]

---

which provided the required correction for ballistic deflection resulting from changes in ship speed along the meridian. The liquid ballistic was comparatively insensitive to quick alternating accelerations initiated by the motion of a rolling ship. Arthur Rawlings, a scientist for the British Admiralty, and Commander G. B. Harrison, R.N., invented the device (U.S. patent issued December 21, 1920 [Patent No. 1,362,940]), and Sperry acquired world-wide commercial rights. In September 1927, the Admiralty established the Philpotts Committee to investigate the origin and development of this control. The committee found that S. G. Brown and Sir James Henderson had the original conception but that Rawlings took issue. See Arthur Rawlings, *The Theory of the Gyroscopic Compass* (London: Macmillan, 1929), pp. 175–84, for an analysis of the Philpotts Committee report.

[64] Tanner to EAS, June 5, 1914 (SP).

[65] The early tests of Guglielmo Marconi's wireless took place in Britain with equipment and personnel provided by the Post Office and the military. By 1903 he had contracts to install wirelesses on thirty-two British ships. In Germany, the military supported two radio pioneers, Adolf Slaby and Ferdinand Braun. See below, for a discussion of the great-powers' support of early aviation.

CHAPTER **VI** Organized Experimentation and Development:
The Sperry Gyroscope Company

It was originally intended that the Sperry Gyroscope Company should specialize in engineering applications of the gyroscope and this intention has been carried out. ... This has been made possible by the large experimental staff of the Company and by the large appropriation which the Company devotes each year to experimental work. REGINALD GILLMOR,
SPERRY COMPANY REPORT, JUNE 1914

BY 1909 ELMER SPERRY VIEWED THE FUTURE OF GYROSCOPIC TECHNOLOGY SO optimistically that he was once again willing to risk forming another company. His latest such venture had been in Cleveland; since that time he had preferred to sell his inventions to others for further development and innovation. Before organizing a gyro company, Sperry found the development of the gyro a complicated and unconventional matter that demanded his close supervision. As the navy's interest and support increased, he also found that he needed a staff to assist him. Rather than part-time outsiders, he wanted his own patent lawyer, experimental shop, manufacturing plant, and sales organization working full-time.

Sperry also had some private reasons for wanting to form a company. He revealed these to his brother-in-law Herbert Goodman when plans for the company were being discussed. Sperry wanted "ample capital" so that he could "take the position with dignity of the inventor and scientific man" and "be relieved of worry, preserve health, and take vacations"—an understandable desire in a fifty-year-old man who drove himself at a pace young engineers found difficult to match. Perhaps he, or Zula, realized that the pace was taking its toll, for he added that he wished to reduce his work "to reasonable limits." This would not, as it turned out, be possible.

Sperry also described the kind of financial backers and managers he wanted in his company. They should be "those who know and appreciate the difficulty of commercially exploiting new inventions; men of experience in other matters; and those interested in scientific research." Remembering, perhaps, Chicago hardships, he sought those who would not burden him with financial problems and debt. He wanted, in short, "to shift the executive load to other shoulders."[1]

i

Sperry turned first to a promoter in an effort to raise capital for his company. In January 1909, after the navy had authorized the testing of his model gyrostabilizer and he had filed patent applications on his automobile and marine stabilizers, Sperry proposed the formation of a gyroscope company to A. P. Lundin. A native of Switzerland who had been a merchant marine officer for several years, Lundin was vice president of the Welin Quadrant Davit Company and an officer of the Lane and De Groot Company, both of New York City. The record does not indicate who may have suggested him to Sperry, but Lundin had probably become known as the successful promoter of the two companies—both of which were in the maritime field, where Sperry saw the largest market for his stabilizer.

The negotiations with Lundin began in December 1908, and extended into early 1909. Although Lundin ultimately declined to promote the company, Sperry's correspondence with him defines the kind of company Sperry wanted. He intended to found the company on inventiveness rather than on natural resources, plant equipment, marketing facilities, or other conventional earning assets. Sperry's own inventive skills and potential would form the company's major asset, although the company was "to control and have exclusive rights in all my future improvements and patents for Gyroscopes for any purpose whatsoever."[2] He proposed to exchange his patents and applications for the $250,000 in capital stock with which the company would be organized. Of this stock, he intended eventually to turn back $150,000 to the treasury for the purpose of raising working capital, although he intended to limit the sale initially to about $25,000. This sum would be used to promote the new company's interest in all countries, to equip vessels with test and demonstration gyrostabilizers and compasses, and to equip one auto with a stabilizer.

For his future services as inventor and chief engineer, Sperry asked that the company negotiate a contract with him that would guarantee either a retainer or a salary equal to at least three percent of the yearly gross income of the company and a minimum of $5000 per year. He also asked that the basic contract provide that, if the capital of the company were increased in the future, at least 25 percent of the increase would be assigned to him. In addition, he wanted the company to reimburse him for prior development of the inventions being turned over to it. Sperry felt that Lundin's "interest as promoter should be recognized" and that he should be, with Sperry, a member of the board of directors.

Lundin was convinced, after several weeks of deliberation (he also engaged an engineer to study Sperry's model gyrostabilizer), that the gyro would eventually find important application. He also believed that Sperry's

[1] An undated memorandum entitled, "For Sperry essentials," in the Herbert Goodman file (SP microfilm).
[2] EAS to Lundin, January 21, 1909 (SP).

proposals were sound and fair. He declined to promote the company, however, because the men to whom he had hoped to sell the Sperry stock decided that they would not subscribe to a device unproved by service. Nor were they willing to finance further experiments and development.[3] In view of Lundin's comments, it seems likely that the financiers that he consulted were not technically trained; it is not surprising, therefore, that they would not accept theory or models as substitutes for performance.

ii

A year later, Sperry, still determined to form a company, had more resources. During 1909 he had surveyed the state of the art in Europe, compared his own work with it, and gained confidence in his own achievement and potential in the gyro field. After returning from Europe, he and Taylor experimented with the model stabilizer, and Taylor's favorable reaction led Sperry to anticipate further support from the navy. These experiments also provided him valuable data on his stabilizer in action. During the year, he also applied for his first gyrocompass patent.

Of special importance, in terms of the establishment of a company, was Sperry's purchase of two basic stabilizer patents that protected his own. Had he not bought them, his active stabilizer might have been found an improvement on earlier designs, in which case the holder of the basic patent might have claimed infringement. One was a patent on a passive gyrostabilizer, for which Thomas C. Forbes of Minneapolis, Minnesota, had applied January 8, 1903. Subsequently, Forbes won an interference case against Schlick and was issued a patent on September 13, 1904 (769,693). When he entered the field and conducted his customary study of existing patents, Sperry learned of the Forbes patent and contacted him in 1908. Although several promoters, realizing the potential litigation value of the Forbes patent, had unsuccessfully attempted to acquire it, Sperry corresponded with him anyway. The two men found that they were kindred spirits, and in May 1909, Forbes assigned his patent to Sperry for "due consideration." In November 1909, Sperry also acquired Rudolf Skutsch's American patent covering the use of two gyros rotating in opposite directions to neutralize reactions.[4]

Late in 1909, Sperry drafted another company prospectus.[5] He offered five patents or applications as company assets. Besides the Forbes and Skutsch patents, Sperry had his 1907 patent on the vehicle (automobile) stabilizer, his application of May 21, 1908, on the ship stabilizer, and the application of September 25, 1909, on the mercury-floated compass. In addition to the patents he controlled, he also listed as assets the calculations, curves, and formulae established by his and Taylor's experiments, and his plans for gyro control and stabilization of airplanes, dirigibles, and submarines. Given these assets, he decided to dispense with a promoter

[3] Lundin to EAS, March 10, 1909 (SP).
[4] Filed March 10, 1906; issued December 17, 1907 (Patent No. 874,255).
[5] Notebook No. 50 (March 10, 1909, to December 31, 1909; SP).

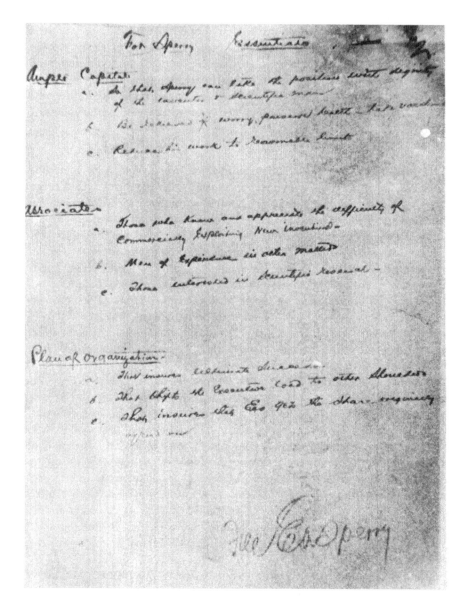

Figure 6.1 Sperry notations of the characteristics he considered essential for the gyroscope company he envisaged in 1909.

THE SPERRY GYROSCOPE COMPANY

MANHATTAN BRIDGE PLAZA

BROOKLYN, NEW YORK

*Figures 6.2
Sperry lists his inventive
ideas as assets as he plans
for the formation
of a gyroscope company.
From Sperry Notebook
No. 50 (1909).*

156

and to raise the needed capital himself. He turned to his brother-in-law Herbert Goodman for advice—and probably for capital—and also to his long-time friend and business associate, Arthur Dana. In addition, C. Horace Conner, a New York businessman, brought in capital and became vice president and business manager of the new enterprise. Sperry drew from his own resources, including the stock he held in the Goodman Manufacturing Company (mining machinery) and the Chicago Fuse Wire and Manufacturing Company (a successor to the Elmer A. Sperry Company in Chicago). According to oral tradition and company publicity, he "decided to mortgage everything he possessed."[6]

On April 14, 1910, Sperry and his associates organized the Sperry Gyroscope Company.[7] At the first meeting of the board of directors, held in the law offices of Jones and Carleton at 40 Wall Street, New York, Sperry was elected president. From these beginnings, the company would become a pioneer research-and-development firm in naval and aviation technology, and, by mid-century, a giant American industrial corporation.[8]

Until the spring of 1913, the company had no plant. Sperry and Hannibal Ford sketched, designed, and experimented, using the New York City office and the rented area at the Pearce shop. Sperry hired several draftsmen to prepare his drawings and blueprints, and models were built by Pearce machinists. Pearce made the gyrocompasses, but heavy work, like the stabilizer for the *Worden,* was subcontracted. By 1913, however, the gyrocompass demand was steady enough to permit Sperry to establish a small manufacturing plant on several floors of the Carey Building at Manhattan Bridge Plaza and Flatbush Avenue in Brooklyn—not far from the Navy Yard. The company acquired the finest machine tools for the precision engineering necessary in the manufacture of gyrocompasses and hired master machinists, some of whom were Pearce employees (Otto Meitzenfeld, who became head of the Sperry machine shop, was a former Pearce machinist).

By March 1913, before moving, the company had acquired tools and equipment valued at $6,000 (compared to a valuation of $475,000 for patents); six months after the move, tools and equipment were valued at almost $40,000 (compared to $350,000 for patents). These figures show that a gradual change from an invention and development enterprise to one of invention, development, and manufacturing was transpiring.[9]

With an enlarged staff and the acquisition of a plant, the company began to articulate its structure and functions, most of which had been performed earlier by Sperry alone. By 1914 the company had been organized so rationally that an analysis of the cost of a gyrocompass could be made. The total cost of a compass was divided among development, patents, manufacture, and sales. The cost of "development" involved experiments

[6] "Twenty-Five Years," *Sperryscope,* VII (1935), pp. 2-3.
[7] Sperry Gyroscope Company, "Annual Report for 1912" (SP).
[8] Sperry Rand was formed in 1955 by a consolidation of the Sperry Corporation and Remington Rand, Inc.
[9] Balance sheets, Sperry Gyroscope Co., March 31, 1913, and December 31, 1913 (SP).

and the "time of inventors and engineers"; sales costs arose from office expenses, agents, publicity, freight, and supervision and maintenance (the sales department supervised the repair and maintenance of installed compasses). Fees for attorneys and patents, specifications, and royalties made up the cost of "patents"; material, labor, drawings, overhead, and assembling and testing comprised the manufacturing costs.[10] The stress on patents and development differentiated this enterprise from a manufacturing concern. Unfortunately, surviving company records do not apportion the cost among the four major divisions of the Sperry enterprise: such an apportionment would probably show even more convincingly a company emphasis on invention and development.

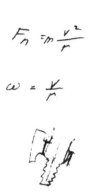

Through invention and development, the company had by 1914 expanded its line of products. The expansion was not diversification but the creation of systems based on the gyro. Beginning with the gyrocompass, the company developed a course recorder which, used with the gyrocompass, provided a graphic record of the ship's course. This instrument evaluated the performance of helmsmen, in navigation, and provided evidence in disputes following upon collisions. Sperry and his inventors and engineers were also developing, before World War I, a naval fire-control system based on the reference line provided by the gyrocompass. They also developed the implications of ship gyrostabilization by developing a roll and pitch recorder and effected a major transition from marine stabilization to the stabilization of airplanes.

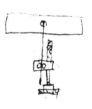

iii

The men in whom Sperry took the keenest interest were his inventor-engineers. As early as 1914, company brochures described this group as a "large experimental staff" supported by "the large appropriation which the Company devotes each year to experimental work." This staff included many engineers of the "highest standing" whose achievements "are well-known in the branches of engineering in which they have specialized."[11] The invention and development staff of the company had, in fact, five experienced engineers. Hannibal Ford, who quickly mastered the gyro and automatic mechanisms, was chief engineer and had responsibility for the development of the compass. Carl Norden, who continued to smoke those "vile black cigars," worked with the development of marine stabilizers. Both of them were graduates of engineering colleges. Harry L. Tanner, a University of Michigan graduate, joined the company in February 1913. Leslie Carter, a graduate of Brooklyn's Pratt Institute, joined the staff in the fall of 1913, as did Elemer Meitner, who would assist Sperry with the development of naval fire-control equipment.

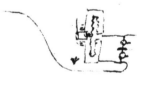

Harry Tanner became the company's chief engineer when Ford left in 1914. After graduation from Michigan in 1908, he had worked for

Figures 6.3
Sperry sketches.

[10] Bulletin of Information No. 1003, Sperry Gyroscope Co. (London), 1913 (SP).
[11] "The Sperry-Curtiss Aeroplane Stabilizer," written in 1914 by R. E. Gillmor of the London Office and mimeographed for distribution at the test of the Sperry-Curtiss airplane stabilizer in June.

General Electric and then returned to Michigan in 1910 as an instructor in electrical engineering. His experience in electrical engineering balanced nicely the mechanical interests of Ford and Norden, but Sperry had other reasons for hiring Tanner. Tanner, like Sperry and Ford, was from Cortland County, and as a college student, he had arranged an interview with the famous Sperry. The young man was a "very fine mechanic" and "was always building when a boy,"[12] and later, when he needed an engineer, Sperry remembered the young man with a background so much like his own.

Sperry could not depend upon Cortland County for his entire staff, but he carefully selected the young men who joined him. Leslie Carter recalled years later that his initial interview with Sperry consisted of a rapid series of questions. Sperry seemed particularly interested to learn that Carter had been building models since boyhood and asked to see Carter's model of a steam-driven airplane. He liked what he saw because he wanted "people around here who can use their hands."[13] (Interestingly, Sperry himself by this time showed little interest in visiting the shop to use his own.) Soon Carter was at work building "breadboard" models of a new two-rotor compass; he took home $10 a week, the prevailing starting rate for engineers. He liked his new job because he thought the small company had a promising future. Sperry's own enthusiasm and optimism had apparently spread to the young men of his staff.

Sperry hired men not only from Cortland County but also from his own family. His eldest son Lawrence became a development engineer and test pilot for the airplane stabilizer in 1912, and Edward and Elmer, Jr., joined the company after 1914. Sperry's daughter Helen could not join the men of the family, but she did research in the New York Public Library on patent matters for her father.

Until the end of World War I, when the company was reorganized, relationships, as one might expect, were informal. The development staff had direct access to Sperry, and he to them, whether at work or at home. Employees—not without pleasure—recalled that a Sunday morning telephone call about an engineering problem might last an hour or more and that the ringing of the telephone in the late evening hours meant "Mr. Sperry." Sperry expected his engineers, as well as other members of his staff, to be able to follow his thought processes and react quickly. He established a frenetic pace, and it was not unusual for him to ask his secretary to accompany him a short distance on the train when he was going out of town so that he could dictate last-minute memos and letters. One of the young engineers who joined Sperry during World War I remembered receiving hurried telephone calls from him:

'Preston, put on your hat and meet me at the elevator.' I would have little idea where we were going. Usually it wasn't far. We crossed the street to his favorite barbershop and I stood by his chair and answered questions while he was getting a haircut and shave. He would become so interested in our discussion that the understanding barber would have to say: 'Now, Mr.

[12] EAS to Helen Willet, December 31, 1912 (SP).
[13] Carter in the *Sperry News*, June 29, 1961.

Sperry, if you'll please not talk for a minute, I will shave your chin.'[14]

In the early years there were no formal meetings of the engineering staff; meetings were held whenever Sperry had an idea that he felt merited discussion. During these meetings, he communicated with constant sketching (he always carried with him a pocket calculator of his own invention and a folding rule). On one occasion, in fact, he made a sketch on the back of an engineer's hand. After outlining his idea, he would suggest how it might be embodied in a model and tested; after a period of discussion, he would decide who among the engineers should "put something together to check it out."[15]

Although he was not quick to anger, Sperry was occasionally exasperated by an engineer "who saw the circus but missed the elephant." He assumed competence in details and looked for the individual who could grasp the essentials. "Each day almost the first thing in the morning, you'd be summoned to his office to tell what had happened and what you had run into. . . . If you had run into a problem, he would listen, make more sketches . . . and have them incorporated as soon as possible."[16]

The engineers brought him problems that they could not solve; if in turn he could not solve them, he often telephoned members of his wide circle of engineering and science friends throughout the country until he got help. His contacts, like his experiences, cut across the arbitrary lines dividing chemical, electrical, and mechanical engineering. Sperry also paid close attention to the ideas of his own staff members. If an idea seemed patentable he might explore it more fully with the young engineer and then have him take the work to the company patent attorney. The result might be a patent in the young man's name, assigned to the company, or joint with Sperry. For assigning a patent, the inventor received a nominal fifteen or twenty-five dollars. One of his engineers, Eric Sparling, remembered at least a dozen patents "which were more or less sparkplugged by Mr. Sperry."[17]

The models embodying the ideas of Sperry and others were built in a shop equipped with the latest general-purpose machine tools. The shop machinists worked directly with the project engineer, a procedure which kept detailed drawings and specifications to a minimum and allowed the machinist to contribute to the realization of the idea. While the model was being tested, Sperry watched the results carefully and made frequent suggestions in light of performance.

Because of Sperry's close contact with the members of his development staff, they adopted his style of invention and development. The engineering college graduates who joined him may have found that he was not as well informed about engineering science as their professors had been, but his ability to maintain a position of leadership suggests that his

[14] Preston Bassett, "A Brief Biography of Elmer Ambrose Sperry," (manuscript), p. 72 (a 104-page memoir, c. 1965; SP).
[15] Interview with Eric Sparling, May 25, 1966.
[16] Ibid.
[17] Ibid. Before 1915 the staff assigned about a dozen patents to the company.

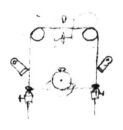

*Figures 6.4
Sperry sketches.*

rich and varied experience created respect. Sperry's characteristics as an inventor and engineer contributed to the growth of a general, "company" approach to solving problems. He believed that the best work was done under pressure, and he therefore often committed himself publicly to the development of a successful invention; he then had to fulfill his commitment or suffer considerable embarrassment. He expected his staff to operate in the same way. He also valued flexibility in the members of his staff because he tended to seek solutions to technical problems by turning from one field to another; if a mechanical solution would not work, he might turn to an electrical or chemical solution. He preferred striking analogies and physical models to mathematical models; but he did not avoid complex problems and solutions, nor did he permit his engineers to do so. A characteristic of his "school of engineers" and of his company was a concern with complex problems and precision manufacturing. Sperry argued that by choosing the difficult, "pretty" problems he could avoid conflicts with mediocre, ruthless, driving competitors, of whom there were too many. He also preferred men of vision and enthusiasm on his staff; his wife knew that the "only way Elmer can work . . . is surrounded by his believers," the young engineers whom he called his "boys."[18]

Sperry never articulated his role as *de facto* director of research and development for his company, but his notebooks and appointment diaries suggest that he preferred to reserve the new problems, those requiring "pure" invention, for himself and to allocate those in the development or post-innovational development phase to specific members of his staff. In 1912–13, for example, when the *Worden* stabilizer installation and the installations of compasses were the company's major activities, the new aerostabilizer interested him most. Presumably, he thought Tanner, Ford, and Norden could carry on the compass and stabilizer work, consulting him when particularly troublesome problems arose. Thus, he reserved for himself the exploration of exciting new fields that attracted his inventive spirit. This preference for the frontier did not wane with time, for the closing years of his life still found him leaving the more routine functions to his staff, while he plunged into promising new territory.

Leaving the more routine development and post-innovational phase of company projects to his engineering staff caused a misunderstanding with Carl Norden. The episode has increased interest because years later Norden developed the famous bombsight, which was similar to the equally famous Sperry sight. The problem arose when Norden decided to leave the company in October 1915, to establish himself in private practice as an engineer. He had come to Sperry without gyrostabilizer experience and during the next four years had become an expert. By 1915 Sperry could write to Norden's uncle, "we are all very fond of your nephew and feel that he is doing excellent work here."[19] Sperry and Norden had worked together for four years "without the slightest misunderstanding"; in fact, Sperry did not believe that "it was possible for reasonable men to have

18 Zula Sperry to Herbert Goodman (no date, SP).
19 EAS to A. Norden, August 11, 1915 (SP).

misunderstandings."[20] When Norden made known his intention of leaving, however, the two men strongly disagreed about the nature of Norden's obligation to Sperry. Sperry contended that he had laid down the general lines of stabilizer development for Norden and that he had cultivated Norden's abilities on the assumption that he would stay with the company until the development produced a successful device. Success meant a stabilizer satisfactorily operating on a battleship. The least Norden could do, Sperry felt, was to continue to develop the stabilizer for Sperry as a paid consultant if he insisted on entering private practice. Furthermore, Sperry thought that Norden should assign future stabilizer patents to Sperry's company for a reasonable compensation.[21] Norden argued that "it was materially through my efforts that the stabilizers are developed to their present stage."[22] He also believed that changes in Sperry's basic design were absolutely necessary and that he had not been given due "recognition" and would not gain "proper compensation" for his contribution to the necessary refinements in design.

The negotiations centered on the disposal of Norden's future patents. Norden would only agree informally to give the company first option on his new gyrostabilizer patents, and as Sperry's patent lawyer later pointed out, "an offer to grant one the first option does not amount to much since all that is necessary to be done is to make the terms so severe that they cannot be accepted."[23] The outcome was that Norden continued to work as a paid consultant on Sperry's stabilizer contracts with the navy until March 1917. During that period, he applied for one stabilizer patent, but Sperry's patent lawyer advised that no issue be raised simply because the patent was substantially the same invention disclosed in a Schlick patent seven years before Norden filed.[24]

iv

After invention, and development, patent matters were closest to Sperry's focus of interest and expertise. Prior to 1914 the company depended upon the services of outside lawyers but then established a patent department headed by patent lawyer Herbert H. Thompson. Sperry knew that the commercial success of an invention depended on the excellence of the patent and the ability to defend it. He also realized the prime role of the patent department in an expanding company whose capital assets were largely patents.

Sperry was a major influence on the policy of his patent department; he also taught his engineers how to make patent applications. "I give the instructions [to the patent attorney] and draw the claims myself," he told a protegé, "because knowing the art as I do I know just where my novelty

[20] "Memorandum of conversation between Mr. Sperry and Mr. Norden, October 11, 1914." The copy carries the date 1914, but other evidence suggests 1915 (SP).
[21] EAS to Norden, November 3, 1915 (SP).
[22] Norden to EAS, November 4, 1915 (SP).
[23] Thompson memorandum to EAS, November 14, 1919 (SP).
[24] "Summary of Patent Situation on Gyroscopic Stabilizers for Ships," memorandum of the patent department (no date but probably about 1919) (SP).

resides and just how generic a claim should be allowed, in view of the state of the art."[25] In criticising another inventor's patent, which the company was considering purchasing, he stressed the naiveté of the claims: not one was generic. Sperry's technique was to give the attorney the principle on which the invention was based—the "metes and bounds" of the invention's contribution—and then demand clearly and simply worded claims. He warned his engineers that if a patent made narrow claims, another inventor, perhaps with the advice of a shrewd attorney, could invent around, or circumvent, the prior patent with alterations in design. Typical of a patent that could be circumvented, Sperry believed, was one whose claims protected only construction and did not claim the principles that governed construction.

H. H. Thompson also helped instruct the Sperry engineers about patents. He stressed in particular the difference between basic patents and dependent patents.[26] A patent, he explained, may merely cover an improvement or modification of a basic patent and therefore be valueless without rights under the basic patent. Yet, he emphasized, the patent on improvements could be more valuable than the basic patent in that the improvement might make the basic device, previously valueless, a commercial success. He offered as an example of the relationship some of Sperry's basic patents and the improvement patents of his engineering staff. The Tanner and Ford patent on the floating ballistic, dependent as it was upon Sperry's broad patent, was, Thompson thought, a good example.

Thompson and the patent department also acquired certain patents of inventors outside the company to supplement and to reinforce Sperry's. By the end of World War I, for example, the company asserted that "through original invention or purchases" it controlled "all important patents covering the art of gyroscopic ship stabilization."[27] Besides the Forbes and Skutsch patents, which Sperry brought into the company initially, it had acquired among others, the basic Schlick patent, a patent on braking the stabilizer of Schlick and Max Wuerl, and the patents of Louis Brennan.[28]

Thompson also advised Sperry about which inventions, patent applications, and patents were sound investments. Because the patenting process itself was expensive and interferences and infringements could add additional expense, the company had to avoid the cost of patenting a device that might be technologically excellent but have a poor commercial potential. Thompson thought, for example, that the Brennan monorail car was a meritorious invention technically, which proved a dismal failure financially. As a general rule, he believed that "probably less than one per cent of the patents granted are very valuable, ten per cent to twenty-five per cent of moderate value, and the remaining percentage of little value."[29] He and Sperry tried to be selective enough in their policy to prevent the company's

25 EAS to R. E. Gillmor, August 25, 1916 (SP).
26 Herbert H. Thompson, "A Patent's Value," *Sperryscope,* I (November and December 1919), pp. 12–14.
27 The Sperry Gyroscope Company, "The Sperry Ship Stabilizer," (1919), pp. 25-27. A typewritten booklet (SP).
28 See above, p. 000.
29 Thompson, "A Patent's Value," p. 12.

patent history falling into "the remaining percentage."

As the company grew, its patenting activity intensified. This meant that Thompson—sometimes Sperry on major inventions—and supporting attorneys spent valuable time shepherding the applications through the Patent Office and the patents, in some instances, through the courts. The patenting process was complex and often cumbersome,[30] as one Sperry patent case suggests. Thompson began processing a patent application for the airplane stabilizer, filed on July 20, 1914. The description of the airplane stabilizer in the application ran close to 10,000 words and was illustrated by six pages of drawings.[31] The total claims in the application numbered sixty-seven; Thompson had examined existing patents to avoid prior claims of other inventors. After these papers had been filed, the exchange between Thompson, in consultation with Sperry, and the patent examiner began. In October 1914, the examiner wrote Thompson that the application included too many different forms of invention and recommended that the application be divided into two or more applications. The examiner also cited inventors and patents that anticipated claims made in the Sperry application. In reply, the next spring, Thompson amended the claims to comply with the division requirements. He also wrote that he had studied the various patents that the examiner believed anticipated Sperry and concluded that the inventions would be practically useless or inoperative on an airplane.

In his next letter, the examiner specified which of Sperry's claims he

Figure 6.5
Page from Sperry notebook.

[30] The application first went to a patent examiner who, in dialogue with the inventor and his attorney, found an acceptable statement of claims: the inventor often claimed too broad an area and the examiner ascertained where he invaded the rights of others. The examiner also insisted on good form in the application. During the course of the examination of the application, the applicant might encounter interference, a rival claim to the same invention. Then, an interference proceeding was instituted to determine priority. The contestants submitted evidence to the Examiner of Interferences, who rendered a decision. Appeal from this went to a Board of Examiners-in-Chief of the Patent Office and from their decision to the Commissioner of Patents. Further appeal was possible from the Commissioner to the Court of Appeals of the District of Columbia. By this time, much money and time would have been expended by both claimants, but the unsuccessful still had recourse to a district court and then to the United States Circuit Court of Appeals. All of this might occur before the patent was issued. The application, as was the rule, included: (1) a preamble stating the name and residence of the applicant, the title of the invention, and if the invention had been patented abroad, the countries in which it had been so patented and the date and number of each patent; (2) a general statement of the object and nature of the invention, except in design applications; (3) a brief description of the several views of the drawings; (4) a detailed description; (5) claims; (6) the signature of inventor; and (7) attestation of two witnesses. This discussion of the patenting process is based on U.S. Patent Office, *Report of the Commissioner of Patents for the Year 1896* (Washington, 1896), pp. viii-xv; Nathan Reingold, "U.S. Patent Office Records as Sources for the History of Invention and Technological Invention," *Technology and Culture*, I (1960), pp. 156–67; and Melville Church, "Needed Reforms in Patent Procedure," *Scientific American*, CV (1911), p. 446ff.

[31] This account of the exchange is drawn from the full case on file in the records of the Patent Office, Arlington, Virginia. The case was on the patent filed July 17, 1914 (851,477), "Aeroplane Stabilizer," issued February 18, 1921 (Patent No. 1,368,226).

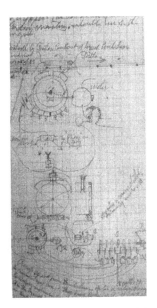

Figure 6.6
Page from Sperry notebook.

rejected outright and which he might accept if the application were further amended. His explanation of his decision consumed four pages of detailed argument illustrated, again, by the patents anticipating Sperry. During the course of the exchange, the examiner discussed no less than sixteen relevant patents. Thompson waited almost a year before cancelling some claims, amending others, and making substitutions. Still not agreeing with the examiner in certain particulars, he again asked for a reconsideration. The examiner, several months later, stood firm on most of his decisions, but accepted Thompson's reasoning and information in some instances. The exchange continued and involved further rejections, substitutions, and allowances. The examiner also declared several interferences which further interrupted the normal routine of patent approval.

Seven years after the application was first filed, the Patent Office and Thompson reached an agreement, and after payment of a $20 fee, the patent was granted. There is no record of how much the process had cost Sperry—and the government—but this case, although drawn out, was not unusual. The possibility of additional expenses arising from infringement litigation still remained. (Zula Sperry, speaking from experience, called a patent a license to litigate.) Another patent holder might bring suit against Sperry after he had begun to manufacture his airplane stabilizer, claiming that the device infringed his patent; Sperry himself might sue another for infringement. In the pre-World War I era, good patent lawyers involved in litigation charged from twenty-five to one hundred dollars a day, and experts who gave testimony, fifty to seventy-five. As a critic of the system wrote, "it can readily be seen how enormous the expenses of patent litigation may become. . . . The system places the poor inventor at the mercy of his rich adversary." Sperry was no poor inventor, but he asked a congressional committee in 1919 to reform the patent law to end the cumbersome and protracted methods of the patent courts. He testified that he doubted his ability to stand the cost of litigation if one of the large financial interests should back an opponent to the full.

Given the expense involved, Sperry considered justified the seventeen years of monopoly granted by a patent. It was a period in which the cost of patenting and development could be recovered. Without the protection of a patent, he found that investors were unwilling to assume the risks, fearing that competitors would take the invention after development.[32] Although often exasperated and bored, therefore, by the process of patenting, Sperry accepted it as an integral part of the professional inventor's life—a part that often meant the difference between ingenious invention (and nothing more) and a successful innovation. He appreciated having at his disposal a patent department in his own company.

v

During its early years, the company grew modestly but steadily. Domestic and foreign sales increased about 50 percent annually from the spring of 1911

[32] On this subject, see Frederick P. Fish, "Patents and Modern Industrial Conditions: The Stimulus of Patent Protection," *Scientific American*, CIX (1913), p. 181.

until the summer of 1914, when the war in Europe brought a dramatic upsurge in foreign sales (see table). Prior to 1914, Sperry had explored the European market and had established a branch of his company abroad. As early as 1911, he hoped to license a European manufacturer for his compass and to commission foreign sales representatives to sell his compass and possibly his stabilizer to the rapidly expanding and technologically advanced European navies. Technically informed representatives could keep him abreast of the needs of the foreign navies, and he could invent and develop to meet these needs. Foreign representatives could also keep him informed of the state of technology there. In addition, a patent lawyer in Europe could be a valuable asset.

Figure 6.7
Page from Sperry notebook.

*Growth of Sperry Gyroscope Company, 1911-14 ***

Year		Quarterly Output	Number of Employees
1911			
2nd	Qr.	$ 338.00	5
3rd	"	2,338.00	5
4th	"	18,750.00	7
		21,426.00	
1912			
1st	"	16,125.00	7
2nd	"	57,475.00	8
3rd	"	9,670.00	12
4th	"	61,031.50	14
		144,301.50	
1913			
1st	"	75,025.00	17
2nd	"	75,071.12	24
3rd	"	54,571.00	77
4th	"	128,692.23	117
		333,359.35	
1914			
1st	"	124,491.60	136
2nd	"	260,564.45	161
3rd	"	416,927.65	199
4th	"	483,173.57	322
		1,285,157.27	

* Source: "Report to the stockholders of the Sperry Gyroscope Company," April 13, 1915.

When he was abroad in 1911, he had negotiated with the British precision-engineering firm, S. G. Brown, about manufacturing under his patent rights. No agreement was reached, but some years later Brown manufactured a compass that competed with the Sperry. Sperry had also explored the British market at the same time and had encouraging conversations with high ranking naval officers about trial installations of the Anschütz compass. Anschütz had a London representative, Elliott Brothers, but Sperry concluded from his conversations that the British Navy might prefer not to depend upon a German compass as the naval armament race continued. He

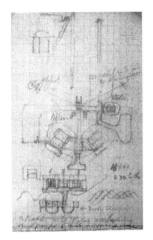

Figure 6.8
Page from Sperry notebook.

also foresaw sales to other naval powers, especially France and Russia.

In June 1913, Sperry established an office at 57 Victoria Street, London, and named the 26-year-old naval officer, Reginald Everett Gillmor, as its head. Gillmor had first come to the inventor's attention as the officer aboard the *Delaware* responsible for the installation and operation of Sperry's first compass. A naval academy graduate, Gillmor went to the U.S. Naval Post Graduate School in 1911 to take an advanced course in electrical engineering. After a year at the school, he resigned from active duty, as a junior grade lieutenant, to join Sperry. Gillmor had not only a naval and an engineering background but also a taste for management. His long letters to Sperry from London reveal an interest in international relations, naval affairs, and managerial theory. He and his family moved with ease in the cultivated naval and business society of London.[33]

Thomas Morgan, who had been an electrician under Gillmor when the gyrocompass was installed on the *Delaware,* also resigned from the navy in the fall of 1912 to join the Sperry company. A few months after Gillmor arrived in London, Morgan was dispatched there to help staff the branch office. Gillmor soon had the London office functioning. Its major activities were the sales and service of compasses. Gillmor and Morgan demonstrated the gyrocompass system to representatives of various navies (the London office had a special room for this purpose), and they also arranged test installations aboard ship. Although the first demonstration compass was assembled and adjusted at the S. G. Brown factory, the London office soon made provisions for assembling, testing, and maintaining compasses shipped from the United States. The London office did not manufacture compasses.

In April 1914, the office was organized as a limited company and named the Sperry Gyroscope Ltd. To expand sales, a number of representatives, who reported to London, were named in other countries. By the outbreak of World War I, representatives were located in Paris, Milan, St. Petersburg, and Constantinople. Gillmor organized the activities of his representatives and drew upon their special knowledge of local conditions. They kept him informed of the contractural requirements of the various navies regarding the purchase of foreign manufactures; they collected testimonials about Sperry equipment from naval officers and officials; and they helped him publish articles on the stabilizer and compass in foreign journals. By the end of 1913, he had articles in *Engineering, Electrical Review* (both of London), *Revue Maritime, Revue Industrielle,* and *Elektroteknisk Tidsskrift* (Copenhagen).[34]

By 1914 the London company was achieving notable successes. Although nineteen Anschütz compasses had been manufactured in Britain by Elliott Brothers and installed in the British fleet in 1912 and 1913, Gillmor predicted, in the fall of 1913, that the outcome of confidential comparative tests being conducted at the Royal Naval College by the British Admiralty

[33] During World War I, when he helped Admiral Sims establish his command in London, Gillmor, so the story has it, established what some considered a first in U.S. naval history: "tea" at headquarters.
[34] Bulletin No. 1001, Sperry Gyroscope Company (London), 1913.

would result in the adoption of the American compass in place of the Anschütz.[35] He had also learned from unofficial reports that sea trials of Sperry compasses on HMS *St. Vincent* and HM Submarine E-1 in 1913 showed the compasses giving "perfect results in every respect." Within a month after the test at the Royal Naval College, the Royal Navy ordered four Sperry compasses and about six months later placed an order for five more complete systems. The Russian navy also ordered a compass for installation on the cruiser *Rurik;* the Italian navy placed an order for installation on the HMS *Roma;* and French officers in the Hydrographic Department were expressing "favourable opinions" about the Sperry. (Before the outbreak of the war, the French ordered at least three compass systems.) As the war approached, the American gyrocompass showed promise of dominating the European market—outside of Germany.

vi

Sperry made an extended trip to Europe in 1914 to survey the activities of his expanding overseas operations and to sell Sperry equipment. During the course of the trip, he visited St. Petersburg, where his sales representative, H. Treeck, arranged for Sperry to have access to the highest naval authorities, who were extremely interested in new developments in naval technology. Sperry found his representative a "strong" man, who owned "a firm which is quite large and important" and had a fortune of about $3 million. Treeck arranged for Professor Kriloff, the "Captain Taylor" of Russia, to spend two days with Sperry helping him prepare for a lecture given before a group that "was simply wonderful."[36] Six admirals and about fifty other high officials of the navy gathered in the large library of the Admiralty building to hear Sperry lecture on a Saturday afternoon (Sperry learned that only Treeck could have assembled so much brass on a Saturday afternoon in St. Petersburg).

Sperry did not find Russia a technological backwater. The Russians, in fact, took keen interest in his marine stabilizer (Sperry believed "we will probably get the imperial yacht and one battleship to start with"),[37] course recorder, artificial horizon, gyrocompass, and fire-control equipment. He found them excited about the possibility of using stabilizers to roll ice-breakers artificially; the prospect would allow them for the first time in history to keep open the port of St. Petersburg during the winter months. Treeck also hoped that Sperry could persuade the government to equip ten giant Sikorsky airplanes with the Sperry airplane stabilizer, shortly to be demonstrated in Paris by young Lawrence Sperry. Treeck promised to broach the question to Sikorsky and anticipated that the designer and the government would request a demonstration. Sperry's letters, however, make no further mention of the plan.

He found the week in St. Petersburg had been too short when he boarded the Nord Express for Berlin in the early evening of June 3. "I

[35] Bulletin No. 1004, Sperry Gyroscope Company (London), October 21, 1913.
[36] EAS to Conner, c. June 3, 1914 (SP).
[37] *Ibid.*

get along with the Russians astonishingly well," he wrote. "I like them and they do me."[38] He was convinced that their officers were fine men of good judgment and had no doubt that "if we work it rightly we can become headquarters for a great quantity of their material for battleships" and generally benefit from their "immense" naval program. He felt so optimistic about the St. Petersburg situation that he began making plans to put Morgan there full-time and to employ another American to take his place in covering other European capitals.

Sperry attributed much of his success to Treeck's commanding position with the navy and his perfect system, "which was not explained to me, of handling an extremely necessary item, which seems here to be the 'right of Kings' and in no wise wrong."[39] Yet the American did have doubts, for he asked rhetorically, "What would we do under these circumstances. . . . There is no possible question but that we must work with him, and we want to. . . ." Treeck obtained orders for Sperry equipment, but his own affairs took a bad turn. With the outbreak of the war two months later, he was arrested as a German collaborator and sent to Siberia. Allowed to move about freely, he escaped from Russia and wrote to Sperry from Finland after the war. Referring to his lost fortune and the miseries he and his family had experienced, Treeck asked for the ten percent commission he thought due him on orders taken when Sperry was in St. Petersburg. Morgan, the Sperry company sales manager in 1919, advised against paying Treeck because the government had asked that the company deal directly with it in these sales. When Treeck appealed to Elmer Sperry, he responded that the matter was closed.[40]

While he was abroad, Sperry also helped plan the defense in the infringement suits brought against him by Anschütz in Britain and Germany. Sperry hoped, with his thorough knowledge of gyrocompass technology and patents, to find in others' patents anticipations of Anschütz where Anschütz claimed infringement. With great determination and ingenuity, he gained permission from the director of the *Conservatoire national des arts et métiers* in Paris to examine the historic Foucault apparatus on display there. Sperry arranged to have photographs and movies made of the Foucault gyroscope in operation and found that the device oscillated exactly 'as a highly sensitive balance on a six-second period and the 'movies' caught it in the act." This proved, he believed, that Foucault had anticipated Anschütz on a cardinal point. "It took little short of an act of Congress and the President of the Republic to put this over," he wrote, "and wasn't it bully that it worked exactly as I anticipated when finally put to the test."[41]

Sperry eventually won the infringement suit in Britain but lost in Germany, where the case dragged on from 1914 to 1917. This experience with the German courts and the conduct of the competitive trials of his

[38] EAS to Conner, June 3, 1914 (on Nord Express to Berlin), (SP).
[39] *Ibid.*
[40] The Treeck correspondence covers a period from October 1914, to November 1919, and the substance of it has been covered here (SP).
[41] EAS to Conner, June 22, 1914 (SP).

compass at Kiel in 1914, reinforced an uncharacteristic prejudice discernible earlier in Sperry. He referred to the Imperial German Patent Office as an institutional embodiment of the inflexible characteristics of the Tsar (Russian, presumedly and unaccountably). Because the German navy would not buy his compass, he assumed that "it would take five or ten years to get into a coppersealed German mind" to demonstrate that his compass was superior. He believed that, in general, the German rejected out of hand "anything that emanated from America." With heavy irony, he wrote a German acquaintance: "You know, we are so aboriginal and crude over here that nothing we do seems to be taken seriously."[42]

Sperry did not allow the infringement and business to occupy all of his time in Europe. Zula had accompanied him, and he wanted her to enjoy the European culture. They traveled in a style suitable for a company president and his wife, crossing the Atlantic first class on the RMS *Oceanic*. In London, they stayed at the Waldorf Hotel in Aldwych, from which Zula could travel easily to Westminster Abbey, Selfridge's, and other places which she found gave her "thrills." In the evening, Gillmor saw that the Sperrys had tickets to the latest Shaw play. Parisian dining she found delightful: "No wonder the French do not travel. . . . How could they ever eat in America. . . . Everything served so beautifully and such service." In St. Petersburg she helped her husband by entertaining officers of the navy and delighted in a visit to the Baltic estate owned by Treeck, who also had a "stable of autos" and one of the country's most luxurious yachts."[43]

In London, Sperry accepted an invitation to attend a *conversazione* of the Royal Society, where he met "a great many celebrities including the President, whom I know personally, and had a very fine time." Another social highlight was an invitation from Captain F. Creagh-Osborne, the Admiralty's expert on compasses, to dine at the United Service Club. The American found it a remarkable club, "a couple of hundred years old," that did not admit officers of the lower ranks. His luncheon with Brennan at the Savage Club was far more business than pleasure, for Brennan's monorail still interested Sperry. He found Brennan "practically down and out, poor fellow," and exhibiting "his soap bubbles" at the annual meeting of the Royal Society. (Sperry and Captain Creagh-Osborne wondered about the relationship between the monorail and soap bubbles.)[44]

Sperry's nature could not permit him to forget the problems of invention, however. While crossing on the *Oceanic*, he experimented with his drift indicator, a new device to show the drift of airplanes from the compass course. After using it to observe the drift of the ship against the lines of its wake, he asked Conner to "please tell Mr. Tanner and Ford that the drift apparatus is a great success." He sent messages to Thompson, instructing him in detail on claims for the indicator in the patent application. On

[42] EAS to Herman Hecht, April 19, March 16, and November 4, 1911 (SP).
[43] Zula to Herbert Goodman, May 11, 1914; and Zula to Helen Sperry, June 19, 1914; and EAS to Conner, June 3, 1914 (SP).
[44] EAS to Conner, May 14, 1914 (SP).

shipboard, Sperry's mind often turned to one of his long-standing development efforts: the compound internal combustion engine. A letter written aboard ship requested that Meitner consult recent articles Sperry was reading on practical thermodynamics and diesel engines. He also sent Meitner a sketch of a transfer valve to test. Also, after learning of the performance of his compass at Kiel, he closely monitored, by transoceanic mail, the efforts of Tanner and Ford to invent a solution to the problems raised in the tests.[45]

Of all the Sperry activities abroad in 1914, the one that best combined business and pleasure was their visit to Paris to watch Lawrence Sperry demonstrate the Sperry airplane stabilizer during the *Concours de la Securité en Aeroplane*, a competition between designers of safety devices for airplanes. Lawrence flew a Curtiss plane equipped with the stabilizer and won not only the prize but also the attention of the Parisians who found the handsome young pilot-engineer a dashing American figure. His triumph was the culmination of several years of Sperry effort to stabilize "a beast of burden obsessed with motion."

[45] EAS to Conner, May 4, May 8, and May 18, 1914; and Conner to EAS, May 29, 1914 (SP).

CHAPTER **VII**   A Beast of Burden Obsessed with Motion:
Stabilizing the Airplane

*Of all vehicles on earth, under the earth and above the earth the airplane is that
particular beast of burden which is obsessed with motions, side pressure, skidding,
acceleration pressures, and strong centrifugal moments, . . . all in endless variety
and endless combination.*   ELMER SPERRY, 1923

ELMER SPERRY'S GYROSCOPIC DEVICES CONTRIBUTED TO THE SOLUTION OF NAVI-
gation and stability problems at sea. In effect, they provided seafaring men
with technological appendages that allowed them to cope with the ocean
environment. After the Wright brothers made their historic flight, increasing numbers of men ventured into the air, where they were even less well
equipped by nature to deal with the environment. Again Sperry provided
technological appendages that helped man function outside of his natural
habitat. Earlier, Sperry had responded with the gyrocompass and stabilizer
when the development of steel-hulled battleships with guns of increased
range and accuracy had revealed the magnetic compass to be an inadequate
component in a developing system. After the early airplane pilots identified
similar inadequacies—reverse salients in an expanding technology—Sperry
again responded to fill the needs of the system.

i

Elmer Sperry became interested in developing an airplane gyroscopic
stabilizer in 1909, about two years after he began work on the ship stabilizer.
Though the Wrights had made their successful flight only six years earlier,
the infant industry attracted considerable attention, and pioneer airplanes
and flights defined the major problems that inventors and developers
needed to solve. In 1908 the era of military aviation began when the Signal
Corps of the U.S. Army awarded a contract to the Wrights for a flying
machine. Abroad, Louis Blériot flew his monoplane across the English

channel in 1909. Glenn Curtiss, in 1908, won the *Scientific American* aviation trophy for the flight of his "June Bug," and a year later he formed the G. H. Curtiss Manufacturing Company to build "Curtiss-Herring" airplanes. The same year he won the speed competition at the meet in Rheims, France. The increasing interest of the military in aviation, the Blériot type of airplane, and the pioneering of Glenn Curtiss would all directly influence Sperry's career.

Figure 7.1
Elmer Sperry in 1914.
From Flying *(August 1914).*

As the pioneer pilots flew—and crashed—the early aircraft, they became increasingly aware of the problem of maintaining stability. Before the Wrights flew at Kitty Hawk in December 1903, the pioneers in aviation, Lilienthal, Maxim, Lanchester, Pilcher, Chanute, and Langley favored an inherently stable aircraft. The success of Wilbur and Orville Wright, where others failed, resulted largely from their rejecting the dogma that aircraft should be inherently stable; their airplane was designed to be dependent upon the pilot to operate control surfaces to maintain the stability of the craft.[1] Their successes, however, depended upon visual contact with the earth and relatively calm days free of the gusty winds that overtaxed the pilot's sense of stability and the airplane's controls.

If man had been content to limit his flying to short hops on fair days, stabilization would not have been a major problem. Because he resisted such limitations, however, an improved stabilization system was needed. Elmer Sperry expressed views widely held in the aviation world when he said:

With the present machines very long flights are nearly beyond the endurance of the aviator [because of fatigue in maintaining stability]. . . . The automatic control of stability of the heavier-than-air machines will do much to decrease the growing list of fatalities. . . . The automatic control of stability will be especially valuable to the military use of the aeroplane, as it will make it possible to fly in almost any condition of weather. . . . In reconnaissance service only one man will be necessary as the machine may be controlled automatically while the aviator makes sketches, records information obtained, or operates the radio set to communicate with his base.[2]

Sperry and other pioneers in stabilization realized that, to maintain stability aloft, a pilot needed a reliable frame of reference and a way to compare his position to it. On earth man had his own reliable reference systems. When there was light, he could use visual frames of reference such as the earth's horizon. In light or dark, he could also rely on his inner ear canals, which correspond roughly "with the principal planes of the body, i.e., front and back, side to side, and transverse or horizontal . . . [and] act as levels."[3] Tactile impressions, especially from the soles of his

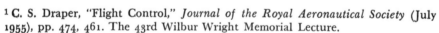

[1] C. S. Draper, "Flight Control," *Journal of the Royal Aeronautical Society* (July 1955), pp. 474, 461. The 43rd Wilbur Wright Memorial Lecture.
[2] Sperry, "Engineering applications of the Gyroscope," *Journal of the Franklin Institute,* CLXXV (1913), pp. 475, 478, and 479.
[3] This analysis of the stabilization problem is based, partially, upon a pre-World War I Sperry mimeograph: Reginald E. Gillmor, "The Sperry-Curtiss Aeroplane Stabilizer," June 1914.

feet, also helped him maintain his stability.

In the air, however, a man's ability to maintain the stability of his airplane is limited. He can depend little upon tactile and muscular impressions, and, when flying at night or in fog, he has no visual impressions. The canals of the inner ear, conditioned by and dependent upon the force of gravity, are confused by acceleration forces, which have the same effect upon the "levels" of the inner ear as gravity. By contrast, animals for whom space is a natural habitat have sensors that supply signals needed to provide the muscular control resulting in stability.

Sperry and the other pioneers saw two possibilities for stabilizing an inherently unstable airplane. They could design instruments to provide a frame of reference which the pilot could use to fly the plane, or they could design a system in which the instruments themselves flew the plane. Sperry would invent some of the aviator's most helpful instruments, but he first took the bolder course and attempted to provide automatic stabilization.

ii

Sperry's first effort to stabilize an airplane gyroscopically was a false start. He attempted to apply the passive gyro to the airplane in the same way he had planned to use it to stabilize the automobile and as Schlick had used it to stabilize the ship. This kind of stabilization depended upon the gyroscope's inertia; the counter force developed when a disturbing force caused the gyro to precess. This design resulted from a request by Stanley Beach, aviation editor of the *Scientific American*, for a gyrostabilizer. In April 1907, Beach began designing an airplane based on the successful Blériot monoplane. He hoped to manufacture a similar type in America for the private pilot interested in transportation and pleasure flying, but he wanted first to find a solution to the stability problem.

Beach may have heard Sperry suggest gyrostabilization. In the fall of 1908, Sperry gave a lecture on gyrostabilization to the New York City Aero Club; afterwards, he was besieged by people building airplanes and airships (dirigibles) who wanted gyroscopic apparatus to stabilize their machines. The lecture probably conveyed his enthusiasm and vision of how things might be, but he certainly had no clearly thought-out design for an airplane stabilizer at that time.

Beach began negotiating with Sperry in the spring of 1909. Since Sperry was then preoccupied with the ship stabilizer and with problems of chemical engineering, Beach provided most of the design specifications. Sperry looked over Beach's calculations and found them generally acceptable, although he suggested that a heavier gyro wheel be used to ensure the counter force needed for the airplane. Beach had proposed ten pounds and Sperry recommended thirty. Beach worked on his plane throughout the summer. His test experiences probably resembled those of many of the pioneers of the time. During a ground test on Labor Day, "one of the wings . . . broke off when caught by a puff of wind." Subsequently, when the plane was running along the ground without wings, the tail became readily airborne and steered satisfactorily; Beach confidently predicted that with suitable wings he would fly. On the same day, however, he read of a fatal accident which made him "anxious to have something that will positively hold the machine on a level keel under all conditions. . . . I want something that will hold the machine level even in a gale, and I believe that you [Sperry] have it."[4]

Late in November, Beach made the first attempt to fly the plane equipped with new wings, although it still lacked a stabilizer. His account vividly portrays his tribulations:

Figure 7.2
Gyro applied to Beach monoplane for automatic stability. From a brochure of the Scientific Aeroplane Company.

It was about 5 P.M. and almost dark, so I could not tell exactly what happened, but I think the machine lifted slightly in front, as otherwise I can't explain its tipping. I had enlarged the tail considerably with the result that it lifted readily. When half way down the track . . . it rose 5 or 6 ft. above ground, and, looking around, I saw it level with my head as I was sitting in the machine . . . suddenly the monoplane swerved to the left, tipping to the right so much that I thought it would upset. I stopped the engine all right, but just as I pulled the switch the right wing struck the ground and was broken back.[5]

He added that "this experience has made me more enthusiastic than ever about your gyroscope." At the end of the year, he expected soon to be flying his stabilized airplane and planned to form a $1 million company that would be "way ahead of Wright Bros."[6] By April, the Sperry gyro-stabilizer had been built at the Pearce shop (at a cost of about $350 for labor and materials). After the gyro was installed on the Beach monoplane, the plane was flight tested on April 10, 1910. Hannibal Ford, who observed the tests, reported that the added weight of the gyro ruined whatever flying ability the plane possessed.[7] Nevertheless, on June 1, Elmer Sperry wrote the Wright brothers that the gyro-equipped plane had made nine different flights. He said that the gyro had been devised to relieve the pilot from the strain of being constantly alert for sudden movements by neutralizing the effects of sudden wind gusts. As a result, tilting was reduced to gradual movements that allowed the pilot ample time to react. "A driver," Sperry reported, "who has operated the aeroplane to which this gyroscope was attached would not go up without it under any consideration."[8] After referring to his successful work for the navy on the ship stabilizer, Sperry proposed attaching stabilizers to one or more Wright airplanes.

Figure 7.3
Gyro for Beach monoplane, mounted on a sawhorse. From a brochure of the Scientific Aeroplane Company.

Following the tests of spring 1910, Sperry began to lose interest in Beach's plans, in part because he did not think that Beach had given him sufficient credit in an interview with the New York *World.* Yet, Stanley Beach's Scientific Aeroplane Co. printed brochures advertising the Beach

---

[4] Beach to EAS, September 22, 1909 (SP).
[5] Beach to EAS, November 29, 1909 (SP).
[6] *Ibid.*
[7] Notes taken from a tape recording of a talk given at the Massachusetts Institute of Technology by Elmer Sperry, Jr., on early airplane instruments, May 1, 1963.
[8] Sperry letter, dated June 1, 1910. This is a carbon copy on which there is no address, but the "Wright brothers" has been penciled in (SP).

gyroscopic monoplane and describing its stabilizer as "a new device, which has been developed after a great deal of experimenting," and has made possible 'machines . . . perfectly stable in flight, and . . . not in a state of unstable equilibrium, transversely, as in the case with all other aeroplanes." The "new device" weighed thirty pounds and, according to claim, produced a resistance to tipping of more than half a ton, or several times the weight of the machine.[9] The Scientific Aeroplane Company predicted that the airplane was here to stay and that "the world will welcome an aeroplane having the greatest of all desiderata—automatic stability." The world would welcome such a plane, but not in 1910 and not the Beach stabilized monoplane. Despite Sperry's sanguine letter to the Wrights and the buoyant advertising of the Beach company, the passive stabilizer proved ineffective in actual use.

iii

Sperry returned to airplane stabilizers in 1912, and this second venture culminated in his becoming the world's foremost pioneer in the field. Sperry, his son Lawrence, and the small staff of engineers at the Sperry Gyroscope Company would not only develop an effective automatic stabilizer before World War I but also transform it, during the war, into the major component of an automatically controlled missile, which would be converted into an automatic pilot after the war. The circumstances that returned Sperry to airplane stabilization arose from the general development of aviation, Glenn Curtiss' contributions to that development, and the navy's growing interest in aviation.

Between 1909 and 1912, aviation had become less a sensational exploit of intrepid pioneers, though it was still that, and more a new form of transportation with obvious commercial and military potential. In 1911 long-distance races and tours replaced short exhibition flights as the major flying activities.[10] The major cross-country races took place in Europe. In America, Harry N. Atwood flew from Boston to Washington with a Wright biplane; later, he flew from St. Louis via Chicago to New York in fourteen days. Galbraith P. Rodgers, starting from Sheepshead Bay, New York, flying a Wright biplane and accompanied by a special train carrying spare parts, made the first trans-continental airplane flight. Long flights tested the pilot's endurance at the manual controls and exposed him and his plane to varying weather conditions. Men and machines often were not equal to the challenge and 76 flyers died in 1911. The increased effort to invent a practical system of stabilization grew with the awareness of the "treacherous and shifty character of the air-ocean."[11]

[9] The stabilizer advertised had a cone-shaped wheel resting on a ball-bearing in the bottom of an aluminum case. The engine of the airplane drove the wheel, by a belt transmission, at 10,000 rpm's using one-fifth horsepower. Using the technique employed in his compass, Sperry ran the wheel in a vacuum made, in this case, by the use of a bicycle pump. Free from air resistance the wheel would continue to revolve long after any engine failure.

[10] "Retrospect of the Year 1911," *Scientific American*, CV (1911), pp. 591, 602.

[11] Grover Cleveland Loening, "Automatic Stability of Aeroplanes," *Scientific American*, CIV (1911), pp. 470, 471, and 488.

The inventors, including Sperry, would have been less interested if aviation had had no industrial side, no commercial promise. The *Scientific American* observed in 1911 that "it seems at first glance somewhat premature to rank the new art of flying among the accepted and recognized branches of industry; but on close examination . . . there is found ample reason for doing so."[12] Sales of airplanes were increasing: Blériot had sold several hundred planes since 1909 and other manufacturers in France were almost as productive; Britain had comparable manufacturers; and the United States had more than a dozen firms, with Wright, Curtiss, and Burgess the leaders. The Wright company had sold between 50 and 100 airplanes. Sales in the U.S. and abroad amounted to about 1,500 up to 1911.[13]

Elmer Sperry met Glenn Curtiss as early as June 1910, and the cooperation of these two pioneers greatly advanced the development of automatic airplane controls. Their cooperation in this early and important venture was fostered by the U.S. Navy and especially encouraged by its head of aviation Captain Washington Irving Chambers. When the two met, Curtiss was already a well-known aviator. Eighteen years younger than Sperry, Curtiss was born in Hammondsport, N. Y., only fifty miles from Cortland, and was similarly gifted with a mechanical aptitude. He progressed from building motorcycles to airplanes, and in 1908 he won the *Scientific American* trophy. After this, he represented the United States in the first international race held August 1909, at Rheims, France, where he defeated Blériot and other leading foreign aviators. "After Rheims, Curtiss was numbered among the world's greatest aviators."[14]

In 1909 he established his own airplane manufacturing company, which concentrated on hydroplanes. The navy became interested in Curtiss in February 1910, when he flew a hydroplane to the USS *Pennsylvania*, landed on the water, was hoisted aboard, and took off again from the water. Captain Chambers predicted that every battleship would soon carry one or two hydroplanes.[15] Curtiss' success spurred the navy to develop an air arm. After Congress made its first appropriation for naval aviation ($25,000), the navy purchased two Curtiss hydroplanes (the A1 and the A2) and a Wright plane converted to a hydroplane (the B1). Three naval officers did temporary duty at the factories to familiarize themselves with the planes and to learn to fly them.[16] Lieutenants T. G. Ellyson and John H. Towers trained with Curtiss at Hammondsport and later tested the Sperry airplane stabilizer.

In 1912, shortly after taking charge of naval aviation, Chambers predicted that airplanes would provide improved communications, facilitate

[12] "The Business Side of Aviation: The Money in Flights and Machines," *Scientific American*, CV (1911), pp. 355, 356.
[13] *Ibid.*
[14] "Curtiss' Career and Experiments," *Scientific American*, CIV (1911), p. 349.
[15] Carl Dienstbach, "The Flying Boat and Its Possibilities," *Scientific American*, CVII (1912), p. 216; "The Hydro-aeroplane," *Scientific American*, CV (1911), p. 492.
[16] *Annual Reports of the Navy Department for the Fiscal Year 1911* (Washington, 1912), p. 59.

reconnaissance, and in warfare be used to bomb submarines, docks, and other facilities. "It has been fully demonstrated," he wrote, "that of two opposing forces, the one which possesses superiority in aerial equipment and skill will surely hold a very great advantage."[17] To achieve this superiority, Chambers advocated the development of an automatic stabilizer for naval aircraft.[18] He thought the day of the "crude efforts of the pioneer inventors" had passed and that scientific engineers were needed to develop flying instruments. He had found that most aviators and manufacturers opposed stabilizers because defective ones had been promoted. This should not deter continued effort, Chambers insisted, for naval aviation could not continue to depend on the acrobatic skill of especially talented pilots. He wanted that skill built into the machine.

iv

Encouraged by the progress of the aviation industry and by developments in naval aviation, Sperry's interest in airplane stabilization was further stimulated by his awareness of the momentum his company had in the gyro field and by his son Lawrence's determination to become an aviator and an aeronautical engineer. Sperry intended to use the techniques and components he had employed in developing the compass and marine stabilizer to improve upon the crude Beach stabilizer. He had spent countless hours analyzing the pitch, roll, and yaw of a vessel upon the surface of the sea; those hours prepared him to analyze the even more complex problem of a vehicle on an ocean of turbulent air. Furthermore, as work on the gyrocompass and ship stabilizer became more routine, his creativity sought expression elsewhere.

In the spring of 1912, Lawrence returned from prep school, where his enthusiasm for flying had been increased by the exploits of the Wrights, Curtiss, and other pioneers. This interest was the latest of his infatuations with machines. He had spent numerous hours working first on bicycles and then on the noisier, dirtier, and more exciting motorcycles; when he had announced his intention to build and fly an airplane, his parents were not surprised.

During the summer of 1910, when the family was away, he had used the Sperry home as an airplane factory. Attic, cellar, and bedrooms served as work rooms, and when dope on the wing fabric did not dry fast enough, he fired the furnace, which had been drained for the summer, to accelerate the process. When the family returned, he triumphantly displayed his plane in the backyard. They also found that the front hall bannisters had been torn out, that a portion of the house wall had been demolished to provide a way for the airplane to be removed, and that the furnace had cracked. Justifying these mishaps in the name of technological advance, Lawrence, with the help of Elmer, Jr., and some neighborhood boys, tested his motor-

[17] Capt. W. I. Chambers, USN, "Report on Aviation," *Annual Report of the Navy Department, 1912* (Washington, 1913), p. 155.
[18] *Ibid.*, p. 160.

less plane at the Sheepshead Bay racetrack by towing it behind the family automobile. But gliding was not enough. During the summer of 1911, Lawrence worked as much as ten hours a day redesigning the plane to accommodate an engine. Using some of the profits from his bicycle repair shop and drawing on the resources of an older friend who shared his interest in aviation, he ordered a second-hand, five-cylinder Anzani radial engine from France. Their scheme was to recover the investment in barnstorming tours with Lawrence as pilot. This was too much for Elmer and Zula. Despite their tolerance of the exploits of their children, they over-ruled Lawrence's plans. Zula was concerned about the dangers of flying and also about Lawrence's health, for she considered him far from robust. Sperry thought Lawrence's work on the plane remarkable but wanted him to study engineering at Cornell ("He needs training if he is to make the man he ought to be").[19] The parents decided that Lawrence in the autumn should enter Professor Evans' Desert School in Arizona, located on a ranch and noted for its outdoor exercises and for its scholastic standing as a college preparatory school.[20]

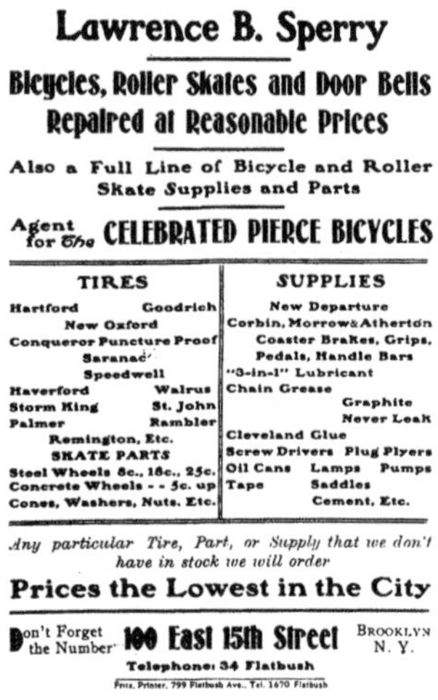

Lawrence B. Sperry

Bicycles, Roller Skates and Door Bells Repaired at Reasonable Prices

Also a Full Line of Bicycle and Roller Skate Supplies and Parts

Agent for the CELEBRATED PIERCE BICYCLES

TIRES

Hartford Goodrich
New Oxford
Conqueror Puncture Proof
Saranac
Speedwell
Haverford Walrus
Storm King St. John
Palmer Rambler
Remington, Etc.

SKATE PARTS

Steel Wheels 8c., 18c., 25c.
Concrete Wheels - - 5c. up
Cones, Washers, Nuts, Etc.

SUPPLIES

New Departure
Corbin, Morrow & Atherton
Coaster Brakes, Grips, Pedals, Handle Bars
"3-in-1" Lubricant
Chain Grease
Graphite
Never Leak
Cleveland Glue
Screw Drivers Plug Plyers
Oil Cans Lamps Pumps
Tape Saddles
Cement, Etc.

Any particular Tire, Part, or Supply that we don't have in stock we will order

Prices the Lowest in the City

Don't Forget the Number 100 East 15th Street Brooklyn N. Y.

Telephone: 34 Flatbush

[illegible], Printer, 799 Flatbush Ave., Tel. 1670 Flatbush

Figure 7.4
Handbill for early Lawrence Sperry venture.

Lawrence's plane was almost ready to fly when he was told of his parents' decision. Lawrence went "wild about it every day or two" during the few weeks before school opened but was somewhat mollified by an agreement that if he did well academically he could decide upon his career the next spring.[21] Lawrence liked the school "pretty well . . . although aeroplanes and their like are sadly lacking."[22] Having improved his health and done well in his studies, he wrote in April:

Dear Mother and Father:

School will be over very soon, as you know.

Almost ever since I came here I have been thinking what I shall want to do after I leave. In making my decision, you can just wager that I have tried to consider your wishes and I know that you will agree with me when I tell you what I want to do. . . .

I want to enter the aeroplane business. I know you have no objections to this save that you think the game has a poor future. As for me, I have utmost confidence in a brilliant one, and I think that is near at hand. . . .

If you doubt the future of aeroplanes, you are simply placing yourselves in the unmagnanimous position of those who doubted the future of the railroad and the automobile. The aeroplane is bound to become safe and practical. Already they know that if they have a certain relation between the rear elevator and the main supporting planes, longitudinal stability would take care of itself.

Where would aeronautics be today if the Wright Brothers and others had not had implicit faith in the future of it? Then why should we, who have seen what can be done with the crude, unscientifically designed machines of today, be afraid to plunge in? . . .

The first thing I want to do is to finish the plane. . . .

You should not condemn aeroplaning as a dangerous pursuit because a

19 EAS to H. H. Wilcox, Newtonville, Massachusetts, August 3, 1911 (SP). Lawrence planned to go into the flying business with Wilcox.

20 Archie Roosevelt, son of Theodore Roosevelt, was preparing for Harvard there.

21 "Zula Goodman," a typescript by Helen Lea (SP).

22 Lawrence to Herbert Goodman, December 22, 1911 (SP).

few reckless and, in fact, ignorant men, wholly incompetent save simply as pilots, are taking reckless, fool chances to make large financial returns. These men . . . bear no relation to sane flight. . . .

I am very determined to go into aeroplanes and I think that you should help me get started, as you promised to do last fall, when you said that if I returned still strong for aeroplanes you would help me.[23]

It was not entirely coincidental that Elmer Sperry approached the navy about developing a stabilizer when his son returned home. If Lawrence was determined to fly, why not let him learn while helping develop a safety device and as a project engineer? Thus began a productive, historic association, for Lawrence, from the spring of 1912 until his tragic death in an airplane accident in 1923, was the individual, besides his father, most responsible for the introduction of the stabilizer and automatic pilot. Colonel (later General) William Mitchell, a pioneer figure in military aviation who came to know Lawrence well, described him as one of aviation's "most brilliant minds, one of its greatest developers and a splendid pilot and air man."[24]

In the spring of 1912, the primary problem was to initiate the project. Informed of Captain Chambers' interest (possibly by a naval friend), Elmer Sperry wrote him in June offering to develop the stabilizer. Thus began more than a decade of invention and development that proved the most frustrating, complicated, original, and perhaps influential of Sperry's gyro projects. After receiving the letter, Captain Chambers traveled to New York to investigate Sperry's offer. He found that there were no detailed drawings, much less a model, but the general plans seemed promising. Chambers would not agree to purchase a stabilizer in advance but offered Sperry the use of a Curtiss plane for testing the device. The plane was to be completed by fall at Curtiss' Hammondsport factory and would be available for a brief period before final acceptance by the navy.[25] Curtiss readily cooperated because his airplanes needed stabilizing.

The stabilizer was built at the Pearce shop because the Sperry company had not yet moved into its Brooklyn factory. Both Lawrence and Edward, on summer vacation, helped with the drawings. Lawrence followed construction closely and experimented at the shop with the prototype. Unfortunately the stabilizer was not ready for installation before the navy plane left Hammondsport. Nevertheless, in November 1912, Lawrence took the partially completed device to Curtiss for trials on another plane. Only that part of the stabilizer intended to maintain lateral stability was ready; work continued at the shop on the longitudinal component.

23 This letter, written in April 1912, is quoted in "Young Men in Aeronautics," a pamphlet published in 1957 as a twenty-year progress report of the "Lawrence B. Sperry Award," administered by the American Institute of Aeronautics and Astronautics, New York, N. Y. This annual award recognized outstanding young men in aeronautics.
24 Colonel William Mitchell, "Lawrence Sperry and the Aerial Torpedo," *U.S. Air Services* (January 1926), p. 16.
25 Captain W. I. Chambers to Chief BuNav, April 2, 1913 (SP).

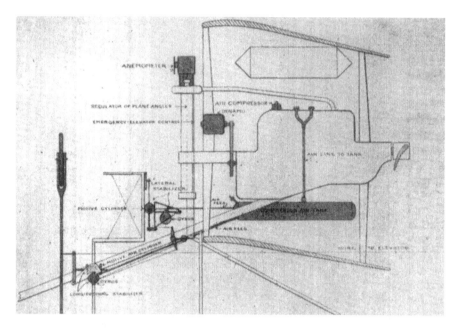

Figure 7.5 Arrangement of the Sperry stabilizer. From the Scientific American, *CVIII (1913).*

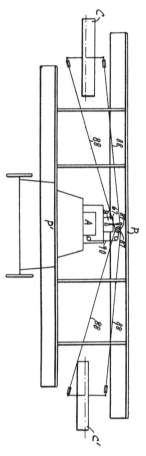

Figure 7.6
The Sperry Airplane
Stabilizer (Patent No.
1,368,226).

v

The stabilizer had two components: one providing lateral stability (about the longitudinal, or roll, axis of the airplane) was mounted behind the pilot; the other, for longitudinal stability, was mounted slightly forward and beneath the pilot's feet.[26] Each unit had a pair of small gyros, the rotors of which turned at the same speed in opposite directions, geared or coupled to precess in opposite directions. The rotational axes of the gyros lay in a horizontal plane: the axes of the lateral unit transverse to, and those of the longitudinal unit parallel to, the longitudinal axis of the airplane. A small alternating-current generator belted to the airplane engine powered the gyro rotors. The unit providing lateral stability was suspended by pivots in a horizontal line along the longitudinal axis of the airplane; the unit for longitudinal stability was suspended from pivots in a horizontal line transverse to the longitudinal axis of the airplane. The purpose of this arrangement was to minimize the transmission of rolling and pitching forces to the lateral stabilizer and the longitudinal stabilizer, respectively. In operation, the gyros acted like long heavy pendulums suspended from the airplane and attracted by gravity. They provided reference lines by which

[26] This description of the 1912–13 model of the stabilizer is based on incomplete descriptions in *The Engineer*, CXV (1913), pp. 600–2; Robert K. Skerrett, "Making the Airplane Safe by the Gyroscopic Stabilizer," *Scientific American*, CVIII (1913), pp. 511–12; Sperry, "Applications of the Gyroscope," pp. 447–82; and "New Device May Rob Aerial Navigation of its Chief Peril," New York *Tribune*, May 18, 1913. The descriptions have been supplemented by information gained from an examination of photographs in the Sperry Papers, which do not illustrate details.

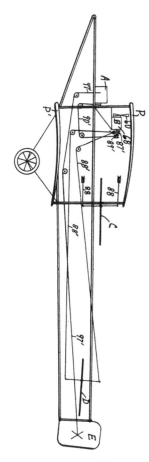

Figure 7.7
The Sperry Airplane
Stabilizer (Patent No. 1,368,
226). (Key: the stabilizer, A;
clutches, B; ailerons, C, C';
elevating rudder, D;
steering rudder, E; wings,
P, P'; windmill power
source, 60.)

the deviations of the airplane from horizontal flight could be determined.[27] The gyros acted like pendulums because they were free to precess and counteract forces caused by the pitching and the rolling of the airplane, forces that tended to disturb the gyros and the reference lines they established. These forces were transmitted by friction at the pivot bearings suspending the stabilizer units. The action of gyros would be more obvious when Sperry remounted them as a single unit in a redesigned airplane stabilizer—the 1914 model—built about a year later.

When the airplane rolled or pitched, its angular displacement from the reference established by the stable gyro units operated the valves of a servomotor.[28] The compressed air for the servos, supplied by an airplane engine cylinder, was stored in a small tank until the gyros opened the intake valve to the servo cylinder and activated the piston of the servomotor. The piston's connecting rod then operated the controls of the airplane normally operated by the pilot. In the Curtiss plane, the pilot, strapped into a shoulder harness, normally controlled the ailerons by swaying his body to right or left; he controlled the lateral elevators with a hand-wheel mounted before him. The stabilizer for lateral stability, therefore, was connected to the shoulder-harness control linkage, and the longitudinal component was connected to the wheel. The pilot was provided a control that allowed him to cut the stabilizer in and out.

Sperry added to his stabilizer system a multi-purpose anemometer which, with a vane mounted forward of the upper wing, measured the force of the wind and translated the force into an air-speed reading. It also transmitted an electric corrective signal to the servomotor of the longitudinal stability unit when the airplane velocity fell below that necessary to maintain flight; the servo responded by adjusting the elevators to turn the airplane into a gliding descent (volplane) until it regained speed and the pilot could once again take control. In 1913 this was a major safety provision because of the low power of the early planes and the tendency of the inexperienced pilot to climb at too steep an angle of incline. An imaginative reporter for the New York *Tribune* graphically described the danger:

The engine was purring away like a contented cat, and in wide circles the machine was climbing heavenward. Below the houses dwindled into toys and the landscape was like a view through a telescope reversed. The aviator was exhilarated by the cooler air and the increased ozone, and in response to sheer buoyancy of spirit set the elevating planes of his machine at a sharper climbing angle. Still, the engine purred its song of deceptive security, and still the air whistled merrily past the ears of the pilot, making him think he was advancing faster than he really was. Suddenly, and without a moment's warning, the aeroplane staggered in its upward course, shivered a second, as if in fearful hesitation, and then started earthward, tail first, at

[27] There are indications that Sperry might have used the precession of the gyros to generate the error signal, but this seems unlikely in view of the design details available.
[28] Skerrett, "Making the Airplane Safe," p. 511. This is an early example of the use of "servomotor." The term was hyphenated in the 1913 article describing the Sperry device.

increasing speed. Before the aviator could make one corrective movement the aeroplane was dropping at a bomb-like velocity. Why dwell upon the rest of the details?[29]

In addition to functioning as a speed indicator and as an automatic safety device, the anemometer also transformed the movements of the stabilizer controls into a function of air speed. Because smaller adjustments of the airplane controls were needed at high speed to correct the attitude of the plane, the anemometer shifted the fulcrum of the plane's control levers so that the resulting angles of the ailerons or elevator suited the speed of the airplane.

vi

Delivering the stabilizer to Curtiss in the fall of 1912 gave Lawrence the exciting opportunity to visit the Hammondsport factory and flying facilities. Located in the village of Hammondsport in upstate New York, the site on Keuka Lake enabled Curtiss to test his hydroplanes easily. Lawrence found the factory unpretentious—a ramshackle set of wooden buildings that somehow managed to produce outstanding airplanes. At Mrs. Mott's boarding house, where the visiting flyers stayed, he talked aviation with others more experienced but no more enthusiastic. He particularly admired Navy Lieutenant T. G. Ellyson, who volunteered (on one occasion) to fly with the stabilizer, and liked the nickname "Gyro" that Ellyson gave him. Lawrence looked forward with high spirits to his own flying lessons.

Because Lawrence had not yet earned his flying license, he was not able to test fly the stabilizer, but he persuaded Curtiss and others to fly the plane on which he installed the lateral component. Lawrence, along as a passenger, closely observed the performance of the device. A flight with Curtiss piloting on Thanksgiving Day, 1912, caused Lawrence to write home enthusiastically that the stabilizer held the plane steady even in a "puffy wind." After the test flights, Lawrence assisted his father in remedying the difficulties that had been encountered. In November, for example, after the valve admitting compressed air from a cylinder of the aircraft's engine to the stabilizer's pneumatic servomotors had malfunctioned, Sperry designed a new check valve which Lawrence installed. After such changes, the younger Sperry had to persuade Curtiss to free the plane and perhaps a pilot to test the modified device.

At the end of December, Elmer Sperry reported to Chambers that the horizontal stabilizer had been "perfected" and attempted to convince him that the navy should order a unit for delivery within a month. Chambers, as cautious as he had been the prior spring, insisted on awaiting the perfection of the longitudinal stabilizer as well, so that the navy could purchase a complete device. He prudently suggested that the navy await the outcome of further tests, which he recommended be conducted at the Washington

[29] New York *Tribune*, May 18, 1913, p. 5.

Figure 7.8 Lawrence Sperry learning to fly at the Curtiss School.

Navy Yard. The Sperrys, however, decided to continue their testing in San Diego, California.[30]

Each winter Curtiss moved his base to San Diego; in the winter of 1912–13, he was working with the army, providing flying instructions. Sperry arranged for Lawrence to continue testing on the army planes with army flyers. After returning the stabilizer to the shop for modifications, Lawrence joined the move to the camp on North Island, next to Coronado Beach. He installed the stabilizer on an army plane, which army pilots flew while Lawrence tended the gyro stabilizer. In March Lt. Harold Geiger gave an exhibition, flying over warships anchored in the bay. Geiger cut figure eights around the masts of cruisers, volplaned "with startling rapidity from dizzy heights," and tried "to careen the machine half-way over on one side when flying low, to thoroughly test the gyro." These flights led an aviation journal to report that the success of the stabilizer was now assured.[31]

But all was not well. First one plane and then two others equipped with the stabilizer crashed, one crash killing a passenger and another causing serious injury. Though the stabilizer was not being used when the

[30] Chambers thought he could monitor the tests better in Washington; the Sperrys preferred San Diego weather. Chambers to EAS, January 2, 1913 (W. I. Chambers Papers, Library of Congress, container 22).

[31] *Aero and Hydro*, VI (1913), p. 50.

Figure 7.9
The Curtiss flying boat.
The Sperry stabilizer
of 1914 was installed
on a similar aircraft.
From the Scientific American.
CVII (1912).

accidents occurred, "the experience did not tend to stimulate interest in the further possibilities of gyroscopic stabilization."[32] Furthermore, reports in late April described the overall results of the tests only a limited success and the stabilizer only "a step in the right direction."[33]

Downcast and pessimistic because of these events, Lawrence returned East. His mood was short-lived, however, for the navy decided to support the experiments. On June 25 it contracted for a stabilizer and provided a newly purchased Curtiss flying boat to be devoted exclusively to two months of tests at Hammondsport. Zula was certain that there would be no more "half-hearted experimenting as at San Diego with the Army." She felt that the navy was more scientific and more mechanically proficient than the army. Furthermore, the Sperrys were far closer to the navy and Glenn Curtiss, who was "such a nice man and so interested in Lawrence and the whole thing."[34]

The navy had designated Lt. Patrick N. L. Bellinger test pilot for the stabilizer-equipped airplane. He first visited the Sperry shop to familiarize himself with the device but could generate little enthusiasm for it. Like so many of the early pilots, he enjoyed the challenge of piloting an airplane and preferred not to turn the controls over to a machine. In Hammondsport, he learned that the other flyers had made a pool and placed bets on whether he would survive the tests.[35] Bellinger wondered, as well, why the navy expended so much effort on an entirely private venture. On many subsequent occasions, both Lawrence and his father encountered pilots who failed to appreciate their efforts with the stabilizer.

When the project took on new life in the spring of 1913, publicity articles appeared in the New York *Tribune* (a full page), the *Engineer* (London), and the *Scientific American*.[36] The publicity stressed the critical importance of airplane stability for the future of aviation; the *Engineer* unequivocally stated that "trite as it may sound, the broadening of aviation

[32] Lawrence Sperry, "The Aerial Torpedo," *U.S. Air Services* (January 1926), p. 18.
[33] *Aero and Hydro*, VI (1913), p. 76.
[34] Zula to Herbert Goodman, May 21, 1913 (SP).
[35] Patrick Bellinger, with Lee Pearson, "Memoirs of Bellinger," (unpublished typescript in the files of the Historian of the Naval Air Systems Command).
[36] "New Device May Rob Aerial Navigation of its Chief Peril," New York *Tribune*, May 18, 1913; *The Engineer*, CXV (1913), pp. 600-2; and Skerrett, "Making the Airplane Safe," pp. 511–12.

in its fields of practical usefulness hinges upon automatic stabilizing."
Though Sperry's stabilizer was not the first in the field and not the only
one being developed in 1913,[37] a writer in the *Scientific American* concluded that "the most successful efforts to deal with these two departments
[lateral and longitudinal] of aeroplane stability have been those of Mr.
Elmer A. Sperry."[38]

Both the lateral and the longitudinal components of the stabilizer had
been tested in San Diego. After studying the results of the test, Elmer Sperry
assured Chambers that the apparatus, "which works almost perfectly," required little if any alteration.[39] Sperry was too optimistic. Almost a decade
later, he observed that this was not "the first time that I have gotten into
a development where I had to swim out and found that the shore was a
'mighty sight' farther away than any of us had ever anticipated." Development expenses mounted so rapidly that the project at times assumed the
character of a "nightmare."[40] With the return to Hammondsport and with
cooperation of the navy, however, the Sperrys were sanguine.

Lawrence Sperry installed the stabilizer on Curtiss' second flying boat
(the "C-2") in the summer of 1913, and between August 31 and October 4,
Bellinger logged 58 flights testing the gyrostabilizer. Lawrence often accompanied him to observe the performance of the stabilizer.[41] With Bel-

---

[37] According to Loening, "Automatic Stabilization," pp. 470ff., stabilizer inventions could be placed in two categories, nongyroscopic and gyroscopic. In the first
classification were devices using light vanes riding in the airplane's air stream.
When the plane deviated from the normal flight position, the vanes briefly retained
their line of drift. The angle between this drift and the new direction of the airplane was transformed into a signal that controlled the movement of the airplane's
control surfaces to return the airplane to the original flight path. Other nongyroscopic devices, including an automatic stabilizer patented by the Wright brothers
in 1913, used a pendulum to establish a frame of reference. As the plane rolled, the
difference between the reference and the airplane's new position generated a signal
operating a pneumatic servomotor that in turn adjusted the control surfaces. The
Sperrys were not the first to invent a gyrostabilizer for an aircraft. Hiram Maxim,
the famous inventor, in 1897 patented a device with components similar to those
in the Sperry stabilizer of 1913. Maxim described a single gyro, a servomotor, and
a follow-up system (English patent 19,228). Herbert Thompson, who later examined
the patent, decided the device could not work because Maxim "did not know how
to build gyros." (H. H. Thompson to EAS, London office, March 24, 1916; SP.)
Before 1912, Louis Marmonier, Paul Regnard, and Paul Victor Arvil all proposed
using the gyro for airplane stabilization; Marmonier and Regnard used the motion
of precession to activate the control surfaces. Perhaps the first patent on a gyrostabilizer for an aircraft was issued in the United States in 1877 for an aerial machine using two gyro wheels, one with a normally vertical, the other with a normally
horizontal, axis; the combination was intended to maintain stability about two
axes (*Scientific American*, CV [1911], p. 232). Sperry conceived of the 1913 stabilizer
as early as 1909 judging by his entry in Notebook No. 50 (March 10, 1909, to December 31, 1909; SP).

[38] Skerrett, "Making the Airplane Safe," p. 511.

[39] EAS to Chambers, April 19, 1913 (SP).

[40] Elmer Sperry testifying, April 30, 1923, in patent interference, 47032, *Sperry* v.
John Hays Hammond, Jr., "Wireless Controlled Aerial Torpedo," U.S. Patent
Office files.

[41] The Sperrys experimented with other combinations while they had the use of

linger at the controls, Lawrence had to sit in a contorted position in order to observe the device, and his concentration was often so intense that on one occasion he remained oblivious as the plane, struck by a sudden gust of wind, plunged into the water. Lawrence sometimes pleaded with Bellinger to continue a particularly rigorous test even though the plane and the occupants seemed in danger. This kind of behavior won the respect of Bellinger, who judged Lawrence brilliant, hard-working, and enthusiastic —and young.[42] Lawrence's enthusiasm and energy increased as the stabilizer approached the level of performance demanded by the Sperrys. His mother noted that by the end of the year he was "going Elmer's pace when he was L's age."

With a new airplane and experimental device, Bellinger and Lawrence never lacked problems. The log of the Curtiss C-2 shows that small adjustments in the stabilizer threw off the setting of the airplane's controls; resetting the controls once again disturbed the functioning of the stabilizer.[43] For example, the nominal neutral setting for the stabilizer, which should have ensured a level flight, resulted in the plane's banking to the left. When the gyrostabilizer was then taken apart and adjusted, the ailerons of the airplane also proved defective and these had to be realigned. Lawrence and Bellinger also found that the vertical rudder of the plane had to be set about five degrees off the neutral position to compensate for the rotation of the propeller wake. These were but a few of the problems; to add to them, Bellinger, on good flying days, preferred the company of other passengers, presumably female.[44]

On hand to evaluate the test was Naval Constructor H. C. Richardson. Richardson, like his senior, David Taylor, and other naval constructors, was thoroughly grounded in engineering and tended to take a more sympathetic view of the mechanized pilot than Bellinger and other aviators. In his report to Chambers, he discussed the essential complexity of the problem: the plane and the atmosphere presented the stabilizer with too many variables. The Sperrys had designed a stabilizer on the basis of the simple flight theory of the day, but reality was far more complicated. Richardson noted, for example, that the tail controls of the C-2 were in the lower half of the propeller race, and, therefore, variations in speed, course, and load altered the response of the airplane to identical commands from the stabilizer, and identical settings of the controls. Despite the complexity of the stabilizer, it did not compensate for such variations. Richardson also found that automatic volplaning was dangerous when the plane flew at less than

the C-2. Among these was a pendulum for both longitudinal and lateral control, "this being the method," Lawrence observed, "employed by fully 95 percent of the inventors." The Sperrys concluded that the pendulum was useless. Lawrence Sperry to the executive committee of the Sperry Gyroscope Company, March 1915 (SP).
42 Bellinger, "Memoirs of Bellinger."
43 H. C. Richardson to Officer-in-Charge of Aviation, Hammondsport, New York, September 12, 1913 (SP).
44 The Sperry Gyroscope Company complained to Captain Chambers that when the wind was good and passengers could be flown Bellinger took friends rather than Lawrence. EAS to W. I. Chambers, September 11, 1913 (SP).

Figure 7.10
From Flying *(August 1914).*

THE SPERRY APPARATUS BEFORE MOUNTING ON THE AEROPLANE: *(a) Servo-motor for longitudinal; (b) Servo-motor for lateral; (c and d) connections to elevating control; (e and f) connections to lateral control or shoulder yoke; (g) The gyroscopic element; (h) the anemometer; (i) the generator with mountings; (j) foot pedal for changing controls from manual to automatic; (k) connections for cylinder for obtaining compressed air.*

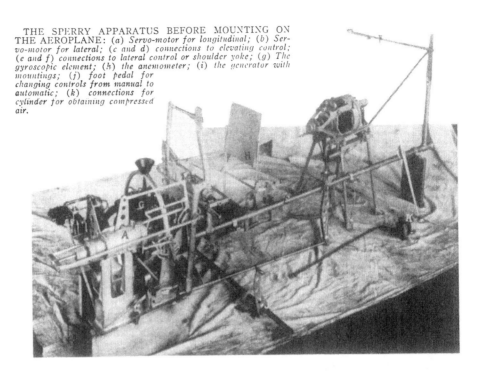

Figure 7.11
Sperry airplane stabilizer in 1914.

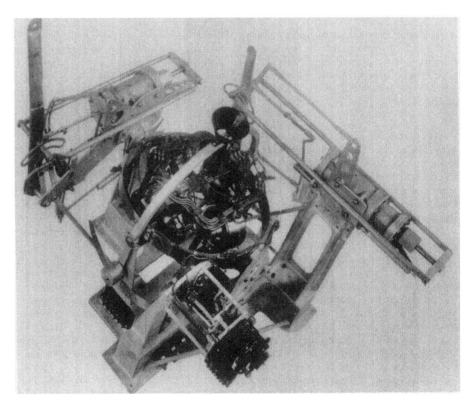

100 feet, for the pilot could not disengage quickly enough to pull out of the automatic dive. In addition, he discovered that during extended use the stabilizer introduced a cumulative deviation from the horizontal. For these and other reasons, he preferred to think of the stabilizer as a supplement to pilot control; he did not consider it a substitute for the judgment and range of response of an experienced pilot.[45]

Despite stabilizer inadequacies, Richardson had confidence in the Sperrys' ability to correct many of the defects and in the fundamental soundness of their concept. He recommended that the experimentation by the navy continue after modifications had been made in the stabilizer on the basis of the test results.[46] The Sperrys shared his desire to improve the design; in the winter of 1913–14 they designed and built a fundamentally improved stabilizer.[47]

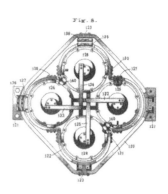

Figure 7.12
Two pairs of gyros used to create a stable platform. From Sperry patent, Gyroscopic Apparatus (Patent No. 1,186,856).

vii

In the 1914 model, as in the 1912–13 model, Sperry used gyros to establish a stable reference and used servos to align the airplane automatically with the reference. But instead of mounting the two stabilizers separately, as he had in 1912, he nested four gyros on a single platform. The stabilized, or stable, platform used as a frame of reference has since become an essential component of guidance systems for missiles, long-range aircraft, and submarines.[48] Sperry described the stable platform in a patent application of 1912:

This invention relates to gyroscopic apparatus and has for its object, among others, to construct such an apparatus which may be used . . . as a reference platform for the control of the stability of unstable objects . . . to provide a reference line or plane which is fixed in space so that it may be used to govern automatically the movements of a body, such as torpedoes, air craft or the like.[49]

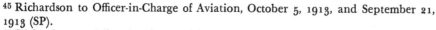

[45] Richardson to Officer-in-Charge of Aviation, October 5, 1913, and September 21, 1913 (SP).
[46] Richardson to Officer-in-Charge of Aviation, October 5, 1913 (SP).
[47] The stabilizer of 1914 was finished in the shop on December 9, 1913; Lawrence Sperry memorandum to the executive committee of the Sperry Gyroscope Company, March 1915. I have designated this the stabilizer of 1914 because it was first installed and flown in an airplane in that year. This designation distinguishes it from the stabilizer publicized in the spring of 1913. The aerial stabilizer made in 1913 and flown at Bezons is at the Smithsonian National Air Museum (accession #1390, catalogue #1963-399).
[48] Robert H. Goddard, "The Use of the Gyroscope in the Balancing and Steering of Aeroplanes," *Scientific American Supplement,* LXIII (June 29, 1907), p. 26330, suggested a gyrostabilized platform as the basis for an airplane stabilizer. Goddard, the pioneer in rocketry, wrote the essay when he was an undergraduate at Worcester Polytechnic Institute. He proposed taking control signals from the platform in order to stabilize the airplane. He also proposed a directional gyro for automatic steering. Goddard supplied few details, probably not reducing his ideas to practice. Later he credited Sperry and Charles Kettering with the invention of a flight-controlled plane; Milton Lehman, *This High Man: The Life of Robert H. Goddard* (New York: Farrar, Straus and Company, 1963), p. 380. In fact, Sperry anticipated Kettering by several years in the development of automatic flight controls.
[49] Filed July 11, 1912 (708,809), "Gyroscopic Apparatus," issued June 13, 1916

*Normal flight.
Roller is on insulated
sector and no signal
is sent to elevator.*

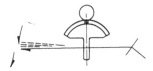

*Plane pitches up,
roller moves into positive
sector, and signal sent
to apply down elevator.
Plane begins to nose down.*

*Plane returns to level flight
and roller moves
on to insulated sector.
Signal for
down elevator cut off.*

*Plane overshoots,
roller moves into negative
sector, and signal for up
elevator sent. Plane noses up,
but again overshoots.
Plane is oscillating since
control has no follow-up.*

*Figures 7.13
Control for
airplane gyrostabilizer
without follow-up
(From Sperry Gyroscope
Company, Inc.,
Automatic Flight Control.)*

The other notable feature of the stabilizer, Sperry stressed, was "the provision of a complete system of automatic control to effect the stabilization of aeroplanes. . . ."[50] He was adding one more link to the long chain of Sperry automatic devices that extended back to his first patent.

To reduce his concept to practice, Sperry used the components with which he had become familiar in making his gyrocompass and his marine stabilizer. The gyro in his gyrocompass was a stable element that retained its alignment along the meridian. He borrowed from the gyrocompass the electromechanical components that kept the phantom automatically aligned with the gyro. In his marine gyrostabilizer, he first used a pendulum, later a sensor gyro, as a stable reference, and employed electromechanical components controlling the stabilizer in order to keep the ship aligned with the stable reference.[51]

In the 1914 airplane stabilizer, the gyros were mounted on a platform suspended in gimbals. The center of gravity of the platform was below the center of suspension so that the platform hung like a pendulum toward the center of the earth. The gimbal suspension allowed the pendulum platform to retain its horizontal alignment despite the motions of the airplane, and the inertia of the gyros maintained the alignment despite the disturbing forces transmitted by the bearing friction of the suspension. The spin axes of all gyros were horizontal, as in the 1912-13 model, with one pair aligned along the longitudinal axis of the airplane and the other pair transversely. One pair maintained the longitudinal axis and the other the lateral axis of the stable, horizontal platform. The gyro pairs had oppositely spinning motors and were coupled for equal and opposite precession.

Sperry used a train of electrical, mechanical, and pneumatic components to align the airplane automatically with the horizontal stable platform. When the airplane deviated from the horizontal—because of strong gusts of wind, for instance—the relative movement between the stabilized platform and the airplane initiated a command signal that operated the servomotors, which moved the airplane's control surfaces. Sperry used a trolley, or roller, and double contact (as he had in the gyrocompass) to send the signal (in the airplane stabilizer the roller and contact were named the "pick-off").[52] The roller was attached to the stabilized platform; the

(Patent No. 1,186,856). This patent also covered the gyro roll-and-pitch recorders. The company successfully built and delivered nine of these in 1911–12. In this device, pairs of coupled, oppositely spinning gyros (one pair for roll, the other pair for pitch) stabilized a heavy pendulum, which carried a pencil that recorded the motions of the ship. Lawrence Sperry to executive board of the Sperry Gyroscope Company, March 1915 (SP).

50 Filed July 17, 1914 (851,477), "Aeroplane Stabilizer," issued February 8, 1921 (Patent No. 1,368,226).

51 It is surprising that Sperry did not see earlier the possibility of dispensing with the heavy gyro wheel and using water fins, like airplane control surfaces, to stabilize the ship.

52 The 1914 model is described as having a roller-contact and sector pick-offs in, Sperry Gyroscope Company, Inc., *"Automatic Flight Control,"* [1947?], p. 21. Charles Keller in "Automatic Pilot, 1913 Version," *Sperry Engineering Review,* XIII (October 1960), pp. 22–23, describes the pick-offs as mechanical (Keller has labeled the stabilizer used at Bezons in 1914 as the 1913 stabilizer; I have desig-

double contact, to the airplane. When the airplane and platform were aligned, the roller rested on a neutral central insulator section between the two contacts. When the roller moved onto one of the contacts, it completed a circuit and sent a signal to the servomotor. If the roller moved on to the sector on one side of neutral, the pneumatic servomotor moved in one direction; if the roller moved in the other direction, so did the servo. In this way the servomotors moved airplane control surfaces to counter the movement of the airplane. There were two pick-offs and two servos—a set for longitudinal stability and a set for lateral stability. The servomotors were of the same pneumatic type used in the stabilizer of 1913.

Other notable characteristics of the stabilizer were the means by which the control surfaces were "eased off" so that the airplane would not pass beyond the horizontal in its response, and the "force impressor," which Sperry and Tanner jointly invented to correct platform deviations from the horizontal during long, steady turns.[53] The easing off was accomplished by a negative feedback signal countering and reducing the error signal (see figure 7.14). Sperry solved a similar problem in a similar way when he designed a gyropilot for ships.

The 1914 stabilizer also included an anemometer and, like the earlier model, took compressed air for the servos from a cylinder of the aircraft engine. The electricity to drive the gyros was supplied by an alternating-current generator driven by the airplane engine. The stabilizer added about forty-five pounds to the weight of the airplane.

Writing of the Sperry airplane stabilizer in 1914, Reginald Gillmor saw an analogy between the device and the sensory motor circuit of human beings and birds. Because birds were well equipped to maintain stability in flight, Gillmor argued that "in the design of a machine for artificially accomplishing what Nature has already accomplished, we cannot do better than to adopt the principles which Nature had adopted. . . ." In the Sperry stabilizer, he continued, "the gyro base-line corresponds to the bird's highly developed semi-circular canals," the anemometer "corresponds to the bird's muscular impressions," and the servomotors "correspond to the bird's muscles and the ailerons and rudders to its wing-tip and tail."[54] Gillmor in 1914 was writing in the language of cybernetics.

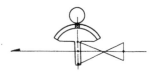

Follow-up arrangement with cables from elevator to sector so that sector will follow the roller.

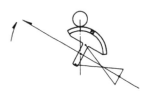

Airplane pitches up; roller moves into positive sector; and signal sent to servo to apply down elevator.

Elevator moves down, plane begins to nose down, and sector follows roller.

---

nated this as the 1914 stabilizer to distinguish it from the stabilizer tested in 1913). Keller believes that the roller contact and the sector (the "electrical") pick-off was introduced on the successor to the Bezons model. The roller and sector pick-off has been described above because it was certainly used on subsequent models of the airplane stabilizer and also on the aerial torpedo.

[53] E. A. Sperry and H. L. Tanner, filed July 14, 1914 (850,874), "Gyroscopic Stabilizer," issued August 14, 1917 (Patent No. 1,236,993). Charles L. Keller has pointed out the importance of the "torquer" in his "Automatic Pilot, 1913 Version," *Sperry Engineering Review*, XIII (1960), pp. 20-23 (Keller's 1913 version is the model that I have designated the 1914 stabilizer.) Elmer Sperry also stressed the value of the force impressor, writing that it would neutralize "outside forces"; EAS to W. I. Chambers, Chambers Papers, Library of Congress, container 22.

[54] Gillmor, "Sperry-Curtiss Aeroplane Stabilizer."

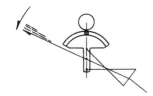

Sector having followed roller, roller rests on insulated sector, and the down movement of elevator is stopped (command signal cut off).

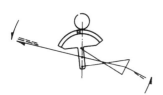

Roller moves into negative sector and elevator begins to move up.

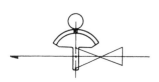

Roller rests on insulated sector as plane resumes level flight.

Figures 7.14 Control for airplane stabilizer with follow-up. (From Sperry Gyroscope Company, Inc., Automatic Flight Control.)

Sperry arranged through the Aero Club of America to enter the 1914 stabilizer in the 400,000 franc airplane safety competition (*Le Concours de la Securité en Aeroplane*), which the Aero Club of France was conducting on behalf of the French war department. Opening January 1, and ending June 30, 1914, the contest featured stabilizers, parachutes, carburetors, self-starters, and other devices thought to contribute to the safety of aviation. France had reason to sponsor such an event. Even though the Wright brothers inaugurated the aviation era, France took the lead in the development of the airplane. By 1914 the French had built more airplanes, had more aviators, and had developed a larger military air arm than any other nation. She also had more flying casualties.

The Sperry stabilizer was displayed for the newspaper press in January before being shipped to France. The New York *Herald* gave the story prominence not only because of interest in the stabilizer and the international character of the competition but also because of the good looks and spirit of twenty-one-year-old Lawrence.[55] Rarely did a newspaper in America or in France carry an illustration of the stabilizer without accompanying it with a picture of Lawrence.

Curtiss agreed to provide a flying boat if Sperry paid all expenses in connection with the undertaking and returned the machine in good order. In case the machine was damaged, Sperry was to pay for the repairs, but Curtiss limited total damage to $3000, $1200 for the motor and $1800 for the "boat and planes." These figures were less than the "actual cost of production."[56] In April, when Lawrence was in France installing the stabilizer in the flying boat, he received a wire from Curtiss recommending that he employ an aviator to fly the machine during the tests. Lawrence wired back, "Think aviator in the way—am flying myself." He had received his flying license on October 5, 1913.

In April the plane was at Bezons, a village on the Seine several miles northwest of Paris. He made practice flights with the aid of a mechanic, who also helped him wrestle the plane in and out of the water. When he encountered mechanical difficulties, Lawrence obtained a delay of his first contest demonstration until June 18. By that time he expected to have completed a smoothly functioning system of pilot, stabilizer, and plane.

This was Lawrence's first visit to Europe. Although his sister, Helen, urged him to live with a French family to learn the language more quickly, Lawrence took up residence at the Hotel Avenida at 41 Rue du Colisée. Staying in the hotel meant a streetcar trip to Bezons, but he found the accommodations in the village "too rank for words"; Paris was far more interesting. Paul Jourdan, son of the Sperry representative in Paris, showed Lawrence many of the sights of the city. He looked at works of art but wished that his sister were there to explain them to him; he attended the

[55] "Ship Aeroplane for Big Safety Contest in France," New York *Herald,* January 20, 1914.
[56] Glenn Curtiss to EAS, February 13, 1914 (SP).

theater and found the actors spoke intelligibly, "instead of sort of mumbling"; and he watched the tumultuous welcome given King George and Queen Mary when they visited Paris. He saw, in short, what most tourists saw in Paris, but he also viewed the city as few tourists or Parisians had then seen it: from the air.

Early in May, having accustomed himself to the controls of the flying boat, he was ready to relinquish them to the stabilizer. Installations brought problems, but on May 13, he turned the plane over to gyro control for the first time. When his father, then traveling through Europe, received word of the event, he wrote, "This is probably the first time in history that an aeroplane has been flown entirely automatically. . . ."[57] Lawrence approached automatic control slowly, first flying with the lateral controls for ten or fifteen minutes, landing, disconnecting these, and then flying with only longitudinal controls. After flying with both connected for a "long time," he found flying under gyro control remarkably pleasant: "Instead of having to be constantly alert, the gyro does it for you."[58]

After the reassuring reports from Lawrence, Elmer Sperry was convinced that the forthcoming tests would provide an excellent opportunity to publicize and promote the stabilizer. He and Lawrence met in England to plan the strategy of the publicity and the presentation. The competition was an opportunity similar to that Edison had when he opened, to gaping thousands, his Menlo Park Laboratories, newly lighted by incandescent bulbs and to that Lee De Forest had when he dispatched reports on the international yacht races off the Atlantic coast to the New York newspapers. The inventors knew the value of publicity. Elmer Sperry intended to make the flight test on June 18 a great occasion, and he wanted all of the American and other naval attaches, as well as his former business associate Ambassador Herrick, on hand.[59]

A few days before the tests, Elmer Sperry went to Bezons to help his son arrange for the jury the most dramatic demonstration possible of the stabilizer's characteristics. It was his idea that Lawrence's French mechanic, Emile Cachin, walk onto the wing to disturb the plane's stability while Lawrence stood up and raised his hands to show that the stabilizer automatically corrected the imbalance. When the feat was performed for Sperry, the ailerons, under gyro control, held the plane in steady horizontal flight. He later wrote: "You know this is one of the dreams that I have had for the last three years, and to actually see it realized was almost too good to be true."[60]

To prepare the stage for the demonstration on June 18, Sperry instructed his representative in France, Commander A. Sauvaire Jourdan, to request that the Aero Club of France send a special delegation to witness and officially record the performance of the stabilizer. To prepare the jury and other prominent observers for the demonstration, the London office

[57] EAS to Conner, May 15, 1914 (SP).
[58] Lawrence Sperry to EAS, May 14, 1914 (SP).
[59] EAS to Conner, May 20, 1914 (SP).
[60] EAS to Conner, June 13, 1914 (SP).

of the Sperry company prepared a twenty-page essay providing a remarkably lucid discussion of the stability problem. It also outlined earlier efforts at stabilization, explained the principles and operation of the Sperry stabilizer, and gave a detailed program for the performance of the stabilizer during the demonstration.[61] This document, prepared by Reginald Gillmor, was a notably forthright statement cast on a sophisticated level of technology.

June 18 was a memorable day for Zula. She and her husband rose early to take a trolley to Bezons. Gillmor, over from London, followed in an automobile with the American naval attachés from Berlin, London, and Paris. Other autos began to roll up to the plane standing on the north bank of the Seine midway between the bridges at Bezons and Argenteuil. These unloaded representatives of other foreign embassies and consulates, the jury (in three automobiles), and representatives of the newspapers and the Pathe and Gaumont newsreels. M. Sauvaire Jourdan's wife, an animated French lady who carried her small dog in her arms, kept Zula company as others crowded around the plane.

Earlier the Sperrys had met their son dressed in the French style with yellow blouse and trousers; Zula thought "he looked happy and picturesque." Lawrence had been up since 4:00 A.M. with his mechanic, Cachin, preparing for the flight. To Zula, Lawrence did not appear concerned about the forthcoming trial, but the reporter for the New York *Herald* found that Mr. Sperry "was rather nervous at the beginning."[62]

After "the wonderful sight" of seeing the airplane skim along the surface of the water and rise over Paris, Zula thought what a grand pilot he was. Frightened as she often was by her son's flying, that day must have been some compensation for her years of contained anxiety. She found his landing beautiful, and she was impressed by his ability to explain in French the intricacies of the apparatus to the crowd gathered around after his successful demonstration. Then there was a lunch with the triumphant young pilot and the Jourdans, including son and daughter, at the simple inn where Lawrence had eaten while working at Bezons. The French proprietress of the Morronier Inn beamed and fussed over Lawrence as if he were her own son and served his family and friends an excellent luncheon of salmon, peas, chicken salad, and big strawberries as they sat in the garden looking down to the Seine. This was indeed a great day for Zula—and for all the Sperrys.

From Lawrence's point of view the day was dominated by the details of the flight. He had taken off toward Paris, turned, and returned in the direction of the spectators. Just over the Argenteuil bridge, he and his mechanic began the first dramatic demonstration. "I disentangled myself from the control of the machine and stood up entirely away from the controls, with both hands over my head . . . my mechanic, Emile Cachin, then stood up in his seat, stepped to the lower wing, thence walked out on the

[61] Gillmor, "Sperry-Curtiss Aeroplane Stabilizer."
[62] "Mr. Lawrence Sperry Gives Splendid Exhibition at Bezons with Automatically Stable Machine," New York *Herald* (European Edition), June 19, 1914, p. 3.

Figure 7.15 The mechanic walks out on the wing as the stabilizer takes control.

Figure 7.16
Father and son at Bezons.

Figure 7.17
Lawrence and mechanic
at Bezons.

Figure 7.18
Observers gather around
the triumphant Lawrence.
Zula Sperry sits
in the background.

wing a distance of about six and one-half feet."[63] This disrupting force had no effect upon the lateral stability of the plane, but the spectators could see the right aileron move automatically to maintain the stability. By this time the machine was approaching the Bezons bridge, and Lawrence made another turn toward the spectators.

A demonstration of the automatic volplaning device followed. Lawrence used the regular foot pedal to throttle back the engine, and the machine dropped below the critical speed for maintaining flight. The automatic volplaning mechanism took over at this point, and the longitudinal stabilizer sent the aircraft into a volplane of 20 degrees. On the third approach to the spectators, after having resumed speed, Lawrence again stood in his seat while the mechanic crawled as far aft as he could without being endangered by the propeller. The plane automatically maintained longitudinal stability. After the mechanic's feat was repeated several times, nearer the spectators, Lawrence landed his plane.

Several days later, June 23, Lawrence flew again for the jury. On this occasion M. René Quinton, President of the *League National Aerienne*, asked to accompany Lawrence. He was a student of stabilizing devices and reportedly had flown in all standard aircraft equipped with stabilizers in France. In Quinton's widely read account of his flight, which appeared in *Le Matin*, he described the weather as unfavorable and the time as poor because midday was thought to bring the most dangerous air currents. "The wind was so strong," he reported, "that there were waves on the surface of the Seine . . . branches of trees were violently shaken . . . smoke from nearby factory chimneys drifted along horizontally . . . there were two distinct aerial currents, one with a downward, the other with an upward current."[64] Under these conditions Lawrence turned over the controls to the automatic pilot and also demonstrated volplaning. He then subjected his passenger to a new experience for which he was not prepared: "a glide with one wing so sharply inclined that it seemed incredible that the apparatus could be working. . . . [it was] over toward the horizon at an angle of 45 degrees."[65] Lawrence did not touch the controls but let the machine govern itself while buffeted in its abnormal position by the wind. Quinton and the watching jurors were impressed.

The Competition for Safety in Aeroplanes ended on June 30, 1914. A 16-man jury nominated by The Union for Safety in Aeroplanes and including five members from the French military, several of whom Lawrence had given flights, had the difficult task of deciding which of the twenty-one competitors had contributed most to safety.[66] They decided not to award

[63] Lawrence's account in *Flying* (August 1914), pp. 197–200. See also, English-language accounts in newspapers, *The Daily Mail*, June 19, 1914, p. 3; New York *Herald* (European Edition), June 19, 1914, p. 3; and a report of the jury's findings published in *L'Aerophile* (July 15, 1914), pp. 320–26 (translation in SP).
[64] René Quinton, "Un aéroplane qui gouverne seul," *Le Matin*, June 24, 1914, p. 1. Translated in *Flying* (August, 1914), pp. 198–99.
[65] *Ibid.*
[66] There were fifty-six entries, but only twenty-one competed; "Result of First Contest for Safety in Aeroplanes, in France," *Scientific American Supplement*, LXXVIII (August 15, 1914), pp. 108–9.

SPERRYS HAVE MADE AEROPLANES SAFE

Stabilizing Device, Tested in France, Work of Father and Son.

USES FOUR GYROSCOPES

Brooklyn Inventors Have Been Experimenting Long, and Now Are Competing for $80,000 Prize.

Cable dispatches from Paris to the effect that Thursday's public tests of the aeroplane stabilizing device invented by Elmer A. Sperry of Brooklyn and son, France, were highly successful caused much comment among persons in the city interested in aviation and aroused enthusiasm at the Sperry plant in Brooklyn yesterday. Albert W. Stringham, who has been associated with Mr. Sperry for several years in the study of the gyroscope, when seen by a TIMES reporter at Mr. Sperry's office at 126 Nassau Street, Brooklyn, said that Thursday's tests were the first public exhibition of the completed stabilizing device.

For the past eighteen months, he said, Mr. Sperry and his son, Lawrence Sperry, 21 years old, have been experimenting with the device in unitary form in this country and have demonstrated at both San Diego, Cal., and at the Curtiss headquarters in Hammondsport that in this form the device would make a flying machine stable either laterally or longitudinally. They had not proven, he said, that when the units were combined, the aeroplane could be stabilized both ways. He said the Sperrys went to France to compete for prizes aggregating $80,000 offered by the Union Cour la Securitie en Aeroplane for the most practicable safety devices exhibited. There will be another public test next week, he said, after which the inventor and his son will return to America.

The test of the device in unitary form in this country have received much attention from followers of aviation and from scientific and technical publications. The Scientific American on June 7 of last year, in commenting on the two units, or departments of aeroplane stability, said:

"The most successful efforts to deal with these two departments of aeroplane stability have been those of Mr. Elmer A. Sperry, who has been experimenting in conjunction with Curtiss machines for more than a year. Recent trials at the Curtiss camp at San Diego have demonstrated the correctness and the efficiency of Mr. Sperry's gyroscopic stabilizer."

Figure 7.19
From the New York Times,
June 20, 1914.

---

the grand prize of 400,000 francs, for the "jury had no apparatus to examine which was of exceptional interest from the point of view of safety."[67] Only two prizes were awarded, the first of 50,000 francs to the Sperry gyroscopic stabilizer and the second of 30,000 francs to the Paul Schmitt biplane with variable incidence planes. The jury also awarded seven small sums ranging from $3,000 to $200 to encourage the recipients.

Perhaps the Sperrys felt disappointment in not winning the grand prize, but there were many compensations. Only their stabilizer fell under the rule reserving to the French Government the option of buying any invention winning a prize of at least 50,000 francs. Their stabilizer won a prize while those of their leading stabilizer competitors—Doutre, Eteve, and Moreau—received only subventions. Lawrence said later that the Sperry stabilizer had been entered principally to compare it with the others representing the achievements of "the very best brains in the world."[68]

Experts also highly praised the Sperry stabilizer and this must have offset any disappointment at not winning the grand prize. Lt. Commander Cayla, the member of the jury from the French navy, who flew with Lawrence, concluded that the Sperry stabilizer, when installed in an airplane with little inherent stability and therefore hard to pilot, increased notably the ease of control and safety.[69] Quinton thought that the Curtiss-Sperry combination was a machine that "controls itself, whether it is rising, flying straight ahead, descending, or battling with the wind. . . . This marvelous machine can recover lost equilibrium unaided by man."

Interest among the military increased the Sperrys' anticipation of a market. Elmer Sperry made an appointment with the "Chief Designer and military authority connected with aviation in Great Britain" to witness a demonstration a few days after the flight before the jury. Naval attachés who had not seen the earlier demonstration were invited to another on June 24. After his father left for London at the end of June, Lawrence received a representative of the German aviation firm, Aviatic A. G. of Mulhouse (Alsace), who said that the German government would buy stabilizers by the hundreds if the performance was as reported. Despite this enthusiastic reception in Europe, Sperry wanted Lawrence to return to the United States to continue tests on the stabilizer, devoting special attention to the perfection of electric servomotors to replace the pneumatic servos, which he believed unreliable. The apparatus needed further development before it could be sold.[70]

An enthusiastic welcome awaited the Sperrys in Brooklyn. The Brooklyn *Eagle*, learning of the demonstration on June 18, carried an editorial predicting that the jury might find that two Brooklynites had made the most important contribution to the science of aviation since the first flights of the Wright brothers. New York City and Brooklyn reporters met the *Olympic* on July 8, and found father and son happy over the triumph and

[67] *L'Aerophile*, (July 15, 1914), p. 320.

[68] *Flying* (August, 1914), p. 198.

[69] *L'Aerophile*, (July 15, 1914), p. 322.

[70] Lawrence Sperry to the executive committee of the Sperry Gyroscope Company, March 1915 (SP).

with plans for supplying the navy with a stabilizer of the latest design. They also told reporters of requests to install a stabilizer on the *America,* the large Curtiss boat that John Cyril Porte intended shortly to fly across the Atlantic. Porte believed that the stabilizer would greatly increase his chances of success.

The Aero Club of New York tendered a luncheon for young Lawrence the day of his arrival. Earlier, the club had wired congratulations to the Sperrys for their signal success with the stabilizer and for greatly advancing the science of aviation. Then Brooklynites expressed their pride with a luncheon for 200 in honor of the Sperrys at the Brooklyn City Club. The official spokesman for the borough compared the prize won with the Nobel prize "for literature, peace, and history," stating that the Sperry achievement had "marked as great an era in the world's development as any of these other prizes." Elmer Sperry, in his speech, moved quickly over the technical description of the stabilizer to note the help received in developing the stabilizer from people in Brooklyn. He also recalled the considerable time and money that it cost the Sperrys to develop the device (later estimated at $8,000).[71] Finally Lawrence told of his flying with the stabilizer, and "the City Club men cheered Lawrence to the echo, for they realized that he, little more than a boy, really 'did the trick.' "[72]

Another honor came to the Sperrys in 1914: Elmer Sperry was awarded the Robert J. Collier Trophy for "gyro controls, the greatest achievement in American aviation in 1914." The three prior winners were Glenn Curtiss (twice) and Orville Wright. Two years later, Sperry again won the trophy for his drift indicator.

[71] *Ibid.* The $10,000 prize covered this.
[72] Brooklyn *Eagle,* July 10, 1914.

CHAPTER **VIII** Brainmill for the Military

This Company, practically from its inception, has been used by the Navy Department and later also by the War Department as nothing short of a 'brain mill' and experimental laboratory . . . for developing . . . intricate and even abstruse instruments of high precision.   SPERRY COMPANY MEMORANDUM, 1918

Successful laboratory directors may be of several types but a militant optimism, contagious enthusiasm, controlled imagination, and quick human sympathy are common to all.   A. D. LITTLE AND H. E. HOWE,
Mechanical Engineering, 1918

ONE OF THE MOST DRAMATIC EVENTS IN THE HISTORY OF THE SPERRY COMPANY occurred in December 1914, when Reginald Gillmor wired Elmer Sperry: "I wish I could be in New York for just one day to shake you by the hand and celebrate the finest little victory known to the history of the Sperry Gyroscope Company." In a forty-eight-hour period, the British Admiralty had ordered fifty-five submarine and ten battleship compasses, and Russia had ordered one submarine and ten battleship compasses. The two orders totaled $832,000, at a time when the Sperry Gyroscope Company had total assets of only $1 million. Gillmor added, "It means that the last stone of our foundation is laid, and that this foundation is strong enough to support any size structure we care to build upon it."[1]

The war in Europe created a demand for gyrocompasses that greatly accelerated the growth of the company. The Allied navies decided that both surface ships and submarines needed the compass and turned to an American, Sperry, to meet the demand. As a result, the company expanded its manufacturing facilities, its sales and service force, and its financial and managerial organization. The sale of compasses continued as the United States embarked upon a defense program in 1915 and entered the war in 1917.

If Elmer Sperry had not been an inventor and an engineer by nature, he might have been satisfied at the age of 54 to watch his company become a profitable and fairly large manufacturing concern. Not content, however, with the prospect of being the world's leading manufacturer of the gyro-

[1] Gillmor to EAS, December 11, 1914 (SP).

compass and the best known of its inventors, Sperry led his company into new fields by invention and development. As a result, the Sperry Gyroscope Company did not evolve primarily as a manufacturing concern but as a company with manufacturing facilities organized around invention and development—or what was increasingly being called research and development.

Circumstances during World War I made it possible for Sperry to continue research and development: profit from the sale of compasses funded additional projects; the policy of the military favored the use of commercial concerns for research and development; the prolongation of the war provided the time necessary to initiate and carry out projects; and the character of the war itself stimulated research and development.[2] The character of the war changed after the Battle of the Marne stopped the German offensive in 1914 and the Battle of Jutland brought a stand-off in the naval war. Both the Germans and the Allies then looked to technological innovation, as well as battles of attrition, to break the deadlock. Facing the possibility of being drawn into the conflict, the Americans had to cultivate military research and development to keep pace with the European innovations. In these circumstances, Sperry and his company became a brainmill for the military.

Among the major innovations of the war were gas, tank, aerial, and submarine warfare. In each case, a number of interrelated components were required to obtain an effective system of offense or defense; some of the components had yet to be invented and developed. Because of his close association with naval officers, Sperry kept well informed of the needs of evolving naval warfare systems, both marine and aerial. The development of aerial warfare exhibited the pattern of related needs, the complex of critical technological problems, to which he had become accustomed and responsive during his career. Aerial reconnaissance and low-level bombing, for example, were challenges that brought more effective anti-aircraft fire as a response. After aviators reacted by flying at night, inventors like Sperry endeavored to provide better instruments for night flying. Night attacks then brought the development by Sperry and others of improved searchlights for the defense. In bombing from high altitudes to avoid the ground fire and the lights, the aviators needed more accurate bomb sights; fighter pilots wanted machine guns that would fire through the propellers and sights that would automatically lead the target. Sperry and his company responded to all of these problems and became deeply involved in developing a system of aerial warfare.

A primary characteristic of the devices resulting from Sperry research and development in aerial and other fields continued to be, in war as in peace, navigation, or guidance and control. Elmer and Lawrence Sperry had entered the aviation field with an automatic stabilizer, but during the war the demand was for instruments permitting a man-machine system of control rather than fully automatic controls. Sperry and his company re-

[2] On the nature of various types of contracts, see Franklin Crowell, *Government War Contracts* (New York: Oxford University Press, 1920), pp. 135–68.

Figure 8.1
Elmer Sperry patent
applications, 1910-1930.

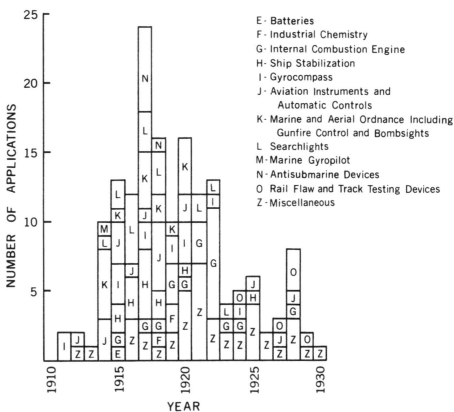

E- Batteries
F- Industrial Chemistry
G- Internal Combustion Engine
H- Ship Stabilization
I- Gyrocompass
J- Aviation Instruments and
    Automatic Controls
K- Marine and Aerial Ordnance Including
    Gunfire Control and Bombsights
L- Searchlights
M- Marine Gyropilot
N- Antisubmarine Devices
O Rail Flaw and Track Testing Devices
Z- Miscellaneous

sponded with an effort to develop a compass for aerial navigation and instruments to provide the pilot with a reference, or horizon, to use in controlling the attitude of his airplane. During the war, Sperry also developed an improved searchlight for aerial warfare in which automatic controls were an important feature. His fire-control system directed with delicate precision the massive guns of a dreadnought whose firepower nearly equalled that of all German army rifles. The Sperry name became associated with sensitivity and precision in the control of ships, airplanes, and other technology. Many of the great inventors and engineers of an earlier era had concentrated upon the control and direction of the forces of the natural world; Sperry controlled and directed the forces generated by technology, or the man-made world.

The increasing number of research and development projects, manufacturing contracts, and company personnel required that Elmer Sperry delegate responsibility. He continued to make the major decisions but delegated the managerial problems of manufacture, sales, and finance to young executives. He remained the director of research and development. Not only did he continue to direct research and development personally, he also continued to be the company's leading inventor, filing more patent applications than any of the bright young men who joined his staff. Throughout the frenzied wartime period, the Sperry Gyroscope Company—especially its engineers and inventors—reflected his professional characteristics.

i

Three of the young engineers whom Elmer Sperry influenced greatly and upon whom he impressed his characteristics were his sons. Lawrence joined him before the war; Edward came to the company in 1915; and Elmer, Jr., the youngest son, joined Sperry the same year. All three held responsible positions in the company before the war ended. Lawrence, as a test pilot and aeronautical engineer, contributed immensely to the development of the automatic stabilizer, the automatic pilot, aerial torpedo, and aircraft instruments. Edward, after graduation from Cornell as a mechanical engineer, assisted his father in the marine stabilizer department. Elmer Sperry, Jr., left Cornell's Sibley College in 1915, after three years of engineering, to act as a special assistant in research and development. Both Edward and Elmer, Jr., had attended Exeter before entering Cornell.

When she was asked why the sons followed so readily in their father's footsteps, their sister Helen recalled that they could imagine no more exciting and satisfying career. Inventors were popular heroes, and the work stirred the boys' imagination. Even before graduation, they chose to work in Sperry's shop, and Elmer, Jr., like the older Lawrence and his father, liked to tinker and build. His subsequent career closely paralleled his father's, while Edward moved toward finance and management.

Sperry's correspondence with Edward shortly after he joined the company sheds light on Elmer Sperry as a director of research and development cultivating the talents of young engineers; it also reveals a patient father giving his sons reasonable responsibility while alerting them to problems and helping to solve those that were too demanding. Most of these problems, judging by the surviving correspondence, were professional, but the attentiveness was both professional and paternal.

Placed by his father in the ship-stabilizer department, Edward was assigned to supervise the building of a stabilizer for the yacht *Widgeon*, which belonged to the wealthy Clevelander, H. M. Hanna, Jr.[3] Elmer Sperry probably thought that young Edward, the most socially sophisticated of his sons, would make a good impression on the prominent Clevelander and perhaps cultivate thereby a market for stabilizers among the wealthy and socially prominent yachtsmen of the day. Before Edward saw Hanna in Cleveland, however, he visited a Sperry subcontractor in Erie, Pennsylvania, working on parts for the *Widgeon* stabilizer. Edward helped install bearings and balance the rotor, both critical for the satisfactory functioning of the stabilizer, and coordinated this work with that being done on the stabilizer elsewhere. Meanwhile, H. M. Hanna waited impatiently for the stabilizer and Edward had to assuage him. The millionaire wanted the device because his yacht had been "rolled around pretty badly of late during storms"; he wanted a speedy installation because he disliked having "his lily white yacht laying at the greasy dock." Edward cajoled the contractors, pleaded

[3] This episode is based on letters between EAS and Edward during the period September 1915, through October 1915 (SP). The quotations are from those letters unless otherwise indicated.

with the shipyard, and reasoned successfully with the engineer aboard Hanna's yacht who, well-informed about steam engines and ill-informed about stabilizers, insisted upon offering advice. Most important, Edward also managed to hold Hanna's confidence.

Elmer Sperry wanted daily progress reports from Edward. He wanted "my dear Ned" to "follow the job hard and stick right to it" but to call for Carl Norden if the installation problems became too complex. Edward did not want Norden and assured his father that "it's pretty simple straight forward kind of work." Although seeking responsibility, he did not hesitate to ask his father about technical details. His questions were always answered patiently. A letter of September 16, 1915, suggests the warmth that permeated a relationship structured by engineering problems:

You must not be disappointed if you do not get along with the work quite as rapidly as you thought, but be very faithful to keep us informed here and also drop a very courteously written note every day to Mr. Hanna, and if you meet with disappointments, tell Mr. Hanna. I notice by your letter that you expected to tell him to have the ship at some certain date which I do not believe you will be able to meet. Go carefully and do not allow your over-enthusiasm to set a date which will in any way hamper Mr. Hanna by depriving him of the use of his yacht before you require it.

After the stabilizer was installed, Edward enthusiastically reported the performance. Even when rough water rolled the yacht twenty degrees, the stabilizer reduced the roll to two degrees. When Edward talked to Hanna after his first trip with the stabilizer, he found him "quite enthusiastic over the plant." In 1916 Edward was made head of the ship-stabilizer department, which in the same year won its eleventh contract.

Elmer, Jr., the youngest, also had a responsible and far from routine position in the company as head of the experimental construction department and special assistant to his father. Elmer, Jr., knew his father's inventive and engineering style well; he had listened attentively to his father's shop talk even before he was old enough to understand much of what he heard. He probably saw more of Sperry than any of the engineers, in part because his job kept him at his father's right hand. Since he had been brought up on his father's sketches, he proved adept at translating these into drawings or instructions for Narveson and the other draftsmen. Elmer, Jr., recalled that he "could read a blueprint before [he] could read anything else." He also enjoyed an advantage over the other engineers because he worked with Sperry during train rides home, at the dinner table, in the evening hours, and on weekends, which were especially busy times for him. He also kept in close contact with those projects in which his father was particularly interested but could not check because of trips to Washington and elsewhere. Elmer, Jr., often worked in the shop making experimental models of new inventions with the precision machine tools, in the drafting room reducing the newest idea to a drawing, and in the field testing new devices. His enthusiasm for engineering and invention was so great that his mother, despite her telephone calls, frequently could not bring the young man home from the plant before eleven in the evening. She worried about his health breaking down.

Elmer Sperry, Jr., delighted later in recalling a wartime episode that not only reveals his own style of engineering but also that of his father.[4] Early in 1917, when the submarine menace was particularly threatening, the American navy learned that the Allies had an effective depth charge. The United States, not yet in the war, could not draw upon the Allies for highly secret information about the depth charge, but the navy was determined to make its own. Under these circumstances, the Chief of the Bureau of Ordnance, Admiral Earle, turned to the "one company in the United States that could work fast and do things for the Navy without any orders or without any contract." Elmer Sperry designated his namesake to represent the company, and the young man took the train to Washington to meet his naval counterpart, Lt. Wilkinson of Naval Ordnance. Admiral Earle told them to get together "and build some of these depth charges in a hell of a hurry." When Sperry and Wilkinson asked him what they looked like he replied, "I don't know, but go find out."

After exploring the problem with Wilkinson and gathering some information from officers who had heard of the depth charges or had seen them, Elmer Sperry, Jr., then returned to New York, drawing the preliminary sketches en route on the train. Back at the Sperry company, the project was discussed and within a few days the first three depth charges were produced. Elmer Sperry, Jr., carried two of the first three models to Washington for testing by the navy. To expedite transportation, he rented a compartment on a pullman, taking the lower himself and giving the upper to the depth charges. After the navy test proved satisfactory, the Sperry company was asked to go into serial production as soon as possible. Wilkinson told Elmer Sperry, Jr., to walk up and down Canal Street in New York subcontracting the containers and non-precision parts to the numerous machine shops there. The Sperry company manufactured the precision detonating device itself. By the time Admiral William Sims sailed for Europe to take command of the U.S. Fleet, 500 of the charges were ready to be shipped. The company went on to supply 10,000 of this early model, but soon an improved hydrostatically fired charge, already being used by the Allies, became standard in the navy.

ii

There was nothing impersonal about the way in which Elmer Sperry recruited his staff of research and development engineers. He not only employed his three sons, but also Cortland men such as Hannibal Ford and Harold Tanner. After he had exhausted the supply of sons and available Cortlanders, he often relied on the good judgment of his sons, frequently recruiting members of his staff from among their fraternity brothers, flying companions, and neighborhood friends. The recommendations of trusted engineering colleagues were also highly valued. The achievements of Preston Bassett, Robert Lea, Mortimer Bates, and other Sperry employees testify

[4] Based on an interview with Elmer Sperry, Jr., February 24, 1966.

Figure 8.2
Elmer Sperry is shown with
Edward Sperry, left;
M. L. Patterson,
searchlight engineer;
and Elmer, Jr., extreme right.

Figure 8.3
Preston Bassett, left,
and Reginald Gillmor, center,
became presidents
of the Sperry Gyroscope
Company. Robert Lea, right,
became a vice president.

Figure 8.4
The engineering department
of the Sperry Gyroscope
Company during World
War I.

to the quality of his judgment.

Sperry was especially interested in the mechanical ability of his young engineers; he wanted his men to have his "mechanical touch." Although he had taught himself engineering, Sperry was not prejudiced against young graduates of the science and engineering schools. Ford and Tanner, his first two chief engineers, had engineering degrees, and Leslie Carter, who became a Sperry engineer in 1913, had a degree from Brooklyn's Pratt Institute. Among the men employed for his engineering staff during the war, Preston Bassett, Eric Sparling, Alexander Schein, William Hight, Raymond Witham, and Edward Sperry had engineering or bachelor of science degrees. At least half of his engineering division in 1918 can be positively placed in this category. Because of the nature of the company's work, these statistics are not surprising and are in harmony with the general trend in the profession.[5] The new century brought to prominence engineers grounded in applied science, and Elmer Sperry, faced by complex engineering problems, did not resist the trend. He did not publicly identify with the self-made and self-educated, as did the irascible Thomas Edison.

The most unorthodox group of young men to join the Sperry company during the war were the young aeronautical engineers. At the time, aeronautical engineering was rarely taught in colleges and universities; this situation, coupled with the desirability of having young inventor-engineers who were also pilots, led Elmer Sperry, on Lawrence's recommendation, to hire young flyers with mechanical aptitude and make aeronautical engineers of them. Mortimer Bates, who contributed greatly to the development of the airplane stabilizer during the war, learned flying at Hammondsport with Lawrence before joining the Sperry company; Morris Titterington, who joined the company in 1914 and worked with the airplane stabilizer and the aerial bomb sight, was also a flyer.

Several young engineers employed during the wartime expansion rose rapidly in the company. Preston Bassett, an Amherst science graduate and subsequently a part-time student at the Brooklyn Polytechnic Institute, began by helping Elmer Sperry with the prolonged white lead project and then became the project engineer in searchlight development. Sperry seems to have been particularly fond of young Bassett, whose family lived not far from the Sperrys, and in 1916 he made Bassett his special assistant on the Naval Consulting Board.[6] Bassett rose through engineering and research to become chief engineer, general manager, and finally president of the company. He attributed much of his approach in engineering and research to his close association with Sperry.

Robert Lea, also destined to rise in the company, joined it in 1915. He had studied engineering at Cornell and was a fraternity brother of Edward and Elmer, Jr. During vacations, the brothers brought him home, and Zula

[5] A third of the American professional engineers born before 1875, like Elmer Sperry, earned no diploma, but more than three-quarters of the group born between 1885 and 1894 had degrees; Edwin T. Layton, "The American Engineering Profession: The Idea of Social Responsibility" (unpublished dissertation, University of California, Los Angeles, 1956), pp. 46–47.

[6] Interview with Preston Bassett, May 27, 1966.

became particularly fond of the young man. Sperry decided that Lea was just "our kind" of person, a fine, conscientious worker, "with splendid push and initiative."[7] He had one more year at Cornell when Elmer Sperry persuaded him to join the company. Lea had admired the inventor since he first heard him speak as a guest lecturer at Cornell. He advanced through engineering sales to become a vice president and, like Bassett, found that the years spent with Elmer Sperry greatly influenced him.[8] Lea was especially welcome at the Sperry home, and in 1921, he married Helen, the Sperrys' only daughter.

One major engineering appointment was exceptional because Elmer Sperry did not pick the man and because he came from outside of the family, outside of Cortland County, and even outside of the Sperry family circle of friends. C. B. Mills was, nevertheless, an excellent choice. He was head of the industrial division of the engineering department of the Westinghouse Electrical and Manufacturing Company and well known to the new factory manager at Sperry, C. S. Doran, who came from a Westinghouse subsidiary. Mills had learned his engineering in night school and on the job, was about forty-three, and had worked for Westinghouse for twenty years, where he was receiving a salary of about $6000 annually. Offered a salary of $7,500 and new challenges at Sperry, Mills accepted and the company had, in the opinion of at least one of the young engineers, "one of the best engineers I ever met."[9] Mills proved himself not only able to coordinate manufacturing with research and development but also to invent.

By the end of the war, Sperry had created a research and development staff numbering about fifteen engineers.[10] The most experienced among these had established areas of special competence for themselves. Tanner, the chief engineer, concentrated on invention and development of electrical automatic-control devices and the compass; a young engineer, G. B. Crouse, also focused on the compass; Lawrence Sperry headed the major projects in the aeronautical field and also made important patent applications on aircraft instruments and aerial torpedoes; C. B. Mills took responsibility for the functional design, the production model, of new devices; Bassett was the leader in searchlight research and development; a foreign-born engineer, Elemer Meitner, worked closely with Elmer Sperry on the invention of fire-control devices; Russian-born engineer A. Schein managed gyro-

[7] EAS to Gillmor, February 25, 1916 (SP).

[8] Interviews with Robert Lea, January and February 1966.

[9] Gillmor to Sperry, October 10, 1918 (SP); Eric Sparling interview, May 25, 1966.

[10] An indication that Elmer Sperry was unable to remain as personally involved in creating his staff when wartime pressures and shortages intervened was his acceptance of a training program to cultivate engineers. The program at the Westinghouse Electric and Manufacturing Company favorably impressed him during a visit to Pittsburgh to address the Westinghouse Club ("the most wonderful group of young engineers that I had ever met"); he readily approved of Tanner's proposal to institute a training program for three or four of the outstanding young men in the Sperry testing department. Like Westinghouse and other large engineering and manufacturing firms, the company also hired engineering students during their vacations. Some of these men were expected to join the company after graduation.

stabilizer development after Edward Sperry assumed more general managerial and financial responsibilities for the company; and Elmer, Jr., as noted, applied himself as his father's assistant. Assisting these men in his position as the chief of the drafting department was Fred Narveson, the senior Sperry employee.

iii

By temperament and experience, Elmer Sperry had little interest in factory management. By the end of the war, however, he had to accept more organization and coordination in his company. The engineering staff had increased, the number of workers in manufacturing had risen from hundreds to thousands, and the sales and capital investment had grown proportionally. The military insisted on careful cost accounting and regularized procedures in research and development and in manufacturing contracts. The Sperry manufacturing department also needed to develop closer cooperation with research and development and sales. The responsibility for the entire enterprise, including the management of research and development, took so much of Sperry's time that he was left too little for his own inventions. In reaction to all of these pressures, and encouraged by Reginald Gillmor, who had returned from London to become general manager of the company, Sperry agreed in 1918 to increased organization and more formalized coordination. Sperry, however, continued to make the major policy decisions about what to invent and develop.

In 1918 the reorganized company (see page 211) consisted of four major divisions, further divided into departments. All divisions, in theory, reported to the general manager, who in turn reported to Sperry; but in practice Sperry continued to deal directly with the engineering division. Within the engineering division, the departmental categories were in accord with the major areas of wartime research and development, with the exceptions of the "research department," the "experimental construction department," and the drafting and the service departments. The research department included the electrical and mechanical laboratories and carried on experimental laboratory testing for the engineers, especially for Sperry and Tanner. The experimental construction department was the bailiwick of Elmer Sperry, Jr., where he and the most skilled mechanics built the test models of newly invented devices.

A major activity of the Sperry company did not appear on the organization chart of 1918. In 1915 Lawrence Sperry had established the aviation department of the company to develop and manufacture the airplane stabilizer. Two years later, he founded the Lawrence Sperry Aircraft Company to develop and market not only the stabilizer but also aircraft instruments. The Sperry Gyroscope Company, however, continued to manufacture the aviation instruments. Charles H. Colvin, who had been an apprentice aeronautical engineer in the aviation department, became superintendent of the instrument manufacturing. Titterington and Bates, the two pilots who had been installation engineers for the aviation department, joined Lawrence under the new arrangement in 1917 and 1918.[11]

[11] This summary is surmised from fragmentary evidence on the aviation department and company from 1915 to 1918.

The new general manager was Reginald Gillmor, whom Sperry had first met as a young naval officer aboard the *Delaware* and made the head of his London office several years later. After the United States entered the war, Gillmor, who had remained a lieutenant in the Naval Reserve, volunteered for temporary active duty and helped Admiral William Sims, who was in command of U.S. naval forces in European waters, organize his headquarters in London.[12] After this duty became routine, he was released and returned to America in January 1918, to help manage and reorganize the Sperry company. Gillmor, long a student of organization and management, drew upon his naval and Sperry experiences. During the war, he often wrote to Sperry his ideas on these subjects. In return, Sperry passed on to the young man his knowledge of invention, development, and entrepreneurship. Their candid letters reveal personal affection and suggest that Sperry early envisaged Gillmor as the leading company executive and perhaps his successor. The correspondence also shows that while the experienced older man wanted spontaneous, informal company relationships, the younger man wanted rational planning and an impersonal, more highly structured organization.

ORGANIZATION OF THE SPERRY GYROSCOPE COMPANY, 1918

PRESIDENT
E. A. Sperry

EXECUTIVE COMMITTEE
E. A. Sperry
H. E. Goodman

VICE PRESIDENT AND GENERAL MANAGER
R. E. Gillmor

Graphic Control
(E. G. Sperry)
Standard Practice

FINANCE DIVISION	ENGINEERING DIVISION	FACTORY DIVISION	CONTRACTS DIVISION
F. Colburn Pinkham	H. L. Tanner	C. S. Doran	T. A. Morgan
Treasurer and Comptroller	Chief Engineer	Factory Manager	Contracts Manager
Treasury Department	Advisory Staff	Production Control Department	Sales Department
Accounting Department	Research Department (2)* (H. L. Tanner)**	Methods Department	Merchandising Department
Estate Department		Purchasing Department	Publication Department
Legal Department (H. H. Thompson)	Experimental Construction Department (1) (E. A. Sperry, Jr.)	Personnel Department	Traffic Department
	Compass Department (4) (G. B. Crouse)	Service & Maintenance Department	
	Searchlight Department (4) (P. R. Bassett)	Factory Department	
		Inspection Department	
	Ship Stabilizer Department (1) (A. Schein)	Gair Building, Factory Department (facilities in a neighboring building.)	
	Drafting Department (F. C. Narveson)		
	Service Department (C. D. Jobson)		

*Number of engineers in department including department head.
**Department heads for engineering shown in parentheses.

Based on *The Sperry Search Light*, August 1918; September 23, 1918.

---

[12] Gillmor to EAS, October 19, 1917 (SP).

The philosophy that guided Gillmor as he reorganized the Sperry company in 1918 was outlined in one of his letters. The products of the Sperry company, according to his analysis, were compounded of three factors: invention, design, and empiricism. Invention he credited to an imponderable quality of genius, especially Elmer Sperry's; design he attributed to the engineers of the research and development section, who were thoroughly immersed in the science and thoroughly experienced in the art of engineering. The empirical factor represented "adaptation to the requirements of the customer" and could best be supplied by the sales and service engineers who worked with the customer and saw the product in use.[13] These factors and personnel, he insisted, had to be coordinated, for they were mutually dependent.

Besides Gillmor, the other executives under the new organization were F. Colburn Pinkham, H. L. Tanner, Tom Morgan, and C. S. Doran. Pinkham, a graduate of Williams College, had organized the National Retail Dry Goods Association before joining Sperry in 1917; he became head of the financial division in 1918. Prior to joining Sperry in 1918, Doran had been the general superintendent of the Krantz Manufacturing Company of Brooklyn, which was owned and operated by the Westinghouse Manufacturing Company. Herbert Thompson continued as company attorney and patent lawyer; and Edward Sperry became a special assistant to Gillmor, specializing in scientific management.

After reorganization, the engineering division, administered in fact by Sperry, continued to have the predominant influence in the company. Its function was primarily research and development, upon which the company's financial, manufacturing, and sales progress largely depended. Engineering had great influence because the president of the company was himself an engineer and the general manager also had an engineering education. In other companies, the research and development division was often responsible to the factory manager or some other executive with relatively little interest in engineering.[14]

There were other circumstances insuring that the Sperry Gyroscope Company, even after reorganization, remained a company organized around research and development. A payroll of a few years later showed about 70 salaried employees, and of these, 26, the highest percentage, were in the engineering division; 24 were in sales and service, 13 in finance and legal, and 8 in the factory division. Furthermore, some of the salaried employees in sales and service were engineers. Of a payroll for salaried employees of about $20,000 per month, more than $7,000 went to the engineers of the engineering division while sales and service received about $6,000, finance and legal $4,000, and factory management $3,000.[15] In average salary, however, the

[13] Gillmor to EAS, December 7, 1916 (SP). Gillmor realized the importance of the service engineer and sales in creating the product because he found that the British environment demanded a different style of product and engineering than the American.

[14] K. C. Mees, "Industrial Research Laboratory Organization," *Mechanical Engineering*, XLI (1919), 667.

[15] These figures are from a payroll of September 1, 1922 (SP). The company organization and size was similar—excepting the labor force in the factory division—to that of 1918.

engineers, earning an average of $280 per month, ranked last behind the salaried employees of the other divisions. The factory management staff was the best paid, each member earning an average of $380 each month.

The factory division's organization showed the influence of new managerial personnel brought to the Sperry company from Westinghouse, a company with an outstanding reputation for rational factory management. Besides C. B. Mills and Charles Doran, I. H. Mills, the new factory supervisor, and G. D. Caspar, head of the production control department in the factory division, were hired from Westinghouse about the time of reorganization.[16] C. B. Mills, who replaced Tanner as chief engineer in 1920, and Doran, who replaced Gillmor when he temporarily left the company because of poor health, emphasized the planning and coordination of factory operations. They also urged engineering and sales to cooperate more closely with the factory by carefully considering factory potential and limitations when inventing and designing.[17]

The functioning of the reorganized company, at least the normative routine, can be outlined by following the development and manufacture of a new product. After Elmer Sperry approved the project, the contracts division drew up, in consultation with the customer, general specifications, insofar as these were predictable, for the desired device. If quantity production was likely, then the production control department of the factory division was consulted. The resulting contract provided, in addition to specifications, the delivery dates and prices. Often in military contracts the price declined with the number of items the customer purchased; the first produced carried much of the development cost. Because the project required research, the research department of the engineering division proceeded with the problem, cooperating with the experimental construction department, which made models for testing. After a successful engineering design was achieved, the appropriate engineering department—the compass department, perhaps—worked in consultation with its corresponding department in manufacturing to produce a design suitable for economical manufacture—a process that often involved many modifications in design. After the drawings were ready, they went again to the production control department, which scheduled the subsequent flow of work, and to the methods department of the factory division, which designed the necessary tools, dies, fixtures, and molds. In the meantime, purchasing would have ordered the materials and scheduled their arrival. Production control then monitored the manufacturing process and inspected each part for precision of size and quality. After the parts were assembled, the finished device was inspected and shipped.

The formal routine occasionally frustrated Elmer Sperry. No longer, for instance, could he simply call in the foremen of a small casting department or machine shop to decide with them the best way to make the new product. No longer could he, without considering manufacturing costs, incorporate into the design aesthetic features that manifested to him the

16 E. M. Herr, president of Westinghouse, was not happy about the loss of these men. Herr to EAS, September 30, 1918 (SP).

17 Both Doran and C. B. Mills wrote articles for the Sperry house organ, *The Sperryscope,* urging close cooperation among the various divisions.

characteristics of his company and his products. It irritated him to have to accept, for example, a housing for a compass that resembled, he believed, an "ashcan," with none of the "dignity" associated with Sperry products. In the case of the "ashcan," Doran persuaded Sperry to accept a design then ready for production, but Sperry insisted upon a more "dignified" design in the next model. These constraints would bring Sperry, after the war, to work increasingly with a small development team that emerged informally from the company structure.

iv

Patent applications reveal the pattern of research and development carried out by Elmer Sperry and his staff from 1914 to 1918 and the impact of war upon the character of technology. Even though a quantitative analysis of patents provides little insight into the quality of the inventiveness, it is a way of comparing the intensity of prewar and wartime inventive activity at Sperry, the focus of research and development in the two periods, and the patenting activity of Elmer Sperry with that of his staff during the war.[18]

Surprisingly, the Sperry Company only acquired patents on 17 inventions before the war. Twelve of these applications were made by its founder; the other five were purchased from inventors not with the company. Three of the seventeen were on the compass, and five were on the marine stabilizer; Elmer Sperry applied for the three on the compass and one on the marine stabilizer. He also contributed an airplane stabilizer and an auto stabilizer patent. The other patent applications acquired by the company were mostly Sperry's and of various descriptions.

The war brought a dramatic increase in inventive activity at the company. Sperry, who was in his fifties, had his most prolific five years, applying for no fewer than seventy-five patents between 1914 and 1918. The only comparable five years in his career came in the early 'nineties and the immediate postwar period. He alone applied for forty percent of the company's patents, despite the increased number of inventive engineers on his research and development staff. His most active year was 1917, when he applied for twenty-four patents, the greatest number in one year of his career. Nineteen-seventeen was also the peak wartime year for his staff, which filed sixty-four patent applications.

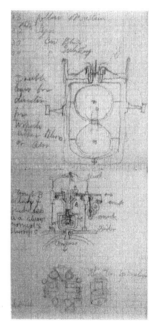

Most of Elmer Sperry's applications of the period fell into five major fields (see chart, page 203): fourteen applications related to searchlights; twelve to ship stabilization; twelve to ordnance, including fire control; thirteen applications to aviation instruments and automatic controls; and five to improvements in the gyrocompass system. Together these applications accounted for about eighty percent of his output during the war years. His applications dominated the searchlight field, where he had twice as many as Bassett, and also the ship stabilizer field, where he had twice the number of the remainder of the staff. Sperry also had more than half of the patent applications in ordnance but could only match his son Lawrence in applications in the aviation field. Harold Tanner dominated in compasses with

Figures 8.5
Pages from Sperry notebooks.

[18] All patent applications made in 1914 and 1918 are taken here as war patents.

more than ten wartime patent applications. The inventive activity of the staff, considered from the point of view of field concentration, resembled that of Elmer Sperry. The staff, too, made about 80 percent of its patent applications in the five fields in which Elmer Sperry concentrated. Staff applications were most numerous in the searchlight, aviation instruments and automatic controls, gyrocompass, ordnance, and ship stabilizer fields, in that order.

Elmer Sperry's patent applications for each of the war years also delineate his special interests. In 1914 the invention of fire-control devices absorbed much of his inventiveness; five of his ten applications were in this area. Some were joint applications, including an application for an automatic gun-pointing device with Admiral Bradley A. Fiske. In 1915 almost half of his patent applications were for improvements in the gyro-stabilizer and gyrocompass, thus demonstrating continuing post-innovational invention and development. In 1915 he also applied for two searchlight patents and for an internal combustion engine patent, a problem in which he never lost interest. The searchlight absorbed much of his inventive energies in 1916; almost half of the applications for the year were in this field. In 1917, at the height of the submarine war, he applied for six antisubmarine devices and for five patents on ship stabilization. In 1918 an internal combustion engine patent appeared again, but in this year, following the peak of 1917, airplane instrumentation would have to be considered the area of special interest.

The efforts of Sperry and his staff in the five fields of wartime concentration were organized into major research and development projects which culminated in major innovations. The quality of the inventive activity can best be conveyed by consideration of some of these major projects and innovations. Because Elmer Sperry seems to have been most interested in the high-intensity searchlight, attention will first be turned to this field.

v

Sperry had not given much thought to arc lamps since the early days in Chicago, and his concentration upon applications of the gyroscope seemed to make unlikely a renewal of interest. A number of factors, however, tended to bring him back into the field in 1914. As always, he wanted to use the technological competence he had acquired, and he had accumulated years of experience during the invention and development of his arc lamp in Cortland and Chicago. He had become an expert in designing automatic mechanisms for the lamp. The most vital of these was a mechanism for feeding the carbon electrodes as they were consumed, the subject of one of his early, important patents (Patent No. 304,966). Sperry was also well-suited to deal with a problem such as the development of a searchlight that involved chemical, mechanical, and electrical experience.

The U.S. Navy, recognizing that the range of existing searchlights did not match that of the improved guns, wanted to correct this imbalance. The navy was interested, therefore, in the high-intensity searchlight designed by Heinrich Beck, manager of the *Physikalisch-technisches Laboratorium* in Meiningen, Germany; navy tests in the fall of 1914 showed that it had a more concentrated beam and an illuminative capacity five times the standard navy searchlight.[19] The navy then asked Beck the cost of converting existing lights and installing new shipboard searchlights. Before Beck and his American agent, L. J. Auerbacher, could negotiate, however, they needed American manufacturing facilities and turned to Sperry as a likely licensee and manufacturer. It seems probable that the navy suggested Sperry. Auerbacher told Sperry that he could anticipate two years of business with profits nearly double the $100,000 Beck asked for the patent rights (including future improvements) to American manufacture and sale.[20]

Sperry was more interested in the patent situation. He was convinced, by his own experience with arcs, that the Beck lamp could be improved and was not convinced that the Beck patent protected the lamp against imitation or competition. Believing as well that the Beck price was unreasonable, Sperry therefore made the counterproposal that a royalty on sales be apportioned among Beck, Sperry, and any other whose patents "really afford protection" for the high-intensity light. Sperry intended to take out his own patents on a high-intensity lamp and acquire other patents that might contribute to the protection of a practical high-intensity lamp. Sperry wanted the kind of protection he had accumulated for his ship stabilizer before investing in the further development and manufacture of the lamp. Because he anticipated a lucrative navy contract, however, Beck was adamant. On the other hand, various "authorities" had confirmed Sperry's opinion that the Beck patent was "utterly wanting and practically worthless."[21] Even though the Beck patents appeared worthless, Sperry was willing to guarantee him a percentage of the royalty, should he buy the patents, because of Beck's early work in the "general line." Sperry noted that his attitude was an unusual one for a manufacturer to take toward an inventor whose patent had been called into question.[22] Nonetheless, Sperry was obviously driving a hard bargain.

Beck replied that "even under the most favorable conditions it is impossible to accept your proposition. I know the value of the invention and am positive that it is fully protected by patents."[23] He then sold the patents to General Electric, a supplier of searchlights to the navy, for $100,000 cash and an additional $35,000 in commission and expenses. For the next two years General Electric supplied the navy with the Beck high-intensity light.[24]

[19] "Report on Tests on Beck Searchlight," signed by Lt. C. S. McDowell, USN (no date but report refers to tests held in August 1914). The Beck lamp was also tested on the USS *Texas;* see "Board Conducting the Tests of the Beck Searchlight to Commanding Officer," October 2, 1914 (SP).
[20] Auerbacher to EAS, October 23, 1914; Admiral Griffin to Auerbacher, October 23, 1914; Auerbacher to EAS, October 26, 1914; and Specifications of Contract, November 1, 1914 (SP).
[21] EAS to Beck, November 13 or 14 (?), 1914 (SP).
[22] *Ibid.*
[23] Beck to EAS, November 14, 1914 (SP).
[24] The agreement between General Electric and Beck was described later when GE and Sperry were contesting the validity of high-intensity patents and seeking a cross-licensing arrangement. H. H. Thompson to EAS, March 29, 1920 (SP).

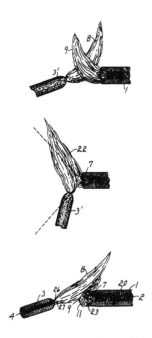

Figures 8.6
Sperry drawings showing
the arc of a searchlight
operating improperly and,
in drawing 6, properly
(Patent No. 1,227,210).

Sperry reacted by investing in the development of his own version of a high-intensity lamp. In Berlin, late in May 1914, before Beck demonstrated his lamp to the navy, Sperry found that "Beck was in some way using alcohol, which was an entirely different method from the one I was pursuing."[25] (This is the only evidence that Sperry was developing his own searchlight before he learned of Beck's.) Not until six months later, however, after learning of the navy tests and of its intention to purchase Beck searchlights, did Sperry inform the navy of his plans for improving the Beck lamp. These included the elimination of the alcohol vapor with which Beck surrounded and cooled the arc, a simplification of the automatic mechanisms in the searchlight, and the use of pure carbon electrodes rather than impregnated carbon.[26]

After deciding to embark on this project, Sperry analyzed Beck's device to discern its weakest points and then invented to improve upon these.[27] Sperry had already filed, on November 11, 1914, an application on a positive-electrode holder that provided "a novel means for rotating, feeding, and supplying current" and cooling the electrode. (The application, written before Beck rejected Sperry's counter contract proposal, could have been used by Sperry to persuade Beck to negotiate on royalties.)

Seeing other possibilities for improving the high-intensity lamp, Sperry assigned the new member of his staff, Preston Bassett, who had majored in chemistry at Amherst, to the project. Bassett worked closely with the National Carbon Company in Cleveland to develop improved carbon electrodes for the high-intensity light. He informed Sperry in detail on the progress of his experiments and Sperry recommended lines of experimentation.[28] Following prior inventors, Bassett designed a positive electrode composed of two parts, a shell and a core. The core was of a material that burned faster than the shell; as a result, the burning core recessed into the shell, forming a deep crater in the positive electrode.[29] Bassett made numerous tests of different designs and various compositions before the proper combination was developed.

Elmer Sperry applied for a basic patent on a high-intensity lamp in June 1915.[30] The patent emphasized the chemistry and design of the carbons and the positioning of the positive and negative electrodes. It also described the formation of the deep crater in the positive electrode, within which the positive-arc flame and luminescent gases were confined by the high velocity discharge and flame from the negative electrode. Sperry explained that the concentration of flame and luminescence in the crater caused the intensity of the searchlight (see figure 8.6). The Sperry high-intensity lamp, like the Beck, was also characterized by a high-current density resulting from the small diameter of the electrodes and from the restriction of flame to the tip of the negative electrode and crater of the

[25] EAS to Gillmor, June 6, 1917 (SP).
[26] EAS to LCDR Herbert Sparrow, November 6, 1914 (SP).
[27] Sperry, "Spirit of Invention," p. 63.
[28] Bassett to EAS, February 16, 1915 (SP).
[29] EAS to George D. Olds, president of Amherst College, March 17, 1926 (SP).
[30] Filed June 28, 1915 (36,615), "Method of Operating Flaming Arc Lights for Projectors," issued May 22, 1917 (Patent No. 1,227,210).

positive electrode. In other arc lamps, the luminous discharge and flame spread on the electrodes when the current flow increased; therefore, the current density remained unchanged. To restrict the flow, both Sperry and Beck cooled the electrodes except at the tip and in the crater. The major difference between the Beck lamp and the Sperry in this respect was that Beck used alcohol vapor to cool the electrodes.[31] Sperry, realizing that apparatus for generating the alcohol fumes and directing them upon the electrodes was clumsy, resorted to air cooling. The use of air cooling strengthened his patent because Beck's patents could be interpreted as limited to an alcohol-cooled, high-intensity arc.[32]

Having developed electrodes and established the character of the arc, Sperry and Bassett then designed the complete searchlight. Problems solved included the ventilation of the searchlight housing, or drum (especially the mirror); the cooling, automatic feeding, and rotation of the electrodes (Sperry had begun on this in his patent of November 1914); and the automatic positioning of the positive electrode crater so that it was at the focal point of the mirror. Sperry cooled the electrodes by circulating air through the metal electrode holders; he also vented the cooling air in the drum so that air currents would not disturb the burning arc. Sperry used an automatic feed for the negative electrode which was similar to the feed on his Cortland arc lamp and designed the automatic mechanism that positioned the arc crater in the focal point of the reflector mirror. To do this, he so placed a small mirror that when the crater was in position light (and heat) focused on a differential thermostat. When the arc was out of position, the thermostat received less light (and heat) and signalled a solenoid to adjust the positive electrode until the crater was at the focal point of the large reflector mirror of the searchlight (see figure 8.7).[33]

In 1916 Sperry was ready to market his high-intensity searchlight. General Electric was selling the Beck lamp to the U.S. Navy, but there were other customers. The army coastal defenses needed an improved searchlight and also recognized the need for searchlights in aerial warfare. Sperry, therefore, entered his lamp in competitive tests held by the Army Coast Artillery Board at Fort Monroe, Virginia, in November and December 1916. General Electric entered a 60″ searchlight equipped with the Beck alcohol cooled high-intensity lamp, and Sperry entered his high-intensity lamp installed

[31] Beck publicity stressed the vapor cooling; "Search Lamp with Vapor-Cooled Electrodes," *Electrical World,* LXIV (1914), p. 181.

[32] Beck's U.S. patent (Patent No. 1,029,787), which he tried to sell to Sperry, was "Method of Burning the Arc in Alcohol." Another Beck patent (Patent No. 1,086,311), also earlier than the Sperry high-intensity patent, was on a method of burning the arc without alcohol, but Herbert Thompson believed that it could be attacked "on the grounds that the phenomena described and claimed therein were present in an arc operated according to the first patent." Herbert Thompson, "Memorandum on the General Electric-Sperry Situation," c. November 1918 (SP).

[33] "High-Intensity Searchlight for Governmental Purposes," *Electrical World,* LXVIII (1916), pp. 611–12. Besides Patent No. 1,227,210, other important Sperry patents were: "Electrode for Searchlights and Method of Making Same," issued June 1, 1920 (Patent No. 1,342,398); and "Feeding Mechanism for Searchlights," issued December 14, 1920 (Patent No. 1,362,575).

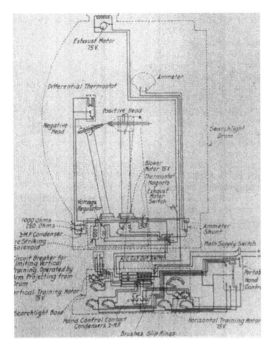

Figure 8.7
Circuit diagram for the
Sperry searchlight. The
differential thermostat
functioned as a sensor to
position the arcs properly.
(From Electrical World,
LXVIII, 1916).

Figure 8.8 The Sperry open-type searchight.

in a General Electric projector. The Sperry lamp included the automatic control to feed the carbons and position the positive crater. On the basis of heat-run, voltage regulation, candlepower, and "practical" illumination tests with ships at sea, the board found that the Sperry lamp was superior in twelve of the thirteen tests. These included light intensity, target illumination, simplicity of operation, and reliability. The Beck matched the Sperry only in carbon consumption. The board recommended that until a more suitable type was developed the Sperry lamp be used.[34] Acting on the report, the army awarded the Sperry company a contract for converting all existing 60" coast-defense searchlights to the Sperry system. The contract kept the Sperry searchlight department busy for over a year.[35]

The introduction of the anti-aircraft searchlight was a part of the general wartime development of aerial warfare. Night flying required searchlight beacons for navigation and searchlight illumination for landings and takeoffs. After night flights and bombing began, searchlights were needed to illuminate the bombers in order to destroy them or force them to fly so high that bombing would be ineffective. Heavy searchlights could be used in the defense of fixed installations, but portable lights were needed to protect temporary field installations and troops. Improvements in searchlight defense also affected other components of the aerial warfare system,

[34] "Report of Competitive Test of Searchlights Held at Fort Monroe, Virginia, November and December 1916," Lt. Col. D. W. Ketchum, test board president (undated; SP).

[35] History of High Intensity Searchlights in the U.S. Army," typescript dated March 9, 1920 (SP).

necessitating, for example, an improvement in bombsights so that airplanes could fly higher and yet bomb accurately.

Lawrence helped develop the aerial searchlight. When he was abroad in 1914 and 1915, he observed the increase in night flights and after his return to the United States made experimental night cross-country flights. In September 1916, he made the first night flight over water in the U.S., flying fifty miles from Moriches to Amityville, Long Island.[36] Lawrence and his father recommended to the army that it promote night flying by using vertical beam searchlights as beacons and searchlights to illuminate air-fields during landing and takeoff. Lawrence used illuminated gyro instruments to establish the horizon during his night flights, and in this way furthered the development of his own instruments while he helped his father with the searchlight.[37]

Elmer Sperry learned more about night warfare needs from the British. Besides the wide publicity given the German air raids on London and the defense organized against these, he had detailed reports from Gillmor and the London office. Zeppelins made their first raids on London in 1915, and the British responded by organizing an anti-aircraft defense under the Royal Navy, which was experienced with searchlights and precision fire control. Searchlights proved particularly effective against the slow-moving Zeppelins. To use fighters at night the British also provided for night flying; this required searchlights for the air fields. With the increased use of night fighters, the army, in 1916, assumed responsibility for the defense of London against air attacks.[38] As Beck searchlights made by a British company were used in anti-aircraft defense, Elmer Sperry wrote Gillmor to inquire about their performance. In the spring of 1916, Sperry was particularly interested in how the light was elevated to ninety degrees to follow targets. In this position the dripping of hot slag from the electrodes was liable to burn and crack the mirror reflectors.[39]

Airplanes proved a more difficult target to illuminate than Zeppelins, and their increased use in 1916 necessitated improved searchlights and better tactics for employing them in conjunction with ground defense and fighters. In May 1917, shortly after the United States entered the war, Sperry conducted a series of tests at the Mineola, Long Island, training field to explore various means of using the searchlight against airplanes. Lawrence did most of the night flying, exhibiting "fearlessness and fine nerve," for night flying was extremely hazardous. Both 60- and 36-inch searchlights were used in the tests. The lights were operated by remote controls located 500 feet away, for anyone working closer than that would have been blinded by the glare. The airplane made various approaches to, and circled, the field to allow the operators to try several techniques of locating it and keeping it in the

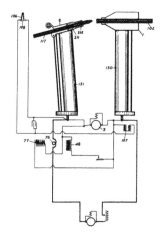

Figure 8.9
Wiring diagram
of feeding mechanism
for searchlight carbons
(Patent No. 1,362,575).

[36] Henry Woodhouse, *Textbook of Naval Aeronautics* (New York: The Century Co., 1917), p. 126.
[37] EAS to Captain Robert Dougherty, Engineer Depot, U.S. Army, September 7, 1916 (SP).
[38] Joseph Morris, *The German Air Raids on Great Britain, 1914–1918* (London: Sampson Low, Marston and Co., n.d.), pp. 63–75, 105.
[39] EAS to Gillmor, March 16, 1916; Gillmor to EAS, June 19, 1916 (SP).

beam. Frank Jewett, chief engineer of the Western Electric Company, lent sound-detection devices, microphones, and megaphones, which were utilized by the searchlight operators to aid in locating the planes.[40]

Development of searchlights at Sperry intensified late in 1917 when the American Expeditionary Force in France found existing searchlight equipment unsuited for operation at the front.[41] The AEF needed light-weight searchlights and more mobile and reliable power plants. It also wanted better tactics for using the searchlight in aircraft detection. The army had decided that the searchlight was the best detection device because "at that time light energy seemed to be the only form of energy which could be projected through space rapidly enough to match the speed of the aircraft."[42] (Radar was still in the future.)

After December 1, 1917, the work at Sperry on the AEF searchlights was directed by the Searchlight Investigation Section of the U.S. Army Corps of Engineers under Captain Chester Lichtenberg. The corps organized the section to establish and carry on an effort. The project lasted 16 months and resulted in unprecedented expenditures on searchlight development.[43] In many ways it anticipated the military research and development projects of World War II, uniting research and development, government and private enterprise, and scientists and engineers. The optical, mechanical, electrical, and chemical properties of searchlights and the psychological and physiological reactions of the men using or exposed to them were investigated in numerous theoretical studies and related experiments. Design problems brought teams of engineers together to work in a common area; development required many tests of newly designed and developed components and systems. The project serves as an excellent example of Sperry involvement in highly-organized, government-directed research and development.

The function of the Searchlight Investigation Section staff was to direct cooperation, which meant defining and assigning research problems and coordinating the scientists, engineers, and manufacturers designing and producing the lights. The section located scientists and engineers through the National Research Council, government agencies, and manufacturers with research and development capability. It also circulated progress reports among all of the collaborators, so that efforts could be more completely coordinated.

Individuals who did research, originated designs, or carried out development for the project were designated collaborators. They were "en-

[40] For a report of these tests, see EAS to Colonel Mason Patrick, Chief of Engineer Depot, Washington, D.C., May 18, 1917 (SP).

[41] This discussion of the Searchlight Investigation Section of the U.S. Army Corps of Engineers is drawn from the *Historical Review* prepared by that section (hereafter cited as *Historical Review*). No date or author is given, but according to a memorandum in the Sperry Papers, the report was published about 1920 by Captain Lichtenberg, the officer in charge of searchlight investigations.

[42] *Ibid.*, p. 2.

[43] "Remarks on U.S. Army Searchlight Investigation Work," an internal report of the Sperry Gyroscope Company, March 3, 1920 (SP).

couraged to make independent and initiatory investigations" on the basis of problems outlined and solutions suggested by the searchlight section. The details of design, construction, and tests were "placed completely in the hands of the individual collaborator who therefore became entirely responsible for the projects."[44] Among the "collaborators" especially commended for their work was Preston Bassett, who helped develop open-type searchlights, light-weight barrel searchlights, simplified searchlight mechanisms, and remote electrical control; and Theodore Hall, another Sperry engineer, who worked on the development of remote electrical controls. Only 11 others among the 106 collaborators were singled out in this way. The searchlight section also designated Bassett as one of five chief design engineers. D. H. Mahood and M. L. Patterson of the Sperry staff also served as collaborators, as did Elmer Sperry—although he mainly contributed by generally supervising the work of his staff. In 1918, Sperry had too many responsibilities as company president and a member of the Naval Consulting Board to commit himself fully to the project.

The project failed to produce a prototype searchlight until April 1918; after that, 17 different kinds of searchlights were partially or wholly completed. Because of the unexpected armistice, only a few searchlights reached the army in France.[45] The 16 months of the development project and an expenditure of $582,764, however, brought "universal acknowledgement that the United States Army had—in 1919—the best mobile searchlight equipment in the world."[46] Furthermore, the development work influenced the design of searchlights manufactured by Sperry, and probably by GE in the postwar decade.

The Sperry Gyroscope Company and General Electric were the major contractors among the 45 concerns that took part in the project. According to a tally made by the Sperry Gyroscope Company, it received $69,700, and General Electric received $281,700, for experimental work. In its report, the searchlight section commended the Sperry company for assigning a large portion of its staff of "investigators" to searchlight work and for producing new designs with "remarkable rapidity." The report also commended General Electric, observing that the unique work of its large staff brought results of far-reaching importance.[47]

The Sperry Gyroscope Company made its own analysis of development work carried out for the Searchlight Investigation Section. According to this typescript report, the army placed duplicate development contracts on various problems with General Electric and Sperry. The companies then proceeded independently. Because of the common problems and the independent approaches, the Sperry company thought the project offered a way of determining the efficiency of the two companies in development work. Efficiency was judged by the success and permanence of contribution in relation to funds expended.

[44] *Historical Review,* p. 27.
[45] The *Historical Review* does not designate the number of searchlights that reached France.
[46] *Ibid.,* p. 75.
[47] *Ibid.,* p. 36.

The Sperry report contended that the company spent $62,540, while General Electric received $200,000 to develop a 60-inch, light-weight drum (enclosed) searchlight, a 60-inch, open-type searchlight, a 500-ampere arc lamp, and distant electric control, and to conduct general tests and investigations. (Sperry also received $5,670 to develop a 36-inch light, but General Electric did not enter this area.) Sperry successfully designed and tested a 60-inch, light-weight drum searchlight, but General Electric abandoned its efforts after the poor performance of its development model. The Sperry company design for the 60-inch open-type searchlight was accepted months before the General Electric. The army subsequently commissioned Sperry to build the high-intensity lamp mechanism for the General Electric searchlight. Sperry successfully tested a 500-amp lamp, while to the best of the Sperry company's knowledge, GE did not have a successful lamp by March 1920, when the report was written. Also, according to the company report, Sperry had devised new kinds of remote control, whereas GE had only modified its existing system of control. General Electric spent $95,700 for tests and investigations alone; Sperry spent only $500. It appears that Sperry achieved the best results for the least amount of money, but it also seems that the company made good use of some of the basic studies and tests made by General Electric.[48] Indicative perhaps of the contrasting styles of the two companies was the comment in the Sperry report that "the bulk of time and expense [of GE's investigation and tests] went into accumulating data with a thoroughness and accuracy hardly warranted in such an experimental state of the field."

vi

Elmer Sperry's wartime patent applications on the ship gyrostabilizer were second in number only to his searchlight applications. The gyrostabilizer applications covered improvements in the design and construction of his active ship stabilizer introduced before the war. As was so often the case with his inventions, the control system was especially interesting. As a sensor, he used a very small pilot gyro that felt the incipient roll of the ship and sent command signals to precess the large gyro stabilizing the ship. He made improvements in the control system to refine the relationship between precession and the roll to gain maximum effect. These improvements, and those in general design and construction, although ingenious, do not reveal much about the inventor or the company; on the other hand, an episode involving the installation of a stabilizer on the *Henderson* proves extremely informative. The Sperry company installed the stabilizer on the transport to test its usefulness as an aid to gunfire control. The project, however, turned out to be a frustrating experience which led to an unusual, acrid exchange between Elmer Sperry and the navy.

The episode began in the fall of 1915 when the general board of the navy again expressed interest in using gyrostabilizers on capital ships to create a stable firing platform. The general board could only recommend, but both the Bureau of Construction and Repair and the Bureau of Ord-

48 *Ibid.*, pp. 2, 32, and 33.

nance thought the project highly promising and decided to have Sperry build a stabilizing gyro and install it on the battleship *Ohio*. Admiral D. W. Taylor looked with favor on the project, although he was concerned that mechanical difficulties might be encountered in scaling-up a stabilizer design intended for the destroyer-class *Worden* into a useful one for a battleship.[49]

The navy contracted for the equipment in July 1915, and Sperry delivered the two massive gyro wheels to the Philadelphia navy yard for machining in May 1916. During the prolonged work at the busy yard, one of the wheels was dropped from a height of about eight inches. Unfortunately, as subsequent events revealed, the wheel was not thoroughly checked for damage, and preparations for installation were hurried along. Then, "on account of the impending crisis of the country," plans were changed, and installation was scheduled for the Navy's first transport, the USS *Henderson*. The stabilizer consisted of two gyro wheels with the axes horizontally mounted in separate housings. A small pilot gyro, used as a controller, could act either as a sensor to feel the movements of the ship and initiate the appropriate response in the giant 25-ton wheels, or as a control to roll the ship. The objective was either complete negation of roll, or a periodic roll created to elevate the guns for increased range. The belief was that fire-control could be as accurate with a predictable roll as with complete stability.[50]

In March 1918, after delays, two Sperry field engineers made an observation trip on the *Henderson* with the stabilizer in place. Edward Sperry was in charge for the Sperry company. He and his assistant had a Sperry pitch-and-roll recorder with which a visual record of the *Henderson's* characteristics could be made. Subsequent analysis of the record would indicate what stabilizer adjustments would be needed. In August 1918, the Bureau of Ordnance installed an improved, centrally directed fire-control system on the *Henderson* and compared its accuracy with and without the stabilizer under various sea conditions. Sperry was convinced from the tests that "it has been demonstrated that we can place the motions of a ship under absolute control by stabilization, so as to produce motions that are so favorable and dependable for gunnery as to double or even quadruple the opportunity for hits. . . ."[51] After reading the full report of the commanding officer of the *Henderson* on the ordnance tests, Admiral Taylor judged them promising but cautioned that a battleship would prove more difficult than the *Henderson,* which was an unusually steady vessel.[52] Admiral Ralph Earle, head of the Bureau of Ordnance, also read the report; although he would not immediately order an installation on a battleship, as Sperry suggested, he asked for plans for installation aboard either the *New Mexico* or the *Delaware*. He wrote Sperry that, given the *Henderson* trials, "we appear to have something now indeed."[53]

[49] From Captain L. H. Chandler, member of General Board, to Secretary of the Navy, Chief of Operations, November 1, 1915 (SP).
[50] EAS to Admiral D. W. Taylor, February 17, 1917 (SP).
[51] Sperry Gyroscope Company to Admiral Ralph Earle, October 1, 1918 (SP).
[52] Taylor to EAS, November 16, 1918 (SP).
[53] Earle to EAS, October 10, 1918 (SP).

Figure 8.10 The gyrostabilizer can be seen, low and amidshships, being slid into the hold of the USS Henderson *at the Philadelphia Navy yard.*

But optimism was somewhat dampened by subsequent events. During the trials on the *Henderson,* there had been repeated trouble with one bearing of a stabilizer wheel. When the ship laid up for extensive overhaul early in 1919, Sperry decided to correct the overheating bearing. A thorough check convinced the Sperry engineers that the trouble could be traced to the wheel that was dropped in the Philadelphia navy yard. Working night and day during the week before the *Henderson* sailed, they repaired the faulty shaft and lapped all of the bearings. Sperry predicted that the stabilizer system could then be run at full speed without overheating the bearings.

To the consternation of Sperry and his engineers, however, in January 1919, the Captain of the *Henderson* placed 1,200 tons of steel billets in her hold as stabilizing ballast; the reasons for the action are not clear, although he may simply have believed the vessel unseaworthy.[54] According to the Sperry engineers, the ballast increased the metacentric height of the ship and made her much stiffer than the ship the stabilizer had been designed to handle. The Sperry engineers subsequently referred to "2 1/3 *Hendersons"* and resigned themselves to being unable to modify the stabilizer. Although trials with the *Henderson* would go on and Sperry engineers would continue to observe, Elmer Sperry informed the Bureau of Construction and Repair that there "remains some doubt as to the stabilizer's ability to stabilize the ship as closely as was originally planned." He hoped that the

[54] EAS to Captain Arthur MacArthur, USN, (a later Captain of the *Henderson),* May 15, 1922 (SP).

ballast would be removed.[55]

The heavily ballasted ship with the repaired gyrostabilizer put to sea again in the spring of 1919 with Alexander Schein, chief gyrostabilizer engineer of the Sperry company, aboard. Schein had been with General Electric before joining Sperry and brought with him experience in steam-turbine design and construction.[56] Because the steam turbine had presented a new order of lubrication and bearings problems to the engineer, Schein's knowledge was especially valuable to Sperry. Schein, however, found that the Captain of the *Henderson* was more of a problem than the bearings of the stabilizer. On the assumption that the gyros were imparting a permanent list to the ship, the Captain would allow them to run for only a few hours at a time. This was the same captain who had ballasted the ship and who reportedly would allow only half of the soldiers on board to "come up for breath" at the same time, keeping the others low in the ship (to provide additional ballast?)[57] Schein reported that the Captain on several occasions ordered that the stabilizer be stopped when in fact it was not running. At other times, Schein ran the stabilizer when the Captain thought it had been stopped.

Despite these limitations, Schein operated the gyros at top speed and full amplitude of precession, one overspeed test extending for almost six hours. After this run, Schein asked the chief engineer of the *Henderson*, who witnessed it, to examine the bearings. He found them in perfect condition and noted that the lubricating oil had not overheated during the extended test. On the basis of this and other runs, Sperry later maintained that the stabilizer plant was in "perfect condition" and operated in a "perfect manner."[58]

The trials of the stabilizer were interrupted when the *Henderson* was ordered to the North Sea. While on station the crew was instructed to run the stabilizer periodically for maintenance purposes. During this cruise, however, a fire in the hold damaged the stabilizer. When the Sperry engineers had another look at the stabilizer in the fall of 1920, they found the strainers in the main-bearing oil system had been punctured during the fire; they assumed that the plant had been run with unstrained, dirty, abrasive oil. They also found the number two gyro wheel badly out of balance.

In the fall of 1920, the navy ordered an overhaul. Although the plant had been superceded by Sperry's newer design, which had only one vertical wheel instead of two horizontal wheels, the navy believed that the *Henderson* plant should be used for further experiments before making another major installation. (Gyrostabilization as an aid to fire-control was still thought promising.) Overhauling the stabilizer was difficult. The cause of the imbalance in the number two wheel proved to be about fifty pounds of solder that had run free from the wiring of the stabilizer when it had been over-

[55] EAS to Bureau of Construction and Repair, May 22, 1919 (SP).
[56] In 1935 Schein became vice president for manufacturing with the Sperry Company.
[57] EAS to Bureau of Construction and Repair, May 17, 1921 (SP).
[58] *Ibid.*

loaded with current by an inexperienced operator. Furthermore, since the journals and bearings had been scored by the dirty, abrasive oil, Sperry removed the main bearings and the navy yard machine shop resurfaced and precisely rebored them.

The balancing of the number two wheel proved a problem. The new captain of the *Henderson*, Captain William Russell White, contracted with another company to balance the wheel. Sperry engineers had intended to use their customary method evolved from years of experience with gyro-compass wheels and improved upon by Schein with his experience in balancing large steam-turbine rotors.[59] Captain White decided to depend upon the Vibration Specialty Company of Philadelphia, headed by a certain Akimoff. In his letter explaining his decision to his superiors, White referred to the "antiquated trial and error method" of the Sperry company and to the "Akimoff scientific method of balancing."[60] To Sperry, the Akimoff balancing procedure seemed unnecessarily complicated, but he did not object, for he and his small staff of engineers were already overloaded with work. The balancing took longer than predicted, delaying the ship's departure from the yard until January 1921. The Sperry service engineer who observed the first tests of the stabilizer reported that the number one wheel showed the same satisfactory balance as when first installed but that the number two wheel—after Akimoff reported it balanced—was far inferior to number one. When the Sperry engineer called Captain White's attention to this, "he agreed." By contrast, the bearings were inspected after the wheels were spun continuously for 8 1/2 hours, with intervals of full precession to create maximum bearing pressures, and were found to be in excellent condition.[61]

Early in February, Captain White reduced the number two gyro's excessive vibration by further balancing; during a subsequent test, it ran satisfactorily for short periods. White, who had been especially assigned for the gyro overhaul and trials, decided, however, that before extended trials could be run modification of the gyro plant was necessary. If this were not done, he and Lt. Cmdr. F. T. Van Auken, also especially assigned, anticipated continued expensive breakdowns and overhauls. In February the officers specifically recommended that the *number one* wheel be balanced by Akimoff's scientific method, that the bearing surfaces be increased in area to reduce the unit pressure, and that the lubrication system for the bearings be improved.[62] Elmer Sperry's reaction to these reports was acrid. He recognized the need for modifications—in fact his new design incorporated many changes—but he took exception to the opinions expressed by the officers about the competence of his company and engineers.

[59] See A. Schein, "Balancing High-Speed Rotors," *American Machinist*, LV (1921), pp. 121–25, for a description of this method.
[60] White to the Secretary of Navy, Operations, November 17, 1920 (SP).
[61] Report of service engineer, W. B. Fletcher, included with letter of Elmer Sperry to Bureau of Construction and Repair, May 17, 1921 (SP).
[62] White to the Secretary of Navy, Operations, February 8, 1921; F. T. Van Auken, the Engineering Officer of the *Henderson*, to the Bureau of Construction and Repair, February 23, 1921 (SP).

White and Van Auken had the notion that the Sperry company used the old empirical approach to engineering problems and that the art of stabilization needed applied science. Having decided that naval officers were not yet thoroughly informed about gyroscopes, White concluded that the navy could not have evaluated adequately the gyroscopic devices supplied by Sperry. The stabilizer equipment on the *Henderson*, White asserted, was poorly, crudely, and unsatisfactorily designed. He added that he and Van Auken had, during the past months, learned much of stabilizer design and were in a position to improve upon it. Furthermore, not limiting himself to stabilizers, White suggested that Sperry was not abreast of science and had supplied the navy with poor compasses.

Captain White concluded that the *Henderson* stabilizer had been designed without scientific research because the Sperry company had not supplied exact design and test figures and information when requested: "The only figures submitted, have been of most elementary school boy character." Captain White reported that he had asked for Commander Van Auken as an assistant because that officer was not only a postgraduate engineer but had studied the intricate mathematics of "the most involved and postgraduate character" night and day in order to analyze the stabilizer scientifically. In passing, the Captain referred to "the phenomenon of nutation" as an example of the kind of stabilizer problem that only scientists, not Sperry personnel, could solve. In conclusion, noting that gyro-stabilization had tremendous potential value on a man-of-war, he strongly recommended that an intensive study be made of "highly involved mathematical phenomena from theoretical, practical and experimental standpoints."[63]

Commander Van Auken, in his separate report, stressed that imbalance was only one of the factors causing failure of the stabilizer. He concurred with Captain White that the controlling factor in the "failure of these machines is bearing trouble" resulting from insufficient size and ineffective lubricating design. With his report, Van Auken sent a nine-page mathematical analysis of the bearing problem. He recommended that in the future the navy conduct careful scientific research on gyroscopically loaded bearings.

In a 22-page response directed to the Bureau of Construction and Repair and to Admiral D. W. Taylor, Elmer Sperry answered the three specific recommendations made by White and Van Auken. Sperry dismissed the first, stating that the *number two* wheel needed balancing, not the *number one*; he rejected the second recommendation, maintaining that the bearing surfaces were adequate; but he accepted the third, concluding that the lubrication system should be modified—a trifling alteration in accord with the company's more recent designs in which the oil would be introduced in the longitudinal center of the journal instead of at the ends.

Following these specific comments Sperry, with the aid of his staff, replied to the remarks about the engineering competence of his company and staff:

---

[63] Quotations are from White to the Secretary of the Navy, February 8, 1921.

It is to be regretted that our naval officers are troubled from time to time with people who are supposed to be abreast of the times in engineering, and who, through a compound of impressive talk, mixed with great amounts of mathematics, make themselves so impressive that they are taken seriously when, in the last analysis, their information is found to be out of date, obsolete and much of their mathematics totally irrelevant, misleading these officers whose minds are taken up in the other lines and who have not the time to disprove their assertions. The upshot of such obsessions is to throw doubt on sound projects and cause serious delays. The officers not being able, offhand, to refute such statements, the engineering staffs of industries to which the matter relates are called upon to turn away from their urgent daily tasks and give attention to properly answering these long drawn out and totally misguiding and misleading masses of statements. This is also to be regretted, but it is doubtless a part of the civilian engineer's burden. The training of naval officers should be such and often does function to detect the fallacies, no matter how plausibly presented they may be, thus saving the civilian from considerable expenditure of time and anxiety.[64]

Sperry regretted that "Mr. Akimoff" considered the gyro calculations "school boy" arithmetic, but he noted "that every effort was made to simplify our formulae, which had been heretofore adversely criticized by the naval officers as being too much involved in higher mathematics." He also confessed that he had not anticipated the "amateurish aspect of Akimoff's attempt at balancing."[65]

A statement, probably by Schein, on journal bearings was appended to Sperry's letter. It argued that the naval officers had confused gyroscope bearings with steam-turbine bearings. Turbine bearings were unusual because their metallic parts never came in contact but were carried upon a film of oil. Because the load-line was constant in these bearings, high temperatures developed owing to the persistence of the location of the pressure; as a result, the oil tended to break down chemically. This limited the unit pressure that could be carried. The gyro, however, offered a different problem, for in precessing it moved the load-line, thereby providing a progressive, or nascent film of oil, always fresh and less likely to overheat. According to Sperry analysis and experience, therefore, the gyro bearings could carry a higher pressure than comparable steam-turbine bearings. Van Auken and White had apparently decided, the memo continued, that the bearings had insufficient surface area because they had analyzed them as though they were turbine bearings. To substantiate the argument, Sperry included a letter from Albert Kingsbury, an expert on bearings, in which Kingsbury noted that his consultation with Sperry over the past several years had convinced him that gyroscopically loaded bearings could carry higher pressures than turbine bearings.[66]

[64] EAS to Bureau of Construction and Repair, May 17, 1921 (SP).
[65] These general observations and other comments were appended to Elmer Sperry's letter of May 17, 1921. The four "exhibits" supplementing the letter are signed in only one case (by the service engineer, W. B. Fletcher, who reported on the overhaul of the fall of 1920). It seems probable that these supplementary statements were prepared jointly by Sperry and his staff.
[66] Kingsbury to EAS, May 16, 1921 (SP).

Sperry added a light touch to his response in the form of a short essay on "nutation." He asserted that he had dispelled the mystery of the phenomenon years earlier, although he did admit that the Egyptians had anticipated him by thousands of years when they found and analyzed nutation in the minute wobbling of the earth's axis. Both to him and to them, it had proved to be of interest, but Sperry had found that it did not greatly affect the design of gyros. He thought that study of nutation had, from the viewpoint of the practical engineer, effectively smothered the whole phenomenon in mathematics.

The exchange between Sperry and the naval officers suggests that the latter, well aware that the navy had long been criticized for its lack of scientific awareness, had made an effort to show that in the case of the *Henderson* stabilizer the navy was more scientifically competent than the civilians supplying complex equipment. In retrospect, the officers' demand for more scientific analysis of such problems seems reasonable. Yet the Sperry response suggests that at the time experience, although based on an empirical approach, was better able to deal with the complex stabilizer problems than the existing theory, which was not yet refined enough to provide a firm basis for analysis and design. Experience running ahead of scientific comprehension has not been unusual in the history of technology, nor has the use of abstruse equations to cover ignorance.

As a sequel to the episode, the navy named a board to evaluate the *Henderson* plant in July 1921. The board ran both gyros and found that the wheel balanced by Akimoff was "still very much out of balance."[67] The board also found that the bearings were in first class condition, despite the strain placed upon them during the test by the unbalanced gyro wheel.[68] Sperry also believed that further trials could be run with the *Henderson* equipment, but he preferred that the *Henderson* chapter be closed and that the navy carry out its future trials on newer Sperry equipment then ready to be installed.

The record does not show that any more trials were run on the *Henderson*. During the next few years, whenever a new skipper took over the ship, Sperry wrote a letter explaining the inactive gyro equipment in the hold of the ship. In these letters, he stressed that the stabilizer was an "old design" and that the *number two* gyro wheel was still unbalanced; the company preferred that the state of the art be judged by its new, single, lighter, vertical stabilizer plants. Yet, the explanation continued, much had been learned from the *Henderson* plant not only about stabilizer design but also about its potential in fire control.[69] The plant had cost the navy $35,000; more than $85,000 had been spent by the company on it.[70]

vii

Half of Sperry's wartime patents in the ordnance field were for gunfire control on naval ships. Together, he and his engineers had by 1916

[67] EAS to D. W. Taylor, July 27, 1921 (SP).
[68] EAS to Taylor, December 15, 1921 (SP).
[69] EAS to Captain Arthur MacArthur, May 1, 1922 (SP).
[70] White to the Secretary of the Navy, Operations, November 17, 1920 (SP).

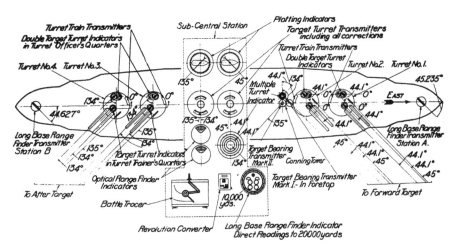

Figure 8.11 The Sperry fire control system for battleships. From the Sperry Gyroscope Company, The Sperry Fire Control System, *Bulletin 301, 1916.*

developed a system of gunfire control for capital ships and had by 1920 installed "Sperry Fire-Control Systems" on nineteen U.S. dreadnoughts, eleven second-line battleships, and nine armored cruisers. The development of these systems led the noted American inventor, former naval officer, and contemporary of Sperry's, Frank Julian Sprague, to describe him as "a man whose work has in the opinion of Naval Officers revolutionized navigation and gun fire."[71]

The gyrocompass was a basic component in the Sperry improved gunfire control system. When Sperry began developing his gyrocompass, the navy needed it to provide a more precise base line or reference line for training the guns. The Chief of the Bureau of Navigation had realized in 1910 that without the improved compass advances in ordnance would to a large extent be nullified. After providing the gyrocompass, Elmer Sperry proceeded to exploit its potential by inventing a system of related components. The system he developed between about 1912 and 1916 increased the responsiveness and firing accuracy of the turret guns, which had become increasingly necessary as improved explosives and metallurgy extended the range of the ten, twelve, and fourteen-inch guns. Before 1900 the maximum range had been under 4000 yards, in 1910 about 10,000 yards, and during the war almost 20,000 yards.[72] Because doubling or tripling the range of the guns also doubled or tripled the effect of the same angular error in training the guns, the increased range made a method of reducing errors essential.

Other improvements in gunnery during the decade or two prior to the war further stimulated a need for better gunfire control. Admiral Bradley A. Fiske, a friend of Sperry's, introduced the telescopic sight for naval guns; the sight was then moved high in the foretop to extend the horizon. The control and plotting room for the senior gunnery officer was established below decks. Admiral William F. Sims, USN, Admiral Sir Percy Scott, RN, and other reform-minded officers introduced, not without opposition, these and other improvements in gunfire in their respective navies. The climate

---

[71] Minutes of the meeting of the Naval Consulting Board, February 9, 1916 (copy in SP).

was right for further improvements of the kind Elmer Sperry could make toward greater precision and control.[73]

Responding to the need and cooperating with progressive officers interested in technology, Sperry had by 1914 contrived around his compass a system for observing and communicating target bearings. His repeater compasses, driven by the master compass below decks, were placed in the foretop and in the plotting room, where the readings were needed for taking bearings or for making calculations. Sperry also provided the communication system needed between observers aloft and plotting-room personnel below.

By 1916 Sperry's target-bearing and turret-control system included a target-bearing transmitter in the foretop and a plotting indicator in the plotting room. After the observer in the foretop sighted the target through the telescope mounted on a repeater compass, the transmitter automatically sent the bearing to the plotting indicator. In the plotting room, the officers and men supplied the necessary deflection, spotters' corrections, and other refinements; the corrected bearing, or train, was then sent by the target turret transmitter to the turret, where a visual display on the target turret indicator gave the information to the gun crew. The train system also included a turret train transmitter, controlled by the rack of the turret when the turret rotated, which showed on the target turret indicator in the turret the position of the turret and on an indicator below decks the position to the gunnery officer. Since both the desired train and the actual train were displayed on the target turret indicator, the turret crew could train the gun by following the pointer. When the gunnery officer saw on his indicator that the turret was correctly trained, he could order fire.[74]

The 1916 system also included an analogue computer, the "battle tracer," that combined the inputs from four electric motors to provide a chart record, or plot, of its own ship's and the target ship's speed and course. This permitted a close prediction of the enemy bearing and range so that hits could be made despite evasive maneuvers.[75] The gyro compass controlled

[72] Herbert K. Weiss, "Influence of the Guidance Designer on Warfare," in *Air, Space, and Instruments,* ed., S. Lees (New York: McGraw-Hill, 1963), p. 37.

[73] For a background to gunfire control see Sir Percy Scott, *Fifty Years in the Royal Navy* (London: John Murray, 1919); Morison, *Admiral Sims;* and the *Scientific American,* which was especially concerned about naval preparedness and gave good coverage to naval matters during the decade before World War I. Description of the naval gunnery can be found in a series of five articles written by J. Bernard Walker, editor of the *Scientific American,* and published in volume CV (1911).

[74] Elmer Sperry's patents on the fire control system of 1916 were: filed August 31, 1914 (859,329), "Multiple Turret Target Indicator," issued March 4, 1919 (Patent No. 1,296,439); filed October 10, 1914 (866,011), "System of Gunfire Control," issued October 19, 1920 (Patent No. 1,356,505); and filed December 18, 1914 (877,953), "Plotting Indicator," issued February 13, 1917 (Patent No. 1,215,425). An Elmer Sperry and Elmer Meitner joint patent, filed April 9, 1917 (160,877), "Director Firing System," issued April 22, 1930 (Patent No. 1,755,340), was a basic wartime fire-control application.

[75] Rear Admiral Bradley A. Fiske, "Sperry's Contributions to the Naval Arts," *Mechanical Engineering,* XLIX (1927), pp. 111–12.

one of the motors and from it the battle tracer received the true compass course of the ship; two revolution counters on the ship's propeller shafts controlled another motor providing the battle tracer an input proportional to the speed of the ship. This input was modified by corrections for currents and tides. The third motor of the battle tracer was controlled by the target bearing transmitter, which gave the target bearing. The fourth motor was controlled by a signal from a range finder.

The battle tracer was about seven inches in diameter and five inches deep. Within it were the four motors and the necessary gearing that combined the input from the motors. The device rode on a wheel, which was pivoted by the motor controlled by the gyrocompass and rotated about its own axis by the motor providing the speed of ship input. Thus the battle tracer moving across the chart simulated the movements of its own ship across the sea. From the battle tracer extended a pivoted arm about a foot-and-a-half long; the third motor turned the arm in azimuth in accord with the signals taken from the target-bearing transmitter. The arm supported a small, movable carriage that was controlled by the range motor. This carriage, making its pencil marks upon the chart, simulated the movements of the target ship.[76] Before the war ended, the navy ordered twenty of the battle tracers and hundreds of the various components making up the Sperry fire control system.[77]

viii

In 1915 Sperry decided to allocate company resources to the development of flying instruments. From 1914 to 1918, Elmer Sperry, Lawrence, Elmer, Jr., Morris M. Titterington, Charles H. Colvin, Francis Champlin, and Omar Whitaker—Sperry inventors and engineers—all applied for patents on aircraft instruments. Elmer Sperry made five applications, Lawrence five, and the remainder of the staff made four successful patent applications in the aircraft instrument field.

This line of research and development was not far afield for Sperry because of his prior work with the automatic airplane stabilizer. The gyros that established a frame of reference for the automatic stabilizer could also be used in instruments. The difference would be that the pilot would take readings from the instruments and operate the controls himself, whereas the automatic stabilizer bypassed the pilot by a communication system, servomechanism, and feedback. For several reasons, the military pilots wanted instruments rather than automatic systems. They desired maneuverability and could obtain this better with the pilot at the controls. The weight of the

[76] The "Sperry Fire Control System" of 1916 is described in the company *Bulletin*, nos. 301 to 304, 1916.

[77] U.S. Bureau of Ordnance, Navy Department, *Naval Ordnance Activities: World War, 1917–1918* (Washington, 1920), p. 152. The young man whom Sperry had cultivated in the gyroscope and instrumentation field, Hannibal Ford, became the developer and supplier to the navy of range-elevation (in contrast to Sperry's bearing-train) fire-control instruments. Ford left Sperry in 1914 and organized the Ford Marine Appliance Corporation, which became the Ford Instrument Company, Inc., in 1915. In 1933 the Sperry company and the Ford company were merged.

stabilizer was also a problem for the planes of World War I, and in the opinion of many of the pilots, the automatic stabilizer could not control as sensitively as the pilot. To use recent terminology, the pilots thought they could close the loop better than a mechanism in a feedback system.[78]

The need for gyro instruments became acute when the pilots ventured into the overcast or the darkness. Because the air corps that could fly under these conditions had an advantage, the American military, after 1914, encouraged Sperry. He saw the possibility of developing a gyro instrument that would provide an artificial horizon with which the pilot could align his plane to achieve the desired attitude of flight and a gyro instrument that would indicate deviations from an established attitude, or line, of flight. The instrument providing an artificial horizon would use the stability characteristic of a pendulous gyro and the one indicating change of state would use the precession characteristic.

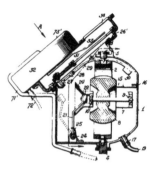

Figure 8.12
Gyroscopic direction
indicator (Patent No.
1,522,924).

In 1915 the pilot had only a few instruments. These included a tachometer to measure engine revolutions, a wind gauge to show air speed, a magnetic compass, a barometer (or altimeter), and inclinometers to show inclinations of the machine fore and aft and from side to side.[79] Some aviators also used a side-slip indicator consisting of about a foot of thick light cord fastened to a strut to indicate the airstream direction and to warn of losing altitude in a side slip.[80] Because the inclinometer, a pendulous nongyroscopic device, could not distinguish accelerations from gravity, because the magnetic compass malfunctioned, and because the sideslip indicator could not determine drift caused by cross winds, Sperry concentrated on developing gyro instruments to solve these problems.

Sperry corresponded with Captain J. A. Hoogewerff at the U.S. Naval Observatory about the need for an improved compass. Initially, Sperry, who was making magnetic compasses for the airplanes of the Allies, argued that the aviators had not learned to use the magnetic compass properly, and he seemed reluctant to embark upon the development of a gyrocompass, "a problem well worth the mettle of the best of us."[81] He wanted, however, to know how much the navy would cooperate in a development project if it determined "that a gryo-compass is really needed."

The navy decided an aerial gyrocompass was needed and proceeded to work out a development contract with Sperry. In January 1916, Captain Mark Bristol, of the Office of Naval Aeronautics, outlined the specifications; he wanted the gyrocompass to establish both a directional and a horizontal plane of reference. Sperry clarified the problem by showing that two instruments were needed: a gyrocompass and a gyro artificial horizon.[82] The navy

[78] Preston R. Bassett, "The Control of Flight," a paper presented at the Fourth Anglo-American Aeronautical Congress, 1953 (SP).

[79] "The Flying Machine and Its Equipment: A Summary of the Air Navigator's Instruments," *Scientific American*, CX (1914), p. 219.

[80] Elmer Sperry, Jr., "Early Airplane Instruments," a Lester D. Gardner Lecture, May 1, 1963 (SP).

[81] EAS to Captain J. A. Hoogewerff, January 11, 1916 (SP).

[82] Captain Mark L. Bristol to EAS, January 13, 1916; EAS to Bristol, January 14, 1916; EAS to J. A. Hoogewerff, February 2, 1916 (SP).

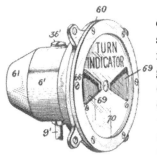

Figure 8.13
Face of
gyroscopic turn indicator.

drew up specifications, with Sperry's advice, and authorized "pecuniary assistance" for the development of the gyrocompass. The navy intended to reimburse Sperry for development by paying a higher price for the first gyrocompass delivered than for subsequent models: $6,000 for the first compass, $2,000 for one additional, $1,800 each for five compasses, and $1,500 each for ten compasses.[83] Sperry contracted for these prices with the stipulation that the compasses could be sold to an unrestricted market. The navy asked that sale be restricted to the U.S. government because development would be supported by the government, but Sperry argued that his company depended on the foreign market to retain his staff of engineers and mechanics and to support his overall development program. He showed that profits from the foreign market for his gyrocompass had made possible a sales-price reduction and continued refinement, and that without the volume made possible by sales abroad, he did not expect that an aerial compass could be similarly improved. Sales abroad were, therefore, of value to the navy.[84]

Development of the gyrocompass proceeded slowly and extended into 1918, for Sperry and his engineers found that the problem did try their mettle. Flight testing an experimental compass convinced them that "the aeroplane has much *greater acceleration* forces than any other craft," compulsively jerking with severe, nonperiodic motions that affected the sensitive parts of the compass.[85] These forces not only damaged the sensitive lightweight device but also confused it; the compass was unable to distinguish them from the gravity force needed to align the compass along the meridian. The disappointing performance of both the magnetic and experimental gyrocompasses led some to say that the magnetic needle had sense but no power and the gyroscope had power but no sense.[86]

By the end of the war, Sperry could write, "It goes without saying that we now know a great deal more about the problem than we did when we started." But the company had not delivered the first compass and Sperry reported that development costs had reached over $9,000. The navy cancelled the contract in January 1919, explaining that the Sperry Company could not ensure delivery of an acceptable compass at a reasonable price.[87] The navy, however, wanted Sperry to continue development of the compass without a contract; Sperry felt that the postwar market would be small and that sales would not repay additional development costs.[88]

In 1918 Sperry and his engineers also worked on an artificial horizon and were again denied success. Similarly, acceleration forces proved the major problem. They built a promising air-driven and air-dampened device which Morris Titterington predicted would "enable an aeroplane to fly on

83 Contract No. 1043, Series 1916, 13 June 1916 (SP).
84 EAS to Captain J. A. Hoogewerff, June 2, 1916 (SP).
85 "Military Aeronautical Gyro Compass," undated Sperry company typescript (SP).
86 Major General George O. Squier, "Aeronautics in the United States, 1918," paper presented at the Meeting of the American Institute of Electrical Engineers, New York City, January 10, 1919, p. 18.
87 T. B. Howard, Inspector of Navigational Material, USN, to the Sperry Gyroscope Company, December 30, 1918 (SP).
88 EAS to LCDR John H. Towers, January 23, 1919 (SP).

the darkest night or through the heaviest fog," but Elmer Sperry reluctantly concluded in September that "we are not getting the results that we anticipated with the simple air spun instrument, and there may be much more delay than any of us expect before a permanent and practical solution of the 'vertical' is attained."[89]

Sperry was not one, however, to give up because of frustrations with the compass and the artificial horizon. He announced in 1918 other means of providing the information desired from a gyrocompass and a gyro horizon. This involved, besides conventional instruments, two "simple" instruments he had recently invented to solve the problem of "fog, night or cloud flying."[90] These were a small air-spun gyro turn-indicator and a side-slip indicator. The gyro device depended upon precession to indicate turns, and the side-slip indicator used a pressure differential to indicate side-slip motion. He described how the pilot would use these:

It will be readily seen that the combined use of these two instruments will at once inform the aviator as to the character of his course. If the gyroscope shows a straight course and the side slip indicator no side slip, the aviator knows that all is well, or, on the other hand, if the gyroscope shows that he is turning and the side slip indicator shows no side slip, he knows that he is banked at the proper angle. If, on the other hand, the gyroscope shows that he is flying a straight course and the slip indicator shows that he is side slipping, it will indicate to the aviator at once that his machine is laterally tilted and sliding downwardly sidewise. Then, again, if the gyroscope shows turning and the side slip indicator side slip, the aviator knows that he is incorrectly banked or that he may be approaching a nose or tail spin.[91]

Sperry also explained how conventional instruments could be used to indicate airplane attitude. To indicate fore and aft deviation from level flight, Sperry recommended that the aviator compare the reading of his tachometer and his air-speed indicator: if the plane was nosed down, the air-speed meter, compared to the tachometer, would read too high; if the plane was nosed up, the air speed read too low. (Pilots were already accustomed to making these comparisons.) Sperry also said that when his turn indicator and side-slip indicator showed level, straight flight, the aviator could then read and depend upon his magnetic compass.[92] Sperry's expedients were ingenious, but the navy, at least, hoped that Sperry would continue to work on the gyrocompass and artificial horizon.[93]

At about the same time that Sperry embarked on the development of the gyrocompass and the gyro horizon, he assigned his staff to work on a bombsight. By 1915 airplane and zeppelin bombing was "lavishly indulged in,"[94] and both sides called for improved bombsights; without them, the pilots

[89] EAS to L. de Florez, September 13, 1918 (SP).
[90] *Ibid.*
[91] Filed September 18, 1918 (254,534), "Position Indicator for Aircraft," issued January 13, 1925 (Patent No. 1,522,924).
[92] Sperry to J. C. Hunsaker, September 13, 1918 (SP).
[93] L. de Florez to EAS, September 28, 1918 (SP).
[94] C. Dienstbach, "The Aeronautical Lessons of the European War," *Scientific American,* CXII (1915), 624.

W. DUDLEY CARLETON
SECRETARY AND TREASURER

NEW YORK May 15, 1911.

40 WALL STREET

THE SPERRY GYROSCOPE COMPANY

ELMER A. SPERRY
PRESIDENT AND CHIEF ENGINEER

MANUFACTURERS OF
SHIP GYROSCOPES
BATTLE COMPASSES
GYROSCOPE COMPASSES
GYROSCOPES FOR AEROPLANES
GYROSCOPIC MONORAIL CARS

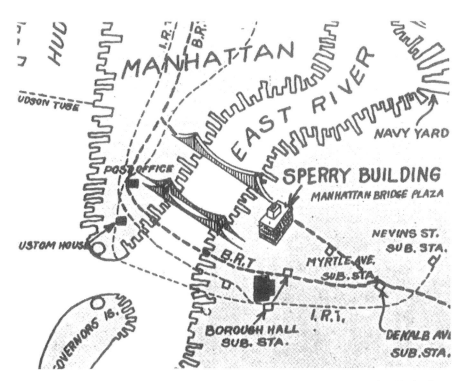

Figure 8.14 Elmer Sperry located his building near the port of New York where ship captains could easily visit his factory to see the gyrocompass. From The Sperryscope, *III (1922).*

were thought remarkable if they dropped their lethal load within the same country as the target.[95] When searchlights and anti-aircraft forced the bombers to fly higher, the need to improve bombsights increased.

Early bombsights consisted of telescopes equipped with simple calculators that showed the pilot how much to lead the target. A major problem was that anything other than level flight threw off the sight. The navy wanted Sperry to stabilize the telescope sight in a horizontal plane. In 1918 a few of the bombsights were "put into service," but the Sperry gyrostabilized bombsight would not become famous until World War II.

The company's overall record of achievement in aircraft instruments at the war's end was a disappointment to Sperry. He agreed that the necessary instruments for aerial navigation did not yet exist, but he would not accept ultimate failure. To his son Lawrence he wrote, "We have absolutely got to solve this problem; if we die in the attempt and could have registered a single notch in advance, it seems to me that it would be well worth while."[96] These would prove to be tragically prophetic words, but in the meantime Sperry turned over to Lawrence the problem of aircraft instruments.

Later, the pioneer Sperry effort yielded a rich harvest. Early in the twenties, pilots who had flown "by the seat of their pants" began to depend

[95] "Bombing Planes and Their Targets: How Aerial Bombing Has Become an Exact Science After Four Years of War," *Scientific American,* CXIX (1918), pp. 86ff.
[96] EAS to Lawrence Sperry, August 10, 1920 (SP).

on the Sperry turn-indicator when visibility was poor.[97] Two Sperry aeronautical engineers, Charles Colvin and Morris Titterington, and another employee, Goldsborough, left him shortly after the war to form the Pioneer Instrument Company, which later acquired licenses to Sperry instruments and redesigned and improved upon them.[98] It became one of the major aircraft instrument makers and later merged with the Bendix Aviation Corporation. Elmer Sperry, Jr., Preston Bassett, and other Sperry engineers again took up the development of gyro instruments for the Sperry company in the late twenties, and their efforts met with admirable success when they provided a directional gyro and an artificial horizon used by Army Lieutenant James H. Doolittle in his epochal blind-flying hop in 1929. Doolittle said that the artificial horizon was "like cutting a port-hole through the fog to look at the real horizon."[99]

The Sperry contribution to aircraft instrumentation during World War I and after was not limited to gyro instruments. Sperry, his engineers, and the company also developed and manufactured nongyroscopic instruments of their own and under licenses from others. The drift set invented and developed by Elmer Sperry and his son won the Collier Trophy for 1915. The company also made a nongyroscopic inclinometer invented by the senior Sperry and manufactured, in addition, a nongyroscopic banking indicator, an angle-of-incidence indicator, an air-speed indicator, and thousands of the Creagh-Osborne magnetic air compasses.[100]

ix

Ventures into new fields and a continuing emphasis on research and development show that the essential character of Sperry and his company did not change during the war, despite the increased activity of the manufacturing division. Because his company[101] grew strong and viable, and

[97] James H. Doolittle, "Early Blind Flying," *Aerospace Engineering* (October 1961), p. 56.

[98] Elmer Sperry, Jr., "Early Airplane Instruments," p. 17 (SP).

[99] *Ibid.*, pp. 21, 22.

[100] There are more than 50 Sperry airplane instruments at the Smithsonian National Air Museum (accession numbers 123 and 1488).

[101] The Sperry Gyroscope Company has been described as Elmer Sperry's company throughout this biography. A copy of an agreement between him and Charles H. Conner, the first business manager and vice president of the company, states that the entire authorized capital stock of the company stood in the name of Elmer Sperry and members of his immediate family before November 1914. Under the agreement, dated November 30, 1914, the stock was reallotted as follows: Elmer A. Sperry (3,000 shares); Conner (1,250); Samuel L. Fuller (250); Arthur D. Dana (250); and Herbert E. Goodman (250). Assuming that the new arrangement went into effect, Fuller, Dana, and Goodman received all the stock as voting trustees for a period of five years. The voting trustees were to select seven directors, Conner having two of these and the other five representing Sperry's interest. The board of directors was also "to elect an Executive Committee to consist of Messrs. Dana, Fuller and Goodman." Copy of agreement between E. A. Sperry and Conner, November 30, 1914 (SP). Fragmentary evidence suggests that during the war, Elmer Sperry bought out both Conner and Dana, and probably Fuller as well.

Figure 8.15 The Sperry building in Brooklyn, N.Y., not far from the Navy Yard.

Figure 8.16 Sperry shown bareheaded at the dedication of his building on December 8, 1916.

because it was cast in his mold, Elmer Sperry was assured a lasting influence upon the course of technological development. A half-century later, the giant corporation still bearing his name would clearly evidence his characteristics, his style, as an inventor and engineer.

The substantial growth of the company was reflected in the increased size and strength of the labor force, the building of a new factory and office building, and the general increase in capital assets. The productive capacity and increased sales of the company manifest this substantial growth. The number of workers—about one-ninth of whom were highly skilled machinists—rose from several hundred before the war to more than two thousand during the war. The proportion of skilled, highly paid machinists remained that low only because of the full exploitation of automated jigs, fixtures, tools, and other devices.[102] Because of wartime shortages in the machine-tool industry, the Sperry company could not add all of the productive facilities needed, but during the war the value of company machinery, tools, and equipment increased from about $100,000 to more than $600,000. Company assets in general grew from about $1 million in 1914 to over $5 million at the end of the war.

The most impressive single material acquisition during the war was a well-designed factory and office building at Manhattan Bridge Plaza. The new eleven-story building, dedicated December 1916, was a striking improvement over the small office in Manhattan and the shop space at Frederick Pearce's with which the company had begun. Harvey Corbett, the new Sperry building's architect, in consultation with Elmer Sperry designed a building with an exterior more attractive than the ugly slab construction then common. The edifice housed, in its 320,000-square-feet of floor space, a foundry, machine shop, finishing and testing department, offices, and other work spaces. Its impressive location dominated the plaza at the Brooklyn end of the Manhattan Bridge and was not far from the navy yard. Sperry's office, with a handsome interior reflecting his daughter Helen's taste, was on the top floor and had projecting bays providing a dramatic view of the skyline. This office, and the building, made a deep impression upon him: his eyes were moist when he first emerged from his new office to receive a silver loving cup and a poetic tribute from his employees. (Gillmor thought it high time that the president have a dignified and quiet office.)

The new plant brought increased production. Sales rose steadily from about $2 million in 1915 to $6 million in 1918, foreign sales comprising for a time a remarkably large part of the total. In 1915 and 1916, foreign sales were about 90 percent of the total; in 1917 and 1918, after the United States entered the war, these dropped to less than 25 percent.[103] Gyrocompass sales formed the single largest bloc in war years total, which included searchlights, aircraft instruments, fire-control equipment, marine stabilizers, and other products. By 1920 the company had equipped 43 U.S. and 66 British battleships, 126 American and 174 British submarines, and 89 U.S. de-

[102] EAS to Herbert Goodman, June 11, 1917 (SP).
[103] Information on sales, assets, and so on are compiled from financial statements and reports in the Sperry Papers.

Figure 8.17
Tribute from
the Sperry employees.

stroyers with compasses. In addition, more than 200 compasses were on the ships of other navies.[104] The total number of Sperry compass installations throughout the world was 726 in 1920. The compass made the company internationally known.

The assets and sales obviously displayed the strength and durability of the company; the spirit of the man was also discernible in the company. It was, like its founder, dedicated to invention and development, not to mass production and marketing. Throughout his career, Elmer Sperry had chosen to solve the difficult technological problems, and his company clearly showed the same preference. Sperry always cultivated the fine mechanical touch in himself and others, and his company manufactured fine precision instruments and other devices. He had a remarkably broad inventive genius ranging over mechanics, electricity, and chemistry; his company also worked in the broad spectrum. He had been drawn throughout his career to automatic controls, incorporating them in a wide variety of machines, some of great subtlety; the Sperry Gyroscope Company research and development engineers had already shown their ability and interest in this line of development—a line that would flourish within decades and cause some to speak of an era of automation. After founding the company in 1910, Elmer Sperry pioneered in providing instruments and controls for ships and airplanes, and the company would continue to lead in marine and aviation instrumentation and control. If society had been willing or able to expend its resources on instruments and controls for airplanes and ships engaged in peaceful pursuits, then the man and the company would have provided these. As it was, the superb inventive and engineering genius of the man and the company were applied to military purposes during the war; the transition to peace would never be quite complete. The Sperry company had become and would become again in time of war a "brainmill for the military." Yet the interwar years gave Elmer Sperry and his company the opportunity to follow, at least for a decade or so, the pursuits of peace.

[104] *The Sperryscope,* II (1920), pp. 14–15. The company monthly periodical, first published May 1919.

CHAPTER **IX** The Assumption of Leadership:
The Naval Consulting Board and the Aerial Torpedo

As a member of the naval consulting board since 1915, Mr. Sperry has rendered invaluable service as chairman of the committees on mines and torpedoes and aids to navigation and as a committee member on aeronautics, internal combustion engines, and special problems. His numerous inventions, including the gyrocompass, plane stabilizer, high intensity searchlight, and his many refinements on apparatus for accurately controlling the fire of our guns, have assisted materially in placing the navy in first class fighting trim. It is safe to say that no one American has contributed so much to our naval technical progress.

SECRETARY OF THE NAVY CHARLES FRANCIS ADAMS,
New York Times, JUNE 17, 1930

ELMER SPERRY, WHO HAD LONG ENJOYED THE RESPECT OF HIS PEERS AND WAS widely known among inventors, engineers, industrialists, and naval officers, rose after 1915 to a position of national prominence and of leadership in the world of technology. In 1915 Secretary of the Navy Josephus Daniels named him to the Naval Consulting Board, which brought Sperry and the other members wide publicity. The *New York Times* considered the board and its members front-page news, as did other newspapers and popular magazines. When the board had its first meeting in the fall, a notable photograph (see figure 9.1) showed the members—a distinguished group of inventors, engineers, scientists, and industrialists—gathered around the navy secretary and Thomas Edison, chairman of the board. Sperry sat on Daniels' left and Edison on his right.

Sperry later proved by his performance as a board member that he merited the prominence given him in this photograph. The subsequent wartime history of the board revealed that Sperry personified the most viable and effective characteristics and functions of the board. Much as the Sperry Gyroscope Company manifested the inventive and engineering characteristics of its founder, the Naval Consulting Board at its best institutionalized the relationship between civilian inventor and the navy that Sperry and the navy had informally established before the war. The board, however, made some false starts and wrong turns before it found how best to fulfill its goals; nevertheless, the experience made it clearer than ever that the military could no longer depend on random and heroic invention but had to look to professionals, like Sperry, who directed their research and development to fulfill the needs of the system.

Deepening American concern about the nation's naval defenses followed the outbreak of war in Europe in 1914. Many Americans reassured themselves that American inventive genius had contributed enormously to the art of naval warfare. The better informed, however, apprehensively considered a troublesome paradox: although the American civilian had invented remarkable devices, the American navy had left to other navies the development and integration of these into systems of naval warfare. The British navy had introduced the dreadnought-class battleship, the German developed submarine warfare, both the German and the British led in complex gunfire control, and the French most fully exploited the airplane for military purposes. Critics charged the American navy with following a "Chinese plan of copying," although defenders cited the support the navy had given Elmer Sperry as an exception to this rule.[1]

American naval officers also became concerned. Wartime secrecy restricted the diffusion of European ideas and made the U.S. navy more dependent upon its own extremely limited resources for research and development. When the *Lusitania* sank and the likelihood that the United States might be drawn into the war increased, this concern deepened. Aware of the problem and the danger, Secretary of the Navy Josephus Daniels announced a plan for cultivating American research and development through a Naval Consulting Board. This civilian board, founded in 1915, and the National Research Council, founded in 1916, became the major wartime efforts to organize America's research and development capability for military purposes.[2]

The sequence of closely linked events culminating in the establishment of the Naval Consulting Board began in May 1915, when Thomas Alva Edison granted an interview to Edward Marshall of the *New York Times*.[3] Edison was already a legendary American, for many believed him to be the sole inventor of the phonograph, the electric lamp, the motion picture, and countless other devices. Of simple stock, plain-spoken, practical-minded, ingenious, and fantastically successful, he symbolized for many the essence of the American dream. Able to express his opinions in bold, lucid concepts, and vividly personable, he was good copy for reporters, who, like the public, viewed him in his old age as a wise man and venerable prophet.

When Edison told Marshall of his "Plan for Preparedness," the *New York Times* spread the interview over two pages of its Sunday magazine section. The source and the subject explained the prominence. Edison had

[1] The quote is from Waldemar Kaempffert who also wrote, "Our navy is but a reproduction of the best to be found abroad," in "The Inventors' Board and the Navy," *American Review of Reviews*, LII (1915), p. 298.
[2] A. Hunter Dupree outlines the history of the NCB and the National Research Council in *Science in the Federal Government* (New York: Harper, 1964), pp. 305–15. Lloyd N. Scott, *Naval Consulting Board of the United States* (Washington, 1920), provides an official history.
[3] Edward Marshall, "Edison's Plan for Preparedness," *New York Times*, May 30, 1915, Section V, pp. 6–7.

First Meeting
of
Naval Consulting Board
and
Heads of Naval Bureaus
Oct. 6. 1915.

Figure 9.1

a plan for creating an invulnerable national defense without raising taxes or introducing militarism into America. Since the war threatened to be long and bloody and the danger of America's being drawn into the conflict was ever more likely, Edison's plan commanded great interest.

Although he had a plan for defense, Edison insisted that he was opposed to war. In an interview with a New York *World* reporter on the same day that the *Times* interview appeared, Edison, "laying a kindly hand upon the reporter's arm explained, 'you see, my boy, the dove is my emblem.' "[4] Protesting that he detested using technology for destruction and being unwilling even to discuss the devices he had in mind, he reluctantly faced the eventuality that, if America were drawn into the war, he would have to invent terrible devices for Americans to use against the forces of despotism. Although Edison feigned moral neutrality, he identified Germany as the despotic nation.

Edison's plan rested on his conviction that war should be mechanized like industry. He called for labor-saving devices for fighting similar to labor-saving devices for working. "Modern warfare," he said, "is more a matter of machines than of men. Most of the machines [of war] are simple matters if we compare them to the machines of industry." Because Americans were excellent mechanics, Edison seemed certain they would make excellent soldiers using the machines of war. His metaphor was the soldier "perspiring in the factory of death at the battle line."[5]

Edison did not favor a large standing army or navy. He suggested that most of the enlisted men be reservists on short tours of duty in peacetime or on emergency service in wartime. He preferred that they receive their basic mechanical training on regular industrial jobs which they would leave temporarily for military service; Edison believed that the men could learn more about machines from the engineers and foremen of industry than from the officers of the military. Officers should continue to be educated at West Point and Annapolis, but they "should be returned to civilian life . . . with annual periods of additional study. . . . They should not be taken permanently from productive and thrust into unproductive effort. . . . What we want," he said, "is a small army [or navy] trained to big knowledge."[6] He not only recommended the preeminence of reservists but he also foresaw preliminary tooling and mothball, or reserve, fleets. "He would," —the reporter found this his most unusual proposal—"build many aeroplanes and submarines, and . . . construct a fleet of cruisers, battleships and other vessels . . . to be kept in drydock, practically in storage . . . until needed."[7]

To insure that the American machines of war would be the most effective, Edison called for a great research laboratory in which the potential of military technology could be exploited "without any vast expense." He thought such a laboratory would give the American military the same advantages that private laboratories gave industry. He envisaged the labora-

[4] New York *World*, May 30, 1915.
[5] *New York Times*, October 16, 1915, p. 4.
[6] *Ibid.*, May 30, 1915, Section V, p. 6.
[7] *Ibid.*

tory doing the research, performing the experimentation, and building and testing the prototypes of armaments to be constructed in mass if America were drawn into the war. Furthermore, he advocated a selective inventory of the nation's manufacturing resources so that within months of the outbreak of war plans could be made for mass producing the prototypes. The laboratory, as time and further elaboration revealed, was his most controversial proposal.

Secretary Daniels responded enthusiastically to the Edison interview. Reflecting upon the transformation technology had wrought in his country, he thought that engineering and invention were the means by which America could counter the force of the military power of a European aggressor. Warm in his adulation of Edison, possessed of great faith in inventive Americans, politically sensitive to the issue of taxes, temperamentally opposed to militarism, and contending with preparedness advocates calling for a larger navy, Daniels found much in Edison's remarks with which he could thoroughly agree.[8] Within two weeks he had asked Edison to be the civilian adviser to the navy on problems of invention and development.[9] "You are recognized," the Secretary wrote, "as the one man who can turn dreams into realities." The dream was to originate within the navy "proper machinery and facilities for utilizing the natural inventive genius of Americans to meet the new conditions of warfare as shown abroad." Daniels knew that if Edison accepted, "the country would feel great relief in these trying times."

Edison responded with a cheerful "aye aye, sir."[10] With his new adviser, the secretary worked out his plan in detail. Early in July, Daniels announced the formation of a civilian advisory board for invention and development to be directed by Edison. The board would consist of "the nation's very greatest civilian experts in machines" and be expected to originate ideas and critically examine those submitted by inventive Americans. Recalling the contributions of civilians such as John Ericsson, Simon Lake, John Holland, and the Wright brothers to modern warfare, Daniels emphasized that his advisory board would help the navy exploit the American genius for mechanical invention. Although he viewed with concern the lack of a national laboratory to develop inventions, he was gratified to find how quickly the concept of the civilian consulting board captured the people's imagination: "The letters have poured in from every sort of people."[11]

At the same time that Daniels issued a call for the ingenious common

---

[8] Daniels had been absorbing most of the blows aimed at the Wilson administration by the preparedness groups. Joseph L. Morrison, *Josephus Daniels: The Small-d Democrat* (Chapel Hill: University of North Carolina Press, 1966), p. 70.
[9] Daniels to Edison, May 31, 1915, in the Josephus Daniels papers, the Library of Congress; hereafter referred to as Daniels Papers.
[10] Daniels at a meeting of the Naval Consulting Board, September 19, 1916, in digest of NCB minutes, Record Group 80, National Archives, Washington, D. C.; hereafter cited as Digest of Board Minutes.
[11] Edward Marshall, interview with Secretary Daniels, *New York Times*, August 8, 1915, magazine section. This article is the basis for the foregoing summary of Daniels' plan.

American to invent, he acknowledged that "sea fighting had developed into a highly specialized science" and that "one chemist, one electrician might be greater in the warfare of the future than Napoleon." The secretary apparently saw no inconsistency in his summons to the grass roots, to the "crank of today" who is the "genius of tomorrow," and his recognition of the growing role of the scientific specialist. Like so many Americans, Daniels could not abandon fond memories of Yankee inventors, even while he recognized the rise of science-based technology cultivated by specialists.

To draw out the "natural inventiveness and genius of Americans," Daniels and Edison had first to name the other members to the civilian advisory board. The newspapers, greatly interested in the plan, speculated about the membership. It was important, the *New York Times* believed, to choose famous men who would inspire the confidence of the American people in the ability of the navy to spend wisely the increased appropriations that would result from Wilson's new preparedness program.[12] Among the inventors and engineers thought likely candidates were Orville Wright, who prematurely, as it turned out, indicated his willingness to serve; Charles Proteus Steinmetz, then an engineer for General Electric; Hudson Maxim; Alexander Graham Bell; Professor R. A. Fessenden, known for his development of wireless; Peter Cooper Hewitt; Nikola Tesla; Henry Ford, whose secretary also said he would serve if asked; Simon Lake; and John Hays Hammond, Jr., wealthy young wireless inventor and entrepreneur.[13] Edison's secretary predicted that the board would "far outrank any body of scientists and experts ever gathered together." Sperry, whose presence on the board was undoubtedly desired by such knowledgeable officers as Admiral D. W. Taylor, was not among those named by the daily press.

Neither the *New York Times* nor Edison's secretary proved accurate. Inventors and engineers of great public renown and rank were notably absent on the board. Instead of choosing the candidates himself, Daniels, so advised by Edison, turned to the professionals to choose the members.[14] Daniels requested that each of eleven professional societies, selected by Edison, name two members. Daniels declared that by this means he would enlist the support of all of the 36,000 engineers, inventors, and scientists in the societies. He asked the president of each society to determine the two representatives "by a poll, by letter of your members, or in whatever ways seem to you most certain of securing the men desired by the majority of your organization."[15]

The societies choosing members of the advisory board and the men chosen were: Matthew Bacon Sellers and Hudson Maxim (Aeronautical Society); Leo Hendrik Baekeland and Willis R. Whitney (American Chemical Society); Benjamin G. Lamme and Frank Julian Sprague (American Institute of Electrical Engineers); Arthur G. Webster and Robert S. Woodward

[12] *New York Times,* July 14, 1915, p. 1.
[13] For biographical sketches of the members of the board, see the *Scientific American,* CXIII (1915), 301ff and 326ff.
[14] Daniels to Edison, December 20, 1916. Daniels Papers.
[15] Letter from Daniels to Benjamin B. Thayer, President of the American Institute of Mining Engineers, July 19, 1915. Daniels Papers.

(American Mathematical Society); William L. Saunders and Benjamin B. Thayer (American Institute of Mining Engineers); Elmer A. Sperry and Henry A. Wise Wood (American Society of Aeronautical Engineers); Howard E. Coffin and Andrew L. Riker (Society of Automobile Engineers); Peter Cooper Hewitt and Thomas Robins (Inventors Guild); A. M. Hunt and Alfred Craven (American Society of Civil Engineers); Lawrence Addicks and Professor Joseph Richards (American Electrochemical Society); and W. L. R. Emmet and Spencer Miller (American Society of Mechanical Engineers). Edison had a special status as the appointee of Daniels, as did M. R. Hutchinson, Edison's chief engineer, who became his personal representative on the board.

The choice of Wood and Sperry generated some criticism because Orville Wright and Glenn Curtiss, far better known to press and public, were not chosen. The Society of Aeronautical Engineers, which named Wood and Sperry, had been newly founded at the suggestion of Edison to select members from the aviation field. The older nonprofessional Aero Club of America had been instrumental in forming the new society, and Wood had been its vice president. Over 80 per cent of the two hundred members of the Society of Aeronautical Engineers voted for Wood and Sperry. In answer to the criticism, Wood pointed out that he and Sperry had named an advisory committee to aid them in their board work and that Wright and Glenn Curtiss were among the members.[16]

In all but three cases, selection of the society representatives was made by the board of directors, the council, or the president of the society. The characteristics of the men chosen suggest that the professionals in America were taking a sophisticated approach to cultivating technology.[17] The *New York Times* decided, "The twenty-three members were chosen for fitness rather than notoriety . . . but in their respective fields they have high standing."[18] Elmer Sperry believed that the members had extensive practical experience and that they knew not only "what to do, but what is also sometimes of very great importance, . . . what not to do."[19] All of the members of the board, except Sperry, Wood, Maxim, and Edison, had college degrees. Although most held only an undergraduate degree in engineering, several had done advanced work abroad. Two members headed major industrial laboratories; nine of the members were company presidents; and more than half were officers in the societies they represented. Also notable was the presence of four naval academy graduates. Members from the scientific or academic communities were Willis R. Whitney, a Massachusetts Institute of Technology professor of chemistry, who had become head of General Electric's research laboratory; Professor Joseph Richards of Lehigh University, who had written a standard work on aluminum metallurgy, a fairly

16 *New York Times,* August 9, p. 7, and August 27, 1915, p. 9.

17 Dupree, *Science in the Federal Government,* pp. 306, 309, writes that the NCB represented only inventors and the engineering societies. He also believes that the screening of the inventions of the public was the major activity of the board; that was not quite the case.

18 *New York Times,* editorial, September 13, 1915, p. 8.

19 EAS to editor of the New York *World,* September 14, 1915.

new field attracting wide interest; Robert Woodward, an authority on mathematical physics and President of the Carnegie Institute; and Professor Arthur Webster of the physics department of Clark University, one of the country's leading science centers. Though best known as an inventor, Baekeland had earlier been a professor at the University of Ghent and at the higher Normal School of Bruges.

At its early meetings, the board established committees to cultivate invention and development in the areas where the navy thought its major technological problems and opportunities lay.[20] Prime problems included a defense against the submarine menace and the design and manufacture of good airplane and submarine engines. At the time America imported the most satisfactory engines.[21] Attention was thus focused on what were already proving in 1915 to be major military innovations of the war, the airplane and the submarine. Sperry was chosen to chair two of the sixteen committees formed to attack the problems; Frank Sprague was the only other member given so much responsibility. Sperry presided over the committees on mines and torpedoes and on aids to navigation. In addition, he sat on the committee for aeronautics and the committee for internal combustion. Later, in 1917, he also joined a special problems committee formed to meet the submarine menace. As chairman of the aids to navigation committee, he headed a group that included Craven, Hunt, Wood, and Woodward; as head of torpedoes and mines, he chaired a group with Baekeland, Maxim, and Hutchinson. Later Sperry also took over the chairmanship of the aeronautics committee. Some of the other board committees were chemistry and physics, wireless and communications, organization, manufacture and standardization, and submarines.

ii

On the second day of meeting, October 7, the board made a fateful decision: It unanimously resolved to recommend the establishment of a "research and development" laboratory.[22] The idea of a naval laboratory appealed because it was thought analogous to the highly successful industrial laboratories. The proponents of the naval laboratory predicted successes similar to those of the General Electric laboratory, where "in the last five years more improvements have been made in electric incandescent lighting than in the previous twenty years," and to those of the industrial chemists, who had created "our magnificent cottonseed and corn products

[20] The problems were summarized for the public in another Edward Marshall interview for the *New York Times*, "Great Experts Discuss the Navy's Needs: A Real Conversation Between the Secretary of the Navy Josephus Daniels, Admiral Taylor, and Thomas A. Edison on How to Meet Some of the Great Problems of the Day, Especially Arranged and Stenographically Reported for the *New York Times*," *New York Times*, October 24, 1915, section IV, p. 1. Marshall called his interview one "of the most remarkable conversations ever occurring anywhere in the world."

[21] Admiral David W. Taylor, Chief of the Bureau of Construction and Repair, quoted in the Marshall interview, *New York Times*, October 24.

[22] Digest of Board Minutes, October 7, 1915.

industries. . . . What may not be expected," the editor of the *Scientific American* asked, "of a Government laboratory in which inventions and discoveries are to be developed for the benefit of the Army and Navy."[23]

Many of the board members, like Sperry, had had experience with industrial laboratories, and they referred to this in their campaign to win congressional support for the proposal. An automobile firm, according to one member, found it advantageous to spend half-a-million dollars annually on testing, research, and experimentation ("How much more important is the business of the United States").[24] Another member stated that even conservative industrial enterprises found it necessary and profitable to spend at least two to five percent of sales on research and experiments. Since the navy contemplated total expenditures of $500 million in five years, logic and analogy warranted at least $5 million for research. Edison was particularly eloquent in drawing analogies with his own laboratory experiences, and initially, the board adopted his plan for a $5 million laboratory. Edison's concept was summarized in a board memorandum:

> Mr. Edison, at his own personal cost, submitted plans of buildings in great detail which if constructed would be of concrete and like a modern manufacturing building of plain construction. The plans submitted showed in detail whereby the various units of machinery for warfare might be produced and standardized, such as aeroplanes, range finders, submarine engines, small guns, and everything relating to war machinery. By these means each unit could be perfected, the gauges produced, and arrangements made with various shops throughout the country, so that on telegraphic notice the various units could be manufactured. It is not intended that a great number of any unit should be manufactured at any time, but that the gauges, etc., be so perfected that the potentiality of having the units made at short notice will be provided for. The laboratory will be equipped with many different tools, but they will be of standard make and capable of making almost any motion. As soon as any unit of war machinery is perfected by this laboratory, specifications can be carefully drawn so that the commercial enterprises of the country can and will know definitely that the work can be done.[25]

Instead of a $5 million laboratory, the House in August 1916, appropriated only $2 million for construction and a half-million for the initial year of operation. Edison may have been affronted by congressional failure to accept his recommendation in full, for he had personally testified on its behalf before the House Committee on Naval Affairs. Other members of the board rallied and came back with a less elaborate plan for a laboratory. Although he supported Edison's bold plan, Elmer Sperry held somewhat different ideas. Sperry circulated among board members a letter from his young protégé, Reginald Gillmor, who was then in London. Gillmor had observed the British effort to organize invention and engineering for the military and had drawn conclusions with which Sperry agreed. Both Sperry

[23] Editorial, *Scientific American*, CXIII (1915), p. 90.
[24] L. H. Baekeland, "The Naval Consulting Board of the United States," *Metallurgical & Chemical Engineering*, December 15, 1915.
[25] Memorandum on the "Experimental Laboratory," prepared by board secretary Thomas Robins (SP).

and Gillmor preferred that the naval laboratory limit itself to basic research and to experimental testing and quality control; they wanted the navy to rely upon private industry for major development projects. Gillmor thought the experience of the British and the Germans had shown the superiority of private industry in large-scale development, citing as an example the superiority of the aircraft engine developed by the Mercedes Motor Company in Germany to the engine developed by the Royal Aircraft Factory in Britain. Mercedes, however, had a "million dollar a year" subsidy.[26]

With Edison's plan frustrated, the board established a special committee to draw up a new proposal. Noting that the original conception was a laboratory in which not only research work but also rapid, heavy construction of all kinds could be carried on, the committee recommended, "because of this great and regrettable reduction in the appropriation," giving up the heavy work and concentrating on light work based upon research and experiment. The laboratory would no longer have a master machine shop, foundry, or forge where, for example, full-scale prototypes of submarines and large guns could be built (Edison's bold concept), but would build only small-size experimental models of equipment. The committee further recommended, again in contrast to Edison's plan, that the laboratory be under a high-ranking officer to whom the navy's bureau chiefs and the board pass ideas and inventions for development. The committee recommended as well that the laboratory be located in Annapolis, on government land at the site of the existing small experimental station operated by the Bureau of Steam Engineering.[27]

Thomas A. Edison, Lawrence Addicks, L. H. Baekeland, Thomas Robins, Frank J. Sprague, and Willis R. Whitney comprised the committee that submitted the report. There was, however, a minority report from Edison. He recommended Sandy Hook, New Jersey, rather than Annapolis, as the site, but the site was only the surface issue of his disagreement. He continued to insist upon the laboratory originally proposed. His minority report called for a laboratory run as a "works" for the rapid construction and testing of experimental machines and devices—a laboratory that depended upon accumulated science. "Research work in every branch of science and industry, costing countless millions of dollars . . . has been on for many years. . . . Only a ridiculously small percentage has yet been applied. . . . It is therefore useless to go on piling up more data. . . ." He also wanted a civilian to manage the proposed laboratory.[28]

The Naval Consulting Board rejected Edison's proposal, and on December 9, 1916, it resolved "by a large majority" to accept the majority report with revision of phraseology.[29] On December 22, Edison wrote to Daniels to inform him that he would not under any circumstances be connected with the proposed Annapolis laboratory. "It is fixed in my mind, whether right or wrong, that the public would look to me to make the Lab-

[26] Gillmor to EAS, November 19, 1917 (SP).

[27] This laboratory concentrated on the photomicroscopic analysis of steel, the quality of fuels, and similar quality control problems.

[28] The majority and minority reports are reprinted in Scott, *Consulting Board,* pp. 225–32.

[29] Robins, Secretary of the Board, to Daniels, December 14, 1916 (SP).

oratory a success, and that I would have to do 90% of the work. Therefore, if I cannot obtain proper conditions to make it a success, I would not undertake it nor be connected with it in the remotest degree. . . ."[30]

A stalemate developed and persisted. Secretary Daniels, although desirous of approving the majority report, would not act without Edison's support and Edison was adamant.[31] Sperry suggested as a compromise building the laboratory in the New York navy yard.[32] The board named, on December 8, 1917, a committee of three to seek unanimity on the site of the proposed laboratory and two weeks later appointed another committee to attempt to persuade Edison to support the board's recommendation.[33] Edison, however, was obdurate. Sperry wrote to Edison, whom he greatly admired, that he regretted exceedingly that the board did not think it wise to follow his recommendations exactly, but "I presume we have to make certain compromises, and if they are wisely made sometimes they are for the best."[34] Sperry assured Edison that despite the board's failure to support him he had been an inspiration to all of them by his devotion to the cause; "the success or failure of this laboratory . . . depends upon you. . . . Its popularity with Congress and with the people . . . rests on you and your name. . . . It is contrary to my habit to offer gratuitous advice, but my great interest in the laboratory and my belief in the great good it may be able to do for the service makes me depart from my custom in this instance."[35] Sperry reasoned that a laboratory in New York could supplement its facilities by using the mechanics, the shops, and dry dock of the New York navy yard, as well as the large ships berthed there, thus answering the need Edison saw for large-scale development work and sea trials. He noted, furthermore, that New York, center of the country's engineering talent and activities, could be drawn upon. After discussing the matter with a close associate of Edison's, whom he did not name, Sperry was so confident that Edison would accept the compromise that he urged the Secretary of the Navy to propose the New York site to Edison.[36]

Edison did not compromise despite the efforts of Sperry and others, and Daniels would not override him. In July 1918, Daniels told the board that he still had not made up his mind about the laboratory's location; he also told them that when built it would be strictly "along the lines first originated."[37] Daniels was not prepared to abandon Edison, whom he had tried to persuade in March to agree to an experimental and research laboratory in Washington "now" and the related manufacturing establishment near New York "possibly" after the war. Daniels wrote that he needed an agreement because members of the board were pressing him to do something.[38]

The war ended, however, before anything was done. Edison remained

30 Edison to Daniels, December 22, 1916. Daniels Papers.
31 *Ibid.*
32 EAS to Edison, January 24, 1917 (SP).
33 Digest of Board Minutes, December 8 and December 22, 1917.
34 EAS to Edison, December 19, 1916 (SP).
35 EAS to Edison, January 24, 1917 (SP).
36 EAS to Daniels, January 29, 1917 (SP).
37 EAS to Daniels, July 27, 1918 (SP).
38 Daniels to Edison, March 1, 1918. Daniels Papers.

as implacable as he had been in 1916. A year after the war ended, he told Daniels his mind had not changed in the least about the location of the laboratory nor his opinion that the laboratory should not be under the control of naval officers "either directly or indirectly." Edison confided to the navy secretary that he did not believe that more than one creative mind was produced at Annapolis in three years, and under the system that one could not possibly exercise its gift. "When you are no longer Secretary," he wrote, "I want to tell you a lot of things about the Navy that you are unaware of."[39]

Two years after the war, when proponents of the laboratory were still seeking to have it built in Washington, Edison wrote to Daniels, "For your own reputation, you should not permit any money to be expended on the laboratory as proposed. . . . While the Naval Consulting Board may have approved something I know nothing about, I have never approved either the design, the location, or the method of administration. . . . I am going to fight this in Congress if necessary."[40] Daniels, however, awarded the laboratory contract, still hoping—although futilely—that Edison would cooperate in working out a plan for organization and management.[41]

iii

Ground was finally broken for the laboratory in December 1920, at the site of the Bellevue Magazine in Washington, and operations were begun in 1923.[42] During the war years, however, the consulting board had to fulfill its functions without the laboratory. One function, especially dear to Daniels, was to solicit and evaluate inventions from amateur inventors. Unfortunately for the reputation of the board, which had already been hurt by its failure to decide on the laboratory, the screening function took an unreasonable amount of time and produced insignificant results. Responding to Secretary Daniels' call, Americans swamped the board with suggestions. Despite the assistance of a preliminary screening group provided by the navy, board members who were committee chairmen found the evaluation of inventions excessively time consuming. Elmer Sperry was "literally" working in every spare minute to examine inventions during the month after the United States entered the war. The board secretary feared that Sperry, who was almost sixty, might break down under such a load. Robins also singled out Maxim and Whitney as men strained by the responsibility.[43] The work was not fruitful. Of the 110,000 "inventions" sent to the board,

[39] Edison to Daniels, November 7, 1919. Daniels Papers.
[40] Edison to Daniels, November 8, 1920. Daniels Papers.
[41] Daniels to Edison, November 19, 1920. Daniels Papers.
[42] Edison did not attend the commissioning, to which he was invited, and never visited the laboratory. Albert Hoyt Taylor, *The First Twenty-Five Years of the Naval Research Laboratory* (Washington, D.C.: Navy Department, 1948), p. 4. Taylor writes that "due to the unusual situation arising from World War I the building of the Laboratory had not been started before the end of that conflict," p. 2. Dr. Hoyt, author of this 75-page commemorative book, joined the laboratory in 1923.
[43] Robins to Saunders, June 29, 1917 (SP).

only 110 were deemed worthy of further development by board committees; of these only one reached production before the war ended.[44]

The board found that the major reason the grass roots could not solve the navy's technological problems was that the common man did not have detailed information defining the navy's problems and was not familiar with the state of the art to be used in solving problems he did not fully understand. Most of the suggestions were either old and already adopted, old and discarded, or could not be worked into the navy system. Some were outrageous, violating, for example, "natural laws as known."[45] Members of the board and the navy learned much about invention and inventors as a result of this effort "to win the war by American inventive genius." If they had not previously appreciated the navy's need for inventors who were both professionals and specialists, they must have after evaluating the deluge of ideas from the public. It was clear that the navy needed professional inventors and specialists acquainted with the needs of naval systems and the state of the art. Hudson Maxim summed it up when he wrote to the board secretary: "This is an age of specialists, and it is necessary for any inventor, scientist, or engineer to devote a large amount of time and attention to the special requirements of naval and military matters before he can qualify himself to be of much use. . . ."[46] He singled out Sperry as an inventor who had developed through experience a special competence in the naval field. Maxim's conclusion was shared by the official historian of the board, a naval officer who had seen the many inventive ideas solicited by the appeal to the grass roots. He wrote that "only those men on the frontiers of scientific development and on the frontiers of development of the various arts concerned in naval warfare would be at all likely to discover anything that was new or could be used by the Navy Department."[47]

iv

Efforts of the board to establish the laboratory and to cultivate American inventive genius were frustrating but well-publicized endeavors. With less publicity, members individually carried out projects with the support of the board. Those who had executive and managerial experience with industrial corporations worked on an industrial preparedness campaign that produced an inventory of the nation's manufacturing resources needed in war. Members with industrial experience also helped establish contacts between the navy and corporations able to solve various technological and supply problems. Inventors and engineers such as Edison, Elmer Sperry,

44 Scott, *Consulting Board*, p. 125. The one innovation from the suggestions was the "orientator," invented by W. Guy Ruggles. This device placed a training airplane pilot in a section of fuselage mounted so that it could move freely in gimbals and thus simulate various flying sensations.
45 The inadequacy of the suggestions was discussed in C. H. Claudy, "Ideas That Will Not Work," *Scientific American*, CXVII (1917), p. 226; and Scott, *Consulting Board*, p. 131. The 17 reasons usually given by the screening committee of the board for rejecting suggestions are listed in its form letter sent to the inventors.
46 Maxim to Robins, November 16, 1916 (SP).
47 Scott, *Consulting Board*, p. 30.

Hudson Maxim, and Frank Sprague worked on research and development projects sponsored by the board but, in the absence of a laboratory, carried out in industrial laboratories and other private facilities.[48]

As chairman of three committees, Sperry received problems from the navy and screened related inventive ideas from the public and elsewhere. If an idea promised the solution of a pressing problem, Sperry, with the advice of his committees, recommended to the Secretary of the Navy the authorization and funding of a research project to develop the idea; a committee member often took charge of the project. On occasion, the committee recommended that Sperry and the Sperry Gyroscope Company carry out a project; conflict of interest was not, it seems, a problem.

In the official history of the consulting board, the contributions of Sperry, through the development projects he headed, are given a prominence second only to those of Edison.[49] As chairman of mines and torpedoes, Sperry helped the navy obtain a much improved steering gyro for its torpedoes and took responsibility for providing the first 10,000 depth charges made in the United States—the project for which Elmer Sperry, Jr., was the field engineer.[50] The development of aviation instruments by Elmer Sperry, Jr., and Lawrence Sperry had the support of his committee on aids to navigation. The board also supported the development by Sperry and his engineers of a bombsight and of an aircraft machine gun sight that automatically provided range and deflection for a moving target. As chairman of the committee on aeronautics, Sperry cooperated with one of his committee members, Matthew Sellers, to investigate and encourage the development of a helicopter for the navy.[51] These are only a few of the Sperry projects initiated or sponsored by the board.

For the special problems committee, established in 1917 to coordinate research and development undertaken to combat the submarine, Sperry tried to develop an underwater searchlight. Judging by the enthusiasm of persons associated with the project, the chances of success were thought good, but after some expensive experimentation, the Sperry group found that despite the enormous power of a Sperry searchlight, the beam could not be projected through water because of the presence of animalcula that exist in water to the depth of over 200 feet.[52] For the committee on special problems, Sperry also experimented with the Elia submarine net, a device that particularly interested him. In 1917 ensnaring or detecting submarines by net in restricted waters was thought promising. Although Sperry had no experience with submarine nets, he became enthusiastic about a system proposed by Commander G. E. Elia of the Italian navy. Commander Elia

[48] For the contributions of the industrial members see *ibid.*, pp. 160–223.
[49] Scott, *Consulting Board;* chapters on the inventive accomplishments of members give Edison thirty-two pages, Sperry five, and other members two or three, at most.
[50] "Report of the chairman of the committee on mines and torpedoes," February 7 and April 1, 1916 (typescript, SP).
[51] On the helicopter, see Thomas Robins to EAS, July 23, 1917; Elihu Thomson to EAS, August 1, 1917; Sellers to EAS, September 14, 1917; and resolution of the NCB, October 6, 1917 (SP).
[52] EAS to Bureau of Construction and Repair, November 7, 1924 (SP).

came to America on the recommendation of the American embassy to aid in an elaborate scheme involving submarine nets, radio-transmitter buoys, patrol boats, and depth charges.[53] The plan called for laying steel nets across the North Sea, the Straits of Gibralter, and similar waters. If a submarine struck a portion of the net, a nearby buoy would be released, rise to the surface, and begin transmission of radio signals. These would be heard by patrol boats that would rush to the area to depth-bomb the submarine.

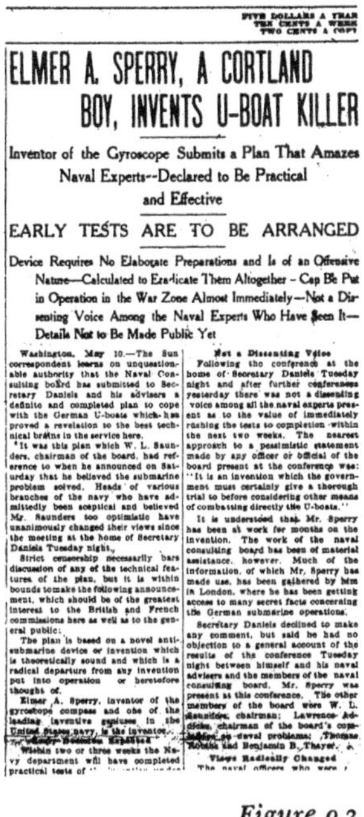

FIVE DOLLARS A YEAR
TEN CENTS A WEEK
TWO CENTS A COPY

ELMER A. SPERRY, A CORTLAND BOY, INVENTS U-BOAT KILLER

Inventor of the Gyroscope Submits a Plan That Amazes Naval Experts--Declared to Be Practical and Effective

EARLY TESTS ARE TO BE ARRANGED

Device Requires No Elaborate Preparations and Is of an Offensive Nature—Calculated to Eradicate Them Altogether – Can Be Put in Operation in the War Zone Almost Immediately—Not a Dissenting Voice Among the Naval Experts Who Have Seen It—Details Not to Be Made Public Yet

Washington, May 10.—The Sun correspondent learns on unquestionable authority that the Naval Consulting board has submitted to Secretary Daniels and his advisers a definite and completed plan to cope with the German U-boats which has proved a revelation to the best technical brains in the service here.

"It was this plan which W. L. Saunders, chairman of the board, had reference to when he announced on Saturday that he believed the submarine problem solved. Heads of various branches of the navy who have admittedly been sceptical and believed Mr. Saunders too optimistic have unanimously changed their views since the meeting at the home of Secretary Daniels Tuesday night.

Strict censorship necessarily bars discussion of any of the technical features of the plan, but it is within bounds to make the following announcement, which should be of the greatest interest to the British and French commissions here as well as to the general public:

The plan is based on a novel anti-submarine device or invention which is theoretically sound and which is a radical departure from any invention put into operation or heretofore thought of.

Elmer A. Sperry, inventor of the gyroscope compass and one of the leading inventive geniuses in the United States navy, is the inventor.

Within two or three weeks the Navy department will have completed practical tests of ...

Not a Dissenting Voice

Following the conference at the home of Secretary Daniels Tuesday night and after further conferences yesterday there was not a dissenting voice among all the naval experts present as to the value of immediately rushing the tests to completion within the next two weeks. The nearest approach to a pessimistic statement made by any officer or official of the board present at the conference was: "It is an invention which the government must certainly give a thorough trial to before considering other means of combatting directly the U-boats."

It is understood that Mr. Sperry has been at work for months on the invention. The work of the naval consulting board has been of material assistance, however. Much of the information, of which Mr. Sperry has made use, has been gathered by him in London, where he has been getting access to many secret facts concerning the German submarine operations.

Secretary Daniels declined to make any comment, but said he had no objection to a general account of the results of the conference Tuesday night between himself and his naval advisers and the members of the naval consulting board. Mr. Sperry was present at this conference. The other members of the board were W. L. Saunders, chairman; Lawrence Addicks, chairman of the board's committee on naval problems; Thomas Robins and Benjamin B. Thayer.

Views Radically Changed

The naval officers who were ...

Figure 9.2
From the Cortland Standard, *May 10, 1917*

A combination of circumstances involving the Elia net and Sperry caused a flurry of excitement in the national press. The net was ready for testing about the time America entered the war and both Americans and Europeans were calling for American inventive genius to defeat the submarine. Stimulated by his enthusiasm for the net—and for another of his projects, the aerial torpedo—Sperry decided, under the pressure of these demands upon American inventiveness, that the net and his aerial torpedo were the solution to the submarine problem. William Saunders, Edison's successor as chairman of the board, persuaded by Sperry's enthusiasm and influenced by the desire both to prove the effectiveness of the board and to raise public morale, wired Secretary Daniels on May 1:

Of the many inventions considered by the Naval Consulting Board two stand out prominently. I am satisfied of the practicability. Great importance and urgency for immediate action in putting into effect abroad these two inventions. Sperry and I are familiar with them and endorse them. We are ready to act. Will you not arrange for us to meet either the Council of National Defense or some instrument of government where we may appear immediately with full particulars and plans.[54]

Saunders' wire referred to the net and the aerial torpedo. He followed the telegram with several well-publicized press announcements. The *New York Times* gave him front page headlines on May 5, when he said that the Naval Consulting Board had the solution for the submarine problem. American inventive genius, Saunders boasted, had responded to the crisis; a special committee of the board had contrived from the antisubmarine devices of American inventors a master antisubmarine plan (presumably, the aerial torpedo and the Elia-Sperry net). Secretary Daniels only expressed faith in American inventiveness and assured the press that Saunders' optimism was not without foundation. Frank Sprague, however, flatly asserted that the board had no solution to the submarine problem and that possibly Saunders had been misunderstood. Other board members agreed with Sprague, although Sperry said that his position was very close to that of Saunders. On May 9, a reporter for the New York *Sun* wrote from Washington that Sperry was the inventor of the new and secret device that would defeat the submarine. By May 11, there were more denials. Robins, secretary of the board, told reporters that Sperry had not invented such a device. He

[53] EAS to Rear Admiral A. W. Grant, May 18, 1917 (SP).
[54] Sperry identifies the net as one of two inventions; EAS to Elia, May 2, 1914 (SP).

attributed the *Sun* story to the fact that Sperry had been in Washington taking part in testing an antisubmarine device of another inventor. Sperry then also denied the *Sun* story, and the flurry of excitement—and confusion—subsided.

The subsequent history of the Elia net was comparatively dull. The Bureau of Ordnance authorized testing the Elia system with two navy submarines and navy ships. The Sperry company constructed the radio signalling buoys, and Roebling & Sons produced the net. The consulting board closely monitored tests that Sperry engineers helped stage in June. The first reports were not unfavorable, for the submarines did not detect the nets and were snared by them. The wireless signal buoys, however, became entangled in the nets and failed to signal. Commander Elia and Sperry looked forward to modifications and to more extensive tests.[55]

A subsequent report was unfavorable. According to it, all of the elements of the Elia net system were already known by the Allies and had been employed—with the exception of the radio buoy, which did not function anyway.[56] The report doubted the value of the radio signal even if it did function because its only advantage over a visual signal (flare) was that the patrol boats could be stationed at greater intervals; at such intervals, however, the boats could not reach the detected submarine before it escaped. After Sperry sent an $8,500 bill to the Bureau of Ordnance for the tests, its head expressed surprise that costs were three times higher than predicted and curtly stated that there would be no further tests.[57] Sperry wrote to explain that the increased costs had come from overtime made necessary when foundry subcontractors had not filled specifications.[58] He noted that Sperry engineers did not charge for their time and several had worked thirty-six hours at a stretch to meet deadlines.

The other device about which Saunders wired the secretary proved of far greater significance. The Naval Consulting Board on April 14, a week after America had entered the war, advised Daniels to support the development of the Sperry aerial torpedo. Saunders' wire of May 1 occasioned a meeting of navy bureau chiefs and consulting board members on May 8 at Daniels' Wyoming Avenue residence. Sperry predicted that, used against submarines, the aerial torpedo could clear the shipping lanes. He also said that launched from a distance of 100 miles it could destroy much of a city such as New York, heavily damage a fortification, and "turn a small town or a large munition plant practically upside down."[59] The group, which included Admiral Earle of the Bureau of Ordnance and Admiral Taylor, formed a special navy committee to report to Daniels on the aerial torpedo. The committee reported favorably in less than two weeks, and the navy allocated $200,000 for the project, half of which came from the Bureau of Ordnance, one-quarter from the Bureau of Construction and Repair,

[55] EAS to Rear Admiral A. W. Grant, June 21, 1917 (SP).
[56] Copy of the Force Commander's comments is in the Sperry Papers.
[57] Admiral Earle to EAS, July 23, 1914 (SP).
[58] Telegram, EAS to Earle, July 24, 1914 (SP).
[59] EAS to Naval Consulting Board, April 11, 1917 (SP); the quotation is from Josephus Daniels, *The Cabinet Diaries* (Lincoln, Nebraska: University of Nebraska Press, 1963), p. 149.

and one-quarter from the Bureau of Steam Engineering. Such were the origins of a research and development project designed to produce a weapon in World War I that anticipated the notorious German V–1 flying bombs of World War II.

The project called for the Sperry Gyroscope Company to begin immediately to construct six sets of automatic controls that would transform a Curtiss flying boat and five Curtiss N–9's into automatically piloted torpedoes.[60] While the controls were being tested, Sperry and Curtiss were to design an inexpensive airplane especially for use as an aerial torpedo. Ten of these airplanes, named the "flying bomb" (FB), were then to be outfitted with improved automatic controls. If tests then proved successful, the special Curtiss FB and Sperry automatic pilot were to be mass produced. It was hoped that production would begin by the following spring. The tests would first have to demonstrate that the aerial torpedo could take off automatically, climb to a pre-set altitude, assume level flight, maintain a stable flight path to a target 50 to 100 miles distant, and then dive, carrying a thousand pounds or so of explosives, into the target. During the tests and development with the N–9's and the flying boat, a pilot would monitor the automatic controls and return the airplane to base. Sand bags or flares would be released when the plane's automatic controls signalled the descent to the target. After the FB was ready, it would be tested without a pilot.

v

Elmer Sperry's enthusiastic advocacy of the aerial torpedo and his prediction of successful innovation within a year can be understood only in light of the advances that he and his staff had made in automatic flight controls between 1914 and 1917. These resulted from efforts to improve the 1914 Bezons stabilizer, the attempt to use a Sperry stabilizer for aerial bombing in England, and experiments in 1915 and 1916 to develop a pilotless airplane. Immediately after Bezons, the Sperrys were determined to improve the airplane stabilizer, despite its prize-winning performance. The new design, ready for production by the end of 1914 and usually referred to as the 1915 model, was often tested for the company by Lawrence in a Curtiss flying boat.[61]

[60] EAS to Chief of Bureau of Operations, May 19, 1914 (SP).
[61] Lawrence, when he became bored by office routine, often flew the Curtiss flying boat equipped with the stabilizer on unscheduled test flights. Employees at the Sperry building were not surprised to see Lawrence, a few hours after his departure, flying along the East River and under the Manhattan bridges. One flight brought considerable newspaper publicity, for he chose to take as a passenger one of New York's most glamorous young society matrons, known among the columnists as the "blue streak" because of her spirited driving through Manhattan in a brightly colored automobile. Flying over Long Island Sound, on that occasion, Lawrence, who never lost the opportunity to demonstrate the dramatic uses of technology, activated the stabilizer and showed the young lady how the invention could facilitate affectionate caresses. Unfortunately, the machine malfunctioned, and the plane plunged into the bay, seriously injuring his companion. Escapades such as these, in conjunction with substantial achievements, led Grover Loening, another young aviation pioneer, to call his friend Lawrence, "a real genius, a terribly hard worker, and equally strenuous in his leisure." Grover Cleveland Loening, *Our Wings Grew Faster* (Garden City, N.Y.: Doubleday, Doran, 1935), p. 93.

Figure 9.3
Sperry gyrostabilizer
and other instruments
installed on a Curtiss flying
boat, February 1915,
at the Brooklyn Navy yard.
The stabilizer
is in the low foreground,
a magnetic compass appears
above it, a searchlight is to
the right and slightly below
a drift indicator,
and a wind-driven
servomotor is mounted
below the engine. The
smaller propeller-driven
instrument to the left
and below the servomotor
is an anemometer.

Figure 9.4
Lawrence Sperry testing
the aerial stabilizer
on the East River
near the Brooklyn Navy yard

The navy also tested the 1915 stabilizer. Between February 15, and March 27, 1915, Lt. (j.g.) R. C. Saufley made 14 flight tests of the device installed on a Curtiss model K flying boat. Perhaps he, like Bellinger, had an aviator's distrust of mechanical pilots, for he found the new airplane well-balanced (stable) but the stabilizer "too delicate" and continually "fighting" the controls. Instead of using the gyro to stabilize the plane, he thought that the gyroscope should be used to stabilize a bombsight (he had learned of Sperry's work on a stabilized bombsight while visiting the factory). In the endorsement to Saufley's report, another officer wrote, "As an example of engineering skill and ingenuity this device is remarkable. . . . Like the majority of aeronautical inventions [however] it is based on principles which are either unnecessary from a practical standpoint, or which neglect the fact that the plane of the horizontal is not the reference for safety in flight, except in a steady horizontal wind."[62] Naval constructor Richardson, who had written a sanguine if critical report on the stabilizer in the fall of 1913, now observed that the device knew nothing about flying and its control required as much skill as the direct control of the plane. Concluding that the defects outweighed the advantages, he recommended indefinite postponement of future tests.[63]

The setback at home was balanced by successes abroad that not only furthered the development of the aerial stabilizer but also led to the aerial torpedo. Sperry's staff of engineers had grown with the outbreak of the war in Europe; two new engineers, Morris M. Titterington and Mortimer F. Bates, were sent abroad to promote the sale of the airplane stabilizer. Titterington concentrated on England and Bates on Italy, both hoping to install stabilizers on giant Curtiss flying boats sold to the Allies. Prospects were so bright that Elmer Sperry decided to send Lawrence to Europe in September 1915, to observe and to help with engineering and sales.

In England Lawrence learned of the effort to use a stabilizer in aerial bombing. The British, with Titterington's help, were testing the stabilizer at the Central Flying School at Upavon. The flying school had the largest airfield in England and, besides sixty machines for flight training, had six machines assigned to a special experimental section staffed by pilots and engineers. This section planned thorough tests of the Sperry stabilizer as a bombing aid. When Lawrence visited Upavon, the British were using the stabilizer with a directional gyro and a telescopic bombsight during bomb runs. The pilot headed the plane toward his target, set his course on the directional gyro, and continually sighted the target through his bombsight. The stabilizer held the plane level while the pilot steered by the gyro and, in this way, quickly ascertained the drift of the airplane. Having ascertained drift, the pilot changed his course, stabilized the airplane, steered the new course by the directional gyro, and kept the bombsight on target until the bomb was released at the point predetermined according to speed and elevation.

62 Naval Observatory Records, National Archives, Record Group 78.
63 Report of May 4, 1915, Naval Observatory Records, National Archives, Record Group 78. I am indebted to Navy Air Systems Historian Lee M. Pearson for this and the previous reference.

As an alternative to the pilot's steering by the directional gyro, it was proposed that the directional gyro, once set, hold the course automatically with the aid of servomotors.[64] After considering the alternatives, Lawrence recommended to his father that the company concentrate on this automatic steering gyro, or "an azimuth stabilizer." During the bombing run the airplane would then be automatically stabilized laterally, longitudinally, and in azimuth by a Sperry system.[65] During the next year, Titterington, in England, and the engineering staff in America worked on the three-way stabilizer. Three-way stabilization would also be essential for the automatic control of the pilotless aircraft, or aerial torpedo.

Lawrence not only participated in the further development of the stabilizer in Britain but also sold, early in 1916, forty of the devices to the French government, the largest of the wartime orders. Both Sperrys, however, were disappointed when the French engineers assigned to test the stabilizers made only a few desultory tests on aircraft that the Sperrys thought unsuitable.[66] The lack of French interest resulted in part from the aviators' conviction that the weight of the stabilizer and the use of automatic controls reduced the maneuverability believed essential for survival in combat. Shortly after the war, the French brought the stabilizers out of storage and used them in the development of their own radio-controlled aerial torpedo.

Experienced with the stabilizer and especially with its use in bombing, Lawrence, back in America in March 1916, became involved in the first phase—a pre-navy and pre-consulting board phase—of the development of the aerial torpedo. Even before Lawrence left for Europe, Elmer Sperry, in conjunction with Peter Cooper Hewitt, had decided to experiment with the development of a flying bomb. Son of Abraham S. Hewitt, the noted iron manufacturer, mayor of New York, and philanthropist, Hewitt was best known as the inventor of the mercury-vapor lamp and a mercury rectifier for converting ac current to dc, but his interests, like Sperry's, ranged wide. When he learned in 1915 that Lawrence and Elmer Sperry were interested in automatic flight, he offered $3,000 to help with aerial torpedo experiments.[67] Hewitt understood from Elmer Sperry that the major problem to be solved before a stabilized plane could become an aerial torpedo was automatic steering. By the terms of the agreement, Sperry was to supply an airplane, a stabilizer, and an "aeronaut"; Hewitt's money was to be used to establish a flying field at Amityville, Long Island, and to support tests there.[68] Hewitt did not want Sperry to give him a receipt for the money, and the return if any that Hewitt would receive was not specified.

A year later the arrangement led to some misunderstanding. According to Hewitt, Sperry had carried on the experiments without keeping him informed of their progress and, furthermore, had made demonstrations for outsiders, although Hewitt had understood that the development would be

[64] Lawrence Sperry to the New York Office, November 2, 1915 (SP).
[65] Lawrence to EAS, November 22, 1915 (SP).
[66] Lawrence Sperry, "The Aerial Torpedo," *U.S. Air Services* (January 1926), p. 18.
[67] Letter written by Hewitt's direction to EAS, July 15, 1915 (SP).
[68] Hewitt to EAS, July 19, 1916; EAS to Hewitt, July 26, 1916 (SP).

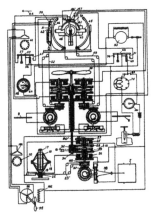

Figure 9.5
Diagrammatic view
of the aerial torpedo
controls. From Lawrence
Burst Sperry patent,
Aerial Torpedo
(Patent No. 1,418,605).
(Key: steering gyro, 17;
windmill for servomotor, 30;
steering servomotor, 27;
electromagnetic clutch, 25,
26, 34; vertical rudder, 7;
barometer, 58; revolution
counter, 75; stabilizing gyro,
41; horizontal rudder, or
elevator, 8, 9.

carried on "as secretly as possible."[69] Sperry replied that Hewitt had not been regularly informed because progress had been slight for most of the year. He acknowledged Hewitt's aid but added that this was nominal when compared to the $15,000 the Sperry Company had put into "this experiment." Sperry wanted it made clear that this was not Hewitt's project. Having been irritated by the coolness of the Hewitt letters, Sperry asked why it was that Hewitt, an inventor and engineer, could not appreciate that what was needed by "an experimenter in such a difficult field as this, where we are practically trying to accomplish the impossible, is not faultfinding and destructive criticism, but sympathy and helpful suggestions."[70]

In the spring and summer of 1916, following Lawrence's return, the Sperrys did make progress. Lawrence, acting as monitor on flights testing the automatic controls installed in a Curtiss flying boat, reported a notable achievement on June 29, 1916, when he allowed the controls to fly the plane automatically for seven miles. Elmer Sperry believed this the first automatic steering of an airplane in history. By the end of July, an automatically controlled flight of thirty miles was recorded and Elmer Sperry's customary enthusiasm asserted itself.[71] On the basis of these performances, he described for Colonel George O. Squier, head of the army's aviation section, to whom Sperry hoped to sell the torpedoes, how the fully automatic aerial torpedo would behave.[72] After course, altitude, and distance were set, the engine started, and the gyros run up to speed, the torpedo, Sperry predicted, would make an "automatic get-away." The AT would then ascend at a predetermined angle, according to the type of airplane and load, and then straighten out to "perfectly" horizontal flight at the desired altitude. "From the first," Elmer Sperry said, "the aeroplane has been held to the proper angles, both laterally and horizontally, perfectly automatically, the flight being a better flight than can be made with a pilot inasmuch as all of the corrections are introduced with extremely slight departures and long before the aviator knows that any correction is necessary." He believed that the steering gyro would operate "for a number of hours with barely a perceptible deviation"; at the exact moment the distance device would dive the plane into its destination "at a tremendous speed."

In 1916, Lawrence filed a patent application on the aerial torpedo. The automatic stabilizer now was a major subsystem in the new automatic pilot system, and to it had been added other automatic subsystems (see figure 9.5). These systems maintained the preset altitude of the torpedo and automatically steered it over a preset course and range into the target. Of these new subsystems, only the steering involved a gyro—one that was nonpendulous with three degrees of freedom and connected to a servomotor that moved the rudder. The altitude of the torpedo was maintained by an aneroid barometer directing a servomotor that operated the ailerons and horizontal rudder (elevator). When the pressure of the atmosphere did not correspond

[69] Hewitt to EAS, July 19, 1916 and August 4, 1916 (SP).
[70] EAS to Hewitt, July 26, 1916 (SP).
[71] EAS wrote to Hewitt of the flights, July 26, 1916 (SP).
[72] EAS to Office of the Chief Signal Officer, U.S. Army, September 2, 1916 (SP).

to that preset in the barometric sensor, the servomotor adjusted the ailerons to correct altitude. In another subsystem, a revolution counter connected to the main propellor shaft of the engine initiated the torpedo's dive after the predetermined distance (number of revolutions) had been flown. The Sperrys powered the servomotors by a windmill; they drove the generator supplying electricity to the gyros by a small propeller in the airstream of the main propeller. To engage and disengage servomotors they used electromagnetic clutches, and to send control signals from sensors to servos they used trolleys and contacts. In this automatic pilot can be found mechanisms that Elmer Sperry had used in his electrical patents in Chicago, as well as many of the devices used in his gyrocompass and ship stabilizer.[73]

Although the aerial torpedo had been patented, the Sperrys still had a difficult road to travel. Test flights of the controls installed on the flying boat had not involved a fully automatic takeoff and were made when weather conditions were favorable. Subsequent development showed that automatic takeoff or launch was a serious problem and that turbulence or unpredicted winds during flight could impose acceleration forces on the gyros and cause substantial deviations from the desired flight path. These post-invention development problems were, however, a commonplace experience in the inventor-engineer's career, and Elmer Sperry had already proved himself persistent and resourceful in dealing with them.

The most discouraging problem in the fall of 1916 was the failure to interest the army—or anyone else—in financing the further development needed to solve those problems. This fact and mounting expenses led Elmer Sperry to call a moratorium. By November Lawrence thought that Hewitt's money should be returned, for "the interest is dead at present in the AT."[74] When the money was returned, Hewitt was told of difficulties underlying the complex and unprecedented problems that had prevented bringing the experiments to a successful conclusion and was thanked for his disinterested action in assuming "without thought of pecuniary consideration" part of the costly experimental burden connected with the trials.[75] An exception to the disappointments of the fall was a successful flight on September 12, 1916, witnessed by navy Lieutenant T. W. Wilkinson, who reported favorably to his superiors.[76]

vi

The torpedo project, dormant in the winter of 1916–17, came to life in April 1917, when the Naval Consulting Board approved the project and the navy awarded a contract to the Sperry Gyroscope Company. When work

[73] This description is based on Lawrence Burst Sperry's patent filed December 22, 1916 (138,485), "Aerial Torpedo," issued June 6, 1922 (Patent No. 1,418,605). Many essentials, the gyros, the servomotors, and the communication and control system among them, are found in earlier Elmer Sperry patents, especially his on the airplane stabilizer (Patent No. 1,368,226). The encompassing system, the aerial torpedo, was the subject of Lawrence's patent.
[74] LBS to F. R. Allen, November 11, 1916 (SP).
[75] F. R. Allen, Sperry company executive, to Hewitt, November 15, 1916 (SP).
[76] Wilkinson to BuOrd, September 13, 1916, and an unsigned letter to Chief of Naval Operations, November 1, 1918 (SP).

began on the revivified project there were three major interrelated problems. The first—and although it preoccupied Elmer Sperry, the one nearest solution—was the perfection of the AT control system. The second problem—and this proved most troublesome—was the coordination of the control system with the new airplane that would carry it. The third problem—and the Sperrys probably did not anticipate that it would be a serious one—was the design of launching gear for a pilotless aircraft. Elmer Sperry, Lawrence, Mortimer Bates, Morris Titterington, the Curtiss Aeroplane Company, Carl Norden, and others were involved in the effort to solve these problems. The navy appointed a liaison officer, to monitor and coordinate the project.

To represent the navy, Admiral Ralph Earle, head of the Bureau of Ordnance, asked a retired officer, Commander Benjamin B. McCormick, to return to active duty.[77] Recommending McCormick was his prior work at E. W. Bliss Company, the torpedo manufacturer in Brooklyn. In July 1917, the first of the Curtiss N-9's arrived, and AT control systems were installed. The N-9's were to be used until the first specially designed aircraft came from Curtiss. The four-gyro stabilizer was used, but at the same time a prototype single-gyro unit was being tested on the flying boat. Lawrence, Titterington, and Bates made many test flights to monitor the automatic controls. The tests revealed a series of problems. The distance meter proved inaccurate and had to be sent to the wind tunnel at the Washington navy yard for calibration. On takeoff, the plane sometimes veered sharply and the resulting acceleration caused the steering gyro to precess and send false guidance signals to the controls. The same problem arose during turbulent flight conditions.[78]

At first, flights with the N-9's were of short range, about five or six nautical miles, but as improvements were made in the control system, the range increased. During these flights, a pilot monitored the automatic controls and returned the plane to base after the distance gear had signalled the descent to target. The pilot also controlled the N-9 until the plane left the ground. Three automatically controlled N-9 flights in November 1917, over a twenty-mile range, resulted in range errors of 275, 375, and 1,100 yards, and deflection errors of 1,400, 1,500, and 2,200 yards.[79] In December the aerial torpedo overshot the target by only 580 yards and erred only 440 yards to the left after a fifty-mile shot.[80] Encouraged, Elmer Sperry pushed Curtiss to complete the specially designed air frames and engines.

Sperry advised Glenn Curtiss on the design of the new airplane. He approved of the two-cylinder engine suggested by Curtiss because it would be free from vibrations likely to disturb the control system,[81] but he urged

[77] Charles L. Keller, "The First Guided Missile Program: The Aerial Torpedo," *Sperry Engineering Review* (March 1961), p. 13. Keller, a research engineer, worked from original sources, and his account has been helpful.
[78] *Ibid.*, pp. 13–14.
[79] B. B. McCormick to EAS January 21, 1919 (SP).
[80] Keller, "Aerial Torpedo," p. 14.
[81] EAS to Glen Curtiss, September 24, 1917; EAS to Admiral Earle, October 4, 1917, and October 5, 1917 (SP).

Figure 9.6 *The flying bomb of 1917-18, which had Sperry controls on a Curtiss specially designed airplane.*

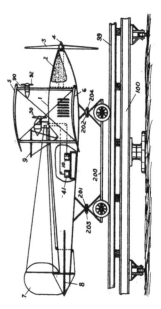

Figure 9.7
Aerial torpedo mounted on a launching track. From Lawrence Burst Sperry patent, Aerial Torpedo (Patent No. 1,418,605).

design changes in the airframe to increase the speed from 70 mph to 90 mph. He knew from his marine torpedo work that the shorter the time of the run, the fewer the accumulated gyro errors. Not only did the airplane have to be fast and relatively free from vibration but it also needed to be light (under 1,000 pounds) so that launching by catapult would not be too difficult. The airplane also had to be capable of carrying a 1,000 pound load of explosives. Ease of assembly was also incorporated into the design because the AT was to be assembled in the field. Because an automatically flown airplane posed special problems that Curtiss had not previously encountered, Elmer Sperry sent Lawrence to the Buffalo plant of the Curtiss Aeroplane & Motor Corporation to assist in the design. By mid-October, specifications had been drawn, and the navy ordered ten of the airplanes, the first to be delivered in November.

For the newly designed airplane, Elmer Sperry and his engineers designed a simpler and less expensive control system. Whereas the 1914 and 1915 stabilizers had the four-gyro unit, the force impressor, "and a number of other complications," the new stabilizer involved one gyro and no force impressor; the servomotors were simplified and mounted on the same base with the gyro. There were no complicating follow-up connections.[82] A heavy, electrically sustained, marine-torpedo gyro was used for the automatic steering.

The project entered a new phase in late November, when controls were installed on the first of these Curtiss "flying bombs." For the first time, a system involving the new airplane, the new control system, and a new launch system would be tested. Three hurried flight tests in December resulted, however, in less than satisfactory results. First, the launching device—a set of

[82] EAS to P. N. L. Bellinger, January 29, 1918 (SP).

heavy cables down which the plane was intended to slide for 400 feet until airborne—wrecked the plane during take off. Next, the repaired airplane failed structurally during the descent on the launching rig; and on the third try, a new catapult—a 150-foot track along which the plane was accelerated by a falling 6000 pound weight—malfunctioned and wrecked the plane during the launch.

Nineteen-eighteen brought efforts to reorganize and redesign. Since the launching device now appeared to be the major problem and the navy wanted a catapult for launching the aerial torpedoes from shipboard, Elmer Sperry and his staff, with the help of Carl Norden, who was brought in as a consultant, concentrated on developing this device. The perfected catapult, which was not ready until July 1918, had a heavy flywheel that was to be spun up to about 2000 rpm by a four-horsepower gasoline engine. The fly-wheel, with its great momentum, was coupled to the launching car, upon which the AT rested, by a tapered drum and cable which accelerated the car.

In the meantime, Lawrence modified a Curtiss "flying bomb" to carry a pilot who could observe its flight characteristics. This became imperative after the second bomb climbed too steeply, stalled, and crashed immediately after launch, and the third crashed after a 200-foot automatically controlled flight. The Sperry staff realized that they knew too little about the flight characteristics of the airplane—much less than they had known about the flying boats and the N-9's, both of which had been flown many times without gyro controls. With the modified "flying bomb," Lawrence made a number of flights from the ice of Great South Bay and was involved in a series of mishaps:

Got one machine all tuned up by flying from the ice all three controls working. This was accomplished only after correcting about twenty different things of the automatic pilot. . . . My fourth smash in one of these little torps resulted in a broken nose—two little bones broken. While the nose has been flattened a little it can be made to grow straight and will not be noticed, although the passages inside are too crowded for breathing and will have to be operated on.[83]

Lawrence's tests, despite the accidents, provided much of the information needed to modify both airplanes and controls.

These efforts brought encouraging progress. In a March 6 test of the launching device, an improved AT rose successfully from its launch under automatic control and made a smooth stabilized flight until the distance control automatically terminated the flight in the water at the preset distance of 1000 yards.[84] This has been judged "the first entirely successful flight of an automatic missile in this country if not in the world."[85] Despite this historic achievement, test flights in the summer and fall of 1918 with the last of the Curtiss FB's were discouraging: the launching device malfunc-

[83] Lawrence to EAS, February 26, 1918 (SP).
[84] The short range had been used because the primary purpose of the flight was to test the launching device.
[85] Rear Admiral Delmar S. Fahrney and Robert Strobell, "America's First Pilotless Aircraft," *Aero Digest*, LXIX (1954), pp. 28ff.

tioned, the airplane failed structurally, and gyro precession during launch caused trouble. The last of the ten Curtiss-built FB's was successfully launched by the catapult on September 26, but only climbed for about 100 yards before spiraling to earth. The prior launch of an FB had ended similarly, and gyro precession during launch was suspected as the source of the trouble. There remained a "special airplane" built by Lawrence Sperry, and this was launched by the catapult on October 29. Having recently witnessed the loss of an aerial torpedo which flew out to sea, Commander McCormick unfortunately ordered all but two gallons of gasoline drained from the tank of this airplane. During launch, acceleration forced the gasoline to the back of the tank, the engine stalled, and the plane wrecked on the catapult.

While the FB's crashed one after the other, test pilots had successfully continued to fly the N-9. Discouraged by the FB's, and wanting to try the new catapult and automatic controls without the complicating factor of the FB airplane, a pilotless launch of an N-9 was ordered on October 17. For this test, a four-unit gyro stabilizer rather than the single unit used on FB's was installed, and the stabilizers were caged, or locked, during launch to be released automatically after the plane became airborne. With the distance gear set for eight miles, the N-9 rose from the launch, climbed steadily, flew in a perfectly straight line over an observer at the Bay Shore Air Station, and then eastward across the Atlantic never to be seen again. The distance control had failed to function.[86] This flight suggests that if the project had not been complicated by the determination to use the inexpensive FB and a simplified control system, a practical but expensive aerial torpedo might have been ready for production much earlier.[87]

The armistice followed a few weeks after the last FB test, and the pressure to develop the aerial torpedo diminished. In January 1919, the contract terminated and the field at Copiague closed down. Although the goal of producing aerial torpedoes in quantity by the spring of 1918 had not been fulfilled, the Bureau of Ordnance thought that an aerial torpedo could be produced by the following spring in quantity at a cost of not over $2,500 per plane. The bureau predicted that the AT would deliver one thousand pounds of explosives up to seventy-five miles with an accuracy of two percent in range and deflection. The weapon was thought not to be accurate enough for use against ships, but in calm weather it could strike a fort at ranges "beyond gun fire."[88] It could prove "very valuable in bombarding cities, such as Wilhem-haven, Heligoland, and possibly Kiel and Berlin" and would have a "strong moral effect" as well. The bureau thought that the major problem remaining was the design of a satisfactory airframe. It judged the catapult that had been designed for the project a striking success,

[86] Keller, "Aerial Torpedo"; and Lee Pearson, "Developing the Flying Bomb," *Naval Aviation News* (May 1968); both have used original reports to describe the various tests.

[87] Rear Admiral Delmar S. Fahrney, an unpublished manuscript on the history of guided missiles, in the custody of the historian, Naval Air Systems Command, Washington, D.C. Hereafter cited as Fahrney, "Missiles."

[88] Bureau of Ordnance to Chief of Naval Operations, November 1, 1918. No author (SP).

with the potential for other applications, and also believed that the gyro-control system had been developed to the level of accuracy thought feasible when the project began.[89]

Most of the difficulty encountered during the project was attributed to the flying-bomb airframe; the success when using the N-9 airplane was believed to support this conclusion.[90] Morris Titterington, in his analysis of the project, stressed the problems encountered with the flying-bomb airframe. He thought the airplane was extremely difficult to control, required excessively exact movements of the control surfaces, and was "critically unstable. . . . If it departs from its attitude of flight beyond a small amount, the controls are not capable of bringing it back."[91] Titterington lamented the fact that the airframe had not been thoroughly flight tested to determine its flight characteristics and that the automatic controls had not been carefully tailored to suit these characteristics. An engineer criticising the project years later thought it suffered from a failure to use the systems approach: the design and development of the airframes, engines, catapults, and automatic controls, all of which critically affected one another, had not been coordinated.[92]

The project might also have progressed more rapidly if a proposal to supplement automatic controls with radio controls had been accepted.[93] Radio signals could have corrected minor errors caused by the automatic controls; postwar achievements by Lawrence Sperry with aerial torpedoes support this conclusion.[94] In his April 1917 proposal to the Naval Consulting Board, Elmer Sperry had suggested "wireless or some other form of radiation . . . as the medium of control." He thought, however, that the increased complications arising from radio control might create an additional and undesirable expense. Throughout the aerial torpedo project, discussions of wireless control persisted.[95] Commander McCormick advocated

---

[89] *Ibid.*
[90] Fahrney, "Missiles," p. 111.
[91] Titterington to McCormick, September 27, 1918 (SP).
[92] Keller, "Aerial Torpedo," p. 11.
[93] Before the war Sperry and others had experimented with wireless control, or "radio-telautomatics." When wireless was still in its infancy, scientists and engineers had taken out patents on wireless control of "dirigible self-propelled vessels." Among those applying for patents in wireless control of marine torpedoes, lighter-than-air craft, or boats, before 1914, were Ernest Wilson (probably the first, with an English patent in 1897), Nikola Tesla, Admiral Sims, Edison, and J. H. Hammond, Jr. (Hammond used some Sperry equipment in his system for wireless control of a boat). Wireless control seemed obvious to inventive minds because the same signal or force used to move the key of a wireless receiver could be used to trip relays and switches to control mechanisms. Benjamin F. Meissner, "The Wirelessly Directed Torpedo: Some New Experiments in an Old Field," *Scientific American,* CVII (1912), p. 53. Anthony Fokker, the Dutch airplane constructor, who built thousands of planes for the Germans during the war, recalled that the German command had asked him to construct a radio-controlled flying bomb. The Ministry of War accepted his plans but did not give a production order until 1918, and Fokker was not ready to begin production until the war ended. Interview in *Le Petit Journal,* September 18, 1919.
[94] See below.
[95] Fahrney, "Missiles," pp. 92–94.

radio control, but he delayed its incorporation in the Sperry system because he thought it prudent to achieve success with a preset automatically controlled aerial torpedo before improving upon its accuracy with radio control. He assumed that radio control could make the corrections that monitoring pilots had sometimes made in the N-9's, when wind or gyro "creep" disturbed the flight. He also hesitated to push radio control development by the Sperry company because there were men of German ancestry—"enemy aliens"—in its radio department. Further, he was of the opinion that the radio control developed at Western Electric was more advanced than that being developed at Sperry.[96] The Sperry-navy phase of the project ended without radio control.

vii

After the armistice, Elmer Sperry left to others the further development of aerial torpedoes. The navy continued experiments with the assistance of two former Sperry employees, Carl Norden and Hannibal Ford, and the army completed a similar project inaugurated during the war under Charles Kettering, the noted inventor and automobile engineer. Later, the army initiated another major aerial torpedo project under the direction of Lawrence Sperry. A short summary of these three endeavors permits comparison with the first of the aerial torpedo projects, and shows how subsequent advances depended on the wartime experiments.

Norden took over further development of the gyro-control system, and Hannibal Ford's company built the system redesigned by Norden. About a week before the armistice, the Bureau of Ordnance had ordered five special airplanes to be built by the Witteman-Lewis Company[97] to replace the Curtiss FB's. The bureau also transferred the entire project to the Naval Proving Grounds at Dahlgren; two Sperry control systems and other usable equipment from Amityville were shipped to the new location. At the same time, Norden, working with Ford, made ready three of his versions of the control system for installation in the new Witteman-Lewis airplane, the other two planes being outfitted with the Sperry controls from Amityville. How Elmer Sperry may have felt about these events is only suggested by a letter to Admiral Earle, head of Bureau of Ordnance, in January 1919. Sperry thought that the aerial torpedo project might best be carried on by shifting to a naval proving ground, where the navy could closely coordinate those, such as Kettering and himself, who had contributed most to the art during the war; he made no mention of Norden or Ford.[98]

Despite the Navy's confidence in Norden—he was considered particularly strong in the mathematical analysis of engineering problems—and despite the new airplane and control system, the tests carried on in 1920 and 1921 brought results no more impressive than those achieved by the

[96] The first of the Sperry patents on radio control was an Elmer Sperry patent filed December 18, 1917 (207,786), issued February 17, 1931 (Patent No. 1,792,937); others followed. For information on Elmer Sperry's inventions, see also interferences between John Hays Hammond and Elmer Sperry, cases 47,032, 47,883, and 56,613, in Patent Office Archives. These interference cases were heard in the 1920's.
[97] Fahrney, "Missiles," p. 113.
[98] Letter of January 2, 1919, quoted in Fahrney, "Missiles," p. 116.

Sperrys and less successful than those obtained with the N–9. The first test of the Witteman-Lewis plane, outfitted by Norden with Sperry controls, terminated in a crash after a flight of 150 yards; the second flight, again with a Sperry control, lasted for about twenty minutes but the plane flew in ascending circles to a high altitude before crashing; and the third, this time with the Norden controls, flew one minute and fifty seconds before crashing. Norden, who had prepared the various tests, attributed this third failure to his setting the controls incorrectly. The navy, however, was losing interest as its funds diminished; in 1922 the last two planes were consigned to the scrap heap.[99]

The wartime army project had been inaugurated after an army delegation in November 1917 studied and inspected the Sperry project. Charles F. Kettering, the inventor and engineer who later became head of research for General Motors, was the most enthusiastic member of the delegation established by General Squier to study the feasibility of a flying delegation established by General Squier to study the feasibility of a flying bomb for the army. In December 1917, Kettering became industrial head of the army's flying bomb project, and Colonel Bion J. Arnold became the military director. Kettering brought in Orville Wright and the Dayton Wright Airplane Company for all of the aerodynamic work. Thus Kettering, Arnold, and Wright were the Army counterparts of the navy's Sperry, McCormick, and Curtiss. Elmer Sperry was consultant for the control system and "invited the Dayton people . . . to get saturated with the details and . . . turned over to Dayton everything we had, not only drawings, but specific pieces of apparatus of all sorts. . . ." Sperry often traveled to Dayton to inform the group there of progress made on the navy project.[100] With the advice of Sperry, Kettering designed the control system, which was cited by Squier as an "achievement of the first magnitude."[101] Also praised was Harold Wills, Henry Ford's chief engineer, who designed the flying bomb's engine and whom, Sperry recalled, Ford described as "the man that the public think I am."[102] The final design in the fall of 1918 was a small biplane with a 15-foot wing span designed to carry about 200 pounds of explosive. Elmer Sperry thought the army-Kettering project simplistic compared to the navy-Sperry.[103]

On October 22, 1918, an army flying bomb, or "bug" as it became known, made a successful test flight after two or three prior tests, also in October, had failed. On this occasion, the "bug" climbed gradually to its predetermined height, flew straight and level, and then dived to the target area 500 yards distant.[104] After the armistice, tests of the army flying bomb

[99] Fahrney, "Missiles," p. 124.

[100] Undated and unaddressed letter, or memorandum, of Sperry, probably written within a year of the war's end (SP). The patent interference case, *Charles F. Kettering* v. *Lawrence Sperry* (51,587) declared in August 1919, provides a comparison of some particulars of the Kettering and Sperry aerial torpedoes.

[101] Fahrney and Strobell, "America's First Pilotless Aircraft." This article has been used in summarizing the Kettering-Army project.

[102] EAS to Commander John H. Towers, October 1, 1918 (SP).

[103] EAS, unaddressed letter, or memorandum, no date (SP).

[104] Colonel Arnold to Elmer Sperry, October 23, 1918 (SP). Arnold congratulated Elmer Sperry, Wright, and Kettering for the work "done in bringing this device here to its present condition" (SP).

continued in Florida. Nine launchings were made in 1919, but only one, covering a distance of 19.5 miles, was completely successful.

Further development of the aerial torpedo came again under Sperry auspices in 1920, but this time under the direction of Lawrence Sperry and with the support of the army rather than the navy. In April 1920, the Lawrence Sperry Aircraft Company received a contract from the Engineering Division of the Army Air Service to provide six aerial torpedoes. The company contracted to install automatic control systems in three standard E–1 advanced training planes and to build five Messenger airplanes, three of which were to be converted to aerial torpedoes. The army agreed to supply three gyro-control units, for the E–1's and the drawings for the Messenger airplane.[105] The Lawrence Sperry Aircraft Company commissioned the Sperry Gyroscope Company to build the gyro-control units for the three Messengers.

The contract was a development one providing for incremental payments as the E–1 planes were converted and the Messenger planes built and converted, and for monthly payments for laboratory, shop, and field or flight experiments necessary to perfect the automatic controls and to prepare for the final tests. The sum for this work was to be $84,000. An unusual feature of the contract was a provision for an additional $40,000 in bonuses for excellent performance of the aerial torpedoes, judged by the number of target hits in twelve tries at thirty, sixty, and ninety-mile ranges.

Lawrence Sperry originally agreed to complete all of the work, development, tests, and deliveries within one year but obtained an extension of the contract to June 29, 1922. After he found that the automatic gyro controls could not hold the aerial torpedoes on target for the ninety-mile range, he obtained permission to use radio control from an accompanying airplane to correct errors in the automatic controls. He was allowed to compete for the bonuses using the radio control because the contract had specified that the torpedoes should function in flight "without manual controls."[106] In a storybook finish, the aerial torpedoes made bonus-winning hits at thirty, sixty, and ninety miles on June 29, 1922, at Mitchell Field.

Throughout the project, Lawrence had the enthusiastic support of many officers of the Air Service, Brigadier General William Mitchell, Assistant Chief of the Air Service, among them. They encouraged him in his efforts to persuade other officers of the promise of the aerial torpedo.[107] In a foreword to an article of Lawrence's published posthumously in January 1926, General Mitchell, referring to the development of the aerial torpedo, described Lawrence as "one of the most brilliant minds and greatest developers in the world of aviation."[108] The project had placed the United

[105] U.S. Department of War, Air Service, Engineering Division, Contract #242, April 14, 1920 (SP).

[106] Lawrence Sperry to the Contracting Officer, McCook Field, Dayton, Ohio, May 1, 1922 (SP).

[107] For example, Lawrence Sperry to General William Mitchell, April 19, 1923 (SP).

[108] Colonel William Mitchell, "Lawrence Sperry and the Aerial Torpedo," *U.S. Air Services*, January 1926, p. 16.

States in the lead in the development of flying bombs, or aerial torpedoes, but subsequent interest languished and the modest expenditures for further flight tests with the existing torpedoes were sharply curtailed with the onset of the depression.

viii

By his wartime service on the Naval Consulting Board and through his aerial torpedo project, Elmer Sperry contributed substantially to the development of twentieth-century technology. The Naval Consulting Board was an advance, if at times a faltering one, in the evolution of the relationship between the military and the American research and development community. The navy learned that it had to depend on the professional inventor, engineer, and scientist well-acquainted with the needs of expanding naval systems. Sperry was highly responsive to the needs of the navy and realized the contribution that the scientifically trained engineer and scientist could make. Possessing a remarkable ability to discover the front edge of major technological trends, he brought this talent to the navy. On the board, he reinforced and accelerated the navy's promotion of aerial warfare—especially the perfection of sensitive instruments for the precise control of massive and complex forces.

In the aerial torpedo project, Sperry gave force, definition, and direction to a major technological trend. His development of the aerial torpedo differed in an important respect from his earlier development of the marine stabilizer and the gyrocompass. In the case of the stabilizer, Sperry had depended on the support of Captain David W. Taylor, who was in a position to act as his patron and to persuade the navy of the need for the Sperry invention. In the second case, the need for a compass was obvious because the advantage that the Anschütz gave the German navy had to be offset. In both cases, Sperry was responding to needs, or reversed salients, in an expanding system. Many of his other inventions and developments had been responses of a similar nature. In the case of the aerial torpedo, however, Sperry was working ahead of obvious need and beyond patrons; he saw in the existing state of technology the possibility of creating the new weapon. He was able to do this in 1917 because of the position of great influence he held on the Naval Consulting Board. From this position, he could create and force the development of a seminal invention that would eventually become the nucleus of a new system.

At the end of World War I, Sperry was close to the successful development of the weapon that would still be considered a major innovation when introduced by others almost two decades later. His aerial torpedo anticipated in general design and in many details the German V–1 flying bomb. Both were winged missiles guided and controlled by preset instruments, and the heart of the instrumentation in both cases was the gyroscope.[109]

109 The German V-1, or FZL 76, was built by the German aircraft firm *Fieseler-Flugzeug-Werke,* Kassel. It had a pulse-jet engine built by Argus Motorenwerke, the firm that conceived of the flying bomb and proposed it to the Air Ministry, which approved of the project in June 1942. The first of the V-1's was launched against Britain two years later. The gyro control for the V-1 was supplied by the

The V–2, though differing from the aerial torpedo and flying bomb in that it was a rocket propelled and a ballistic missile, also operated under gyro guidance. Considerable publicity has been given to such men as Robert Goddard in America, who pioneered in the design of the rocket, but very little to such pioneers as Sperry, who contributed so greatly to the guidance and control of pilotless vehicles.

Although obviously of a sensational nature, the aerial torpedo project received no publicity until eight years after the war. Only in 1926, when Elmer Sperry received the John Fritz medal, did William L. Saunders, chairman of the Naval Consulting Board, publicly describe the wartime project. The result was a flurry of headlines: "Deadly Air Torpedo Ready at War's End: Elmer Sperry's Invention Is Told" (*New York Times*, December 8, 1926); "Sperry Aerial Torpedo Deadly," (New York *Sun*, December 8, 1926); and "Aerial Torpedo Explained: Steered 100 Miles Accurately by Gyroscope Compass, 10,000 Would Win a War," (*Brooklyn Daily Eagle*, December 8, 1926).

Elmer Sperry, of course, realized the tremendous potential of the aerial torpedo. In 1918 he wrote, "We have gone a long way toward completing the development of an extremely significant engine of war, it being nothing short of the coming gun." Yet, like so many other inventors and advocates of destructive weapons, he believed—or hoped—that the aerial torpedo would make "war so extremely hazardous and expensive that no nation will dare go into it." He recalled that the "great engineering savant [Professor Robert H.] Thurston of Cornell was wont to say in this connection ... if the pen was ever mightier than the sword it would be the draftsman's pen."[110]

Even though he did not receive publicity for his most significant research and development project during the war and did not receive nearly as much publicity as Edison for his Naval Consulting Board work, Sperry had established himself in the opinion of his peers as a leader in the world of invention and engineering. The positions of honor and responsibility given him by his fellow engineers and scientists after the war emphatically demonstrated this fact.

---

firm of *Askania*. Like the Sperry AT, the V-1 was designed to carry about 1,000 pounds of explosives. The wingspan of the V-1 was 16 feet and the overall length 25 feet. With fuel and explosives it weighed about two tons, of which about one-half ton was fuel. The range was increased to 250 miles before the war ended and the speed to about 500 miles per hour. Of the 9,300 launched against England in 80 days after June 13, 1944, 2,000 failed shortly after takeoff for technical reasons, and about 800 fell in England short of target because of the failure of the control system. Despite these failures and the success of British fighters and anti-aircraft, 2,400 reached London. About 6,000 persons were killed and 23,000 houses totally destroyed. Basil Collier, *The Battle of the V-Weapons, 1944–45* (London: Hodder and Stoughton, 1964), pp. 179–80; and Rudolf Lusar, *Die deutschen Waffen und Geheimwaffen des 2. Weltkrieges und ihre Weiterentwicklung* (Munich: J. F. Lehmanns Verlag, 1964), pp. 173–81. Collier briefly discusses precedents but is apparently unaware of Sperry's (and Kettering's) early projects; Lusar does not discuss flying bomb projects other than the German of World War II. Generally, Sperry's work in the flying bomb, or guided missile, field is not appreciated or understood by historians.

[110] EAS to Admiral Earle, December 19, 1918 (SP).

CHAPTER X Fulfillment and Legacy

The problem that seems the most insurmountable, like Achilles, has a vulnerable point, and in fact 'the thing that couldn't' has been found very simple of solution after it has been fully analyzed and when attacked from the proper angle.

ELMER SPERRY, BEFORE THE FRANKLIN INSTITUTE,
JANUARY 15, 1913

THE END OF THE WAR LEFT SPERRY AND HIS COMPANY THE PROBLEM OF TRANSITION to peacetime projects. During the conflict, Sperry had applied his research and development capabilities and his company's manufacturing facilities almost entirely to naval and military needs. As Charles Doran, the general manager of the company, later observed, the Sperry Gyroscope Company "was a very important part of the National Military Establishment."[1] Peace dramatically changed the environment in which Sperry and the company had to flourish. Because the armed forces reduced spending drastically, the company had to cultivate new outlets for research and development and its products.

During the war, Sperry and the company were concerned with effectiveness more than economy, for under the extreme conditions of war, the primary objective of the belligerents was victory. With the coming of peace, however, economic efficiency, safety, and comfort were factors to be given more weight in designing and producing devices.

The transition was not as difficult for Sperry, who was primarily concerned with research and development, as it was for manufacturing firms. Presiding over change, in fact, was Sperry's genius: he was an inventor-entrepreneur, not a manager of the *status quo*. The strategy of the company after the war reflects his determination to emphasize invention, development, and innovation. The labor force of the company was reduced by about half, but the size of the research and development staff remained about the same. The highest percentage of salaried employees remained engineers.[2] The sales staff was expanded in order "to reach out," as Sperry

[1] Charles Doran, "The Past Four Years," *The Sperryscope*, IV (1923), pp. 2ff.
[2] See above, p. 209.

often said, to find new outlets for the creative potential of the company. The service and maintenance staff was enlarged because the personnel now using the high-quality Sperry products would not be so thoroughly trained and supervised as the naval and military. Sperry held onto his superb force of skilled labor, especially machinists, and the excellent machine tools and test facilities with which they worked (These were also invaluable assets to the research and development staff.) Robert Lea, one of the most successful Sperry salesmen, recalled that he seldom had to face competition in selling Sperry instruments, such as the compass, because the complexity, quality, and service left would-be competitors behind. The major problem for the salesman was to make a market where one had not previously existed—not to oust a competitor.[3]

Elmer Sperry had several approaches to the peacetime market. A major approach was to modify military products to fill peacetime needs. His outstanding success in modifying a wartime product was the gyrocompass. In order to introduce the compass in the merchant marine, Sperry's engineers developed a compass that was simple in operation, easy to maintain, and comparatively low priced. The result, after several false starts, including one with the "twin compass," was the Mark VI.[4] Salesmen were successful in persuading ship captains, both in the American and foreign merchant marines, that the Sperry compass, although more expensive than the magnetic, offered economies in navigation and safety that the magnetic could not match. By 1932 Sperry gyrocompasses were installed on more than one thousand merchant-marine vessels, the first having been installed on the *Bergensfjord* of the Norwegian-American Line and the *Aquitania* of the Cunard Line.[5] These sales not only helped keep the company solvent but, more important to Elmer Sperry, provided profits that could be reinvested in research and development.

The gyrostabilizer was also modified for the peacetime market. Before America entered the war, Elmer Sperry had supported Edward Sperry's efforts to interest millionaire yacht owners in the stabilizer. After the war, the Sperry sales staff contracted with the United States Shipping Board in 1922 to install a gyrostabilizer on the *Hawk Eye State,* renamed the *President Wilson;* the Westinghouse company built the Sperry-designed stabilizer for the 20,000-ton liner.[6] A notable installation was on the 48,000-ton Italian luxury liner, the *Conte Di Savoia,* in 1930. After this, however, the company, with Elmer Sperry no longer at the head, did not continue to promote sales of the gyrostabilizer; a simpler fin stabilizer with a small gyro control was subsequently introduced.

Sperry and his engineers also found peacetime uses for the searchlight. As commercial aviation developed in the twenties and the need to perfect night flying increased, the market for Sperry beacons at airports and on air

[3] Robert Lea to author, April 1, 1966.
[4] Interview with Eric Sparling, May 25, 1966.
[5] Sperry Gyroscope Company, *The Sperry Register,* 1932.
[6] "Gyro-Stabilizer as Cure for Seasickness," *Scientific American,* CXXVII (1922), p. 16.

NAVAL VESSELS OF THE SPERRY GYRO

276

WORLD EQUIPPED WITH
COMPASSES

Figure 10.1
From the Sperryscope,
II, 1920.

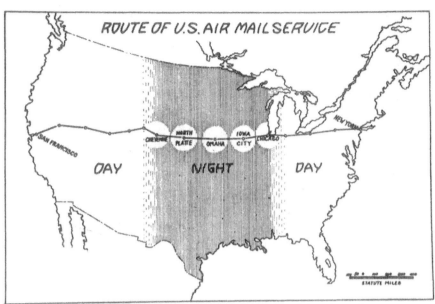

Figure 10.2 Sperry searchlight beacons were needed for flying the mail across country. The map shows the locations of five beacons and the area in which each beacon would be visible to a pilot flying at 5,000 feet. From the Sperryscope, *IV, 1923.*

routes expanded. Sperry beacons guided the early mail pilots as they made dangerous flights along the "night airway." In 1924 five Sperry beacons and landing lights were used on the Chicago-Cheyenne leg of the first Chicago-to-coast airmail service; other routes were soon utilizing Sperry beacons. The Sperry staff also adapted the Sperry high-intensity arc for the motion-picture industry for illuminating sets and for film projectors.

In finding these and other solutions to the transition problem, Elmer Sperry continued to school young engineers in his style. One of them, who later formed his own successful instrument company, said, "The education which I received through my five years' association with him I regard as the most valuable part of my experience."[7] Although Sperry turned over many interesting problems to his staff in the postwar years, he never retired. During the war years, he had received about fifteen patents a year; in the twenties he continued to file patents at the rate of about eleven a year, higher than his average of seven annually during his career of a half-century of patenting.[8] Despite the size and ingenuity of his staff, he produced, in the postwar decade, more than forty percent of the patents issuing from the Sperry company, a percentage slightly higher than that during wartime.

i

The introduction of the automatic pilot for ships was the first major postwar achievement of Sperry and his staff. It was both a culmination of his guidance and control achievements and a manifestation of the ability

[7] Charles Colvin of Pioneer Instrument Company to ?, June 19, 1930 (SP).
[8] Not all of Sperry's applications matured and only those maturing have been counted here. His lifetime total of 355 was acquired from 1880 to 1930.

of his staff to carry his work forward. Although he assigned development responsibility for the pilot to his engineers, the basic invention was Sperry's. Furthermore, the problems encountered in developing the device were solved with components and subsystems that he had used earlier in electrical machinery, streetcars, automobiles, and other gyro-related inventions.

Elmer Sperry began designing the ship's pilot in 1912, not long after successfully placing his compass on the *Delaware*. Having developed the practical gyrocompass, he was seeking ways to apply its directive power and its reference line to other purposes and thus increase the compass' usefulness for the customer. His intention was to use the compass as the nucleus of an automatic-steering system.[9] The problems of developing such a system resembled those he was then encountering in the airplane stabilizer, and while working on the airplane stabilizer, he jotted down sketches and notes that could subsequently be used when drawing up a patent application and further developing a ship pilot.[10] In 1913 the inventor John Hays Hammond suggested that Sperry build the automatic-steering apparatus for use with Hammond's small torpedo boat. Hammond was to use radio signals to change the course of the boat.[11] In the spring of 1914, Sperry and his engineer Elemer Meitner, developed the invention further, but the war interrupted their work; "things that were looked upon as important in July were discarded in August, and we had to right-about-face on many of our propositions."[12] He found time, however, to file a basic patent application for the gyropilot.[13]

After the war, the increased number of compasses being placed on merchant marine ships by the company's energetic sales staff presented Sperry with an opportunity to resume the development of the gyropilot. In May 1921, Robert Lea, who had also been probing the commercial market for new research and development needs, sent Sperry a memorandum in which he asked "just what your plans are with regard to developing the Sperry Automatic Steering."[14] Lea was concerned because he had heard that the Anschütz company was working on an automatic ship's pilot; he also believed that a compass and pilot would reinforce each other on the commercial market.[15]

After consulting with his general manager Charles Doran, Sperry as-

[9] EAS testimony in *Sperry* v. *Henderson*, Interference 50,874 ("Automatic Steering"), September 1927.

[10] These sketches are in his Notebook No. 54 (February 15, 1912, to December 1, 1912; SP).

[11] EAS to Elemer Meitner, June 30, 1926. Sperry wanted Meitner to recall, in connection with an interference suit, his work with the pilot. Sperry had evidence of this work in his Notebook No. 55.

[12] *Sperry* v. *Henderson*, Interference 50,874. EAS to Meitner, May 18, 1914 (SP).

[13] Filed November 13, 1914 (871,885), "Navigational Apparatus," issued November 30, 1920 (Patent No. 1,360,694).

[14] R. B. Lea to EAS, May 26 and May 31, 1921 (SP).

[15] For an early description of the Anschütz, see Alfred Gradenwitz, "Self-Steering Vessels: Recent German Developments in Automatic Gyro Practice," *Scientific American*, CXXVII (1922), p. 96. An Anschütz automatic pilot had been given a sea trial on the *Hansa* in December 1921.

signed C. B. Mills, Elmer Sperry, Jr., Fred Hodgman, and several other development engineers to revitalize the project. They built an experimental model, and Elmer Sperry arranged for a test installation on the Standard Oil Company tanker *John O. Archibold*. In April 1922, Elmer Sperry, Jr., who was present for the trial voyage in the Gulf of Mexico, judged the performance of the pilot remarkable, considering the complexity of the problem and the experimental nature of the apparatus. After modifications, another test installation was placed in 1922 on the Standard Oil Tanker *J. E. Moffett, Jr.*, the largest ship of the tanker fleet. On the first tests, on the *Moffet*, steaming from New York to Galveston, F. S. Hodgman served as test engineer and later recalled that the hand-built pilot "was a rugged looking collection of spare parts which covered the ship like a circus tent," involving "a huge drive motor with yards of open shafting" and a mass "of clapper switches and other noise makers." The ship, strewn with about 200 fathoms of a "sorry" collection of repeater compass cable, "was truly a floating experimental laboratory."[16] The pilot, however, performed well, and the project team moved confidently on to the next installation of the device, which had been named "Metal Mike" by the officers of the *Moffett*.

The next installation was the first permanent one. In October, the Munson liner *Munargo* made an autopilot, roundtrip voyage from New York to Cuba, with Elmer, Jr., and Hodgman aboard. Upon its arrival in New York, Elmer Sperry publicly announced that "Metal Mike" was perfected; Captain Andrew Asborn of the *Munargo* characterized its performance as "a triumph of matter over mind"—an enigmatic but nonetheless enthusiastic comment. The captain reported that he set "Metal Mike" on his course off Barnegat and never put his hand to the wheel again until the vessel reached Nassau. "Metal Mike," the captain noted with satisfaction, "doesn't drink, smoke or kick about working overtime. . . ."[17] (In view of Elmer Sperry's views on prohibition, it is inconceivable that he would have invented a helmsman that *did* smoke or drink.)

After the success of the *Munargo* installation, the number of gyropilot installations increased steadily until more than 400 merchant ships throughout the world had the device.[18] Once a ship owner decided to shift to the gyrocompass—and 1000 had by 1932—the opportunity to sell the gyropilot automatically increased because the gyrocompass was a major component of the pilot system. Not only did the sales department exploit this system relationship but also persuaded owners to take the Sperry helm indicator and course recorder, which made a permanent record of the course steered by either the helmsman or "Metal Mike."

The research and development department made several post-innovational improvements in the design of the pilot. Having found, for example, that ships were likely to have old, mechanically inferior, telemotors to transmit the commands from the autopilot to the steering engines, the Sperry

16 F. S. Hodgman in talk before Sperry Engineers' Club.
17 *New York Sun,* November 9, 1922.
18 Sperry Gyroscope Company, *The Sperry Register* (1932 [?]).

staff designed a two-unit pilot for sending the commands directly and electrically from the gyropilot to the steering engine. The need to improve upon the telemotor so that the gyropilot could be used effectively was the kind of problem frequently encountered by the Sperry company: the state of existing technology had to be brought up to the level of the innovation. The company also often had to improve the electrical system of older ships before the Sperry equipment could be installed.

"Metal Mike" caught the public's imagination, and the press provided considerable publicity. In his satirical drama *R.U.R.*, the Czech playwright Karel Capek used the term "robot" to describe fabricated humans; the term was aptly applied to "Metal Mike."[19] The president of the Massachusetts Institute of Technology, S. W. Stratton, termed "Metal Mike" a most exciting robot, placing it on a par with the "Great Brass Brain" (a computer that predicted tides), the "continuous-product integraph" (a computer that solved second-order differential equations), and the "Televox" (a device that provided telephone control for distant apparatus).[20] Elmer Sperry realized that the performance of the gyropilot seemed uncanny, for he had been told that no one could possibly invent a device incorporating the intuition of an expert quartermaster. He had in fact analyzed the "intuition" problem when he first began work on the pilot in 1912 and had decided that the helmsman's intuition was, in essence, the ability to "ease off" and "meet" the helm.[21] Easing off involved lessening the rudder angle after the rudder had been put over and the ship had responded by swinging toward the desired heading; meeting was putting the rudder over to the other side (overthrow) to counter the tendency of the ship's angular momentum to carry it past the desired heading. If the helmsman did not perform these functions with skill, the ship continued to yaw. Sperry set out to devise a way of performing these functions mechanically and automatically.[22]

Sperry decided to feed into the rudder an overthrow that would be determined, as he put it, "by the left over expression of the amplitude of yaw." The left-over amplitude was the amount the ship would swing past its course if there were no overthrow function. To solve the problem, Sperry needed a mechanical means of measuring the full amplitude of the yaw and of making the overthrow of the rudder a function of it. The more the ship yawed, the greater would be the overthrow necessary to counter the momentum and dampen the yaw.

In 1913 Sperry sketched a mechanism to solve the problem.[23] The ingenious solution involved a device resembling shears. The device was constructed so that the opening of the shears' prongs depended on the amplitude and period of the yaw. This dependence was achieved by using a shaft

Figure 10.3
Page from Sperry Notebook No. 54 (February 15, 1912, to December 1, 1912), on which Sperry sketched his roller and cam, or roller and drum, device for solving the "easing-off" and "meeting" problem of the marine gyropilot.

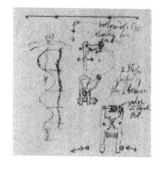

Figure 10.4
Sperry's sketch of his anticipator, or anti-yaw mechanism, for his marine gyropilot. From Notebook No. 55 (December 1, 1912, to January 15, 1914). Note the similarity to the drawing in his patent applied for about a year later (see figure 10.5).

[19] S. W. Stratton, in collaboration with Frank P. Stockbridge, "Robots," *Saturday Evening Post* (January 21, 1928), p. 23.

[20] *Ibid.*, p. 23ff.

[21] Elmer A. Sperry, "Automatic Steering," paper read at the Society of Naval Architects and Marine Engineers, New York, November 8 and 9, 1922.

[22] Typescript dated June 25, 1926, of EAS testimony prepared for *Sperry v. Henderson*.

[23] Notebook No. 55 (December 1, 1912, to January 15, 1914; SP).

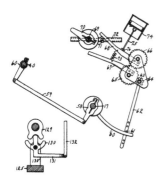

Figure 10.5
Drawing of the anticipator
from the Sperry marine
gyropilot. From Navigational
Apparatus (Patent No.
1,360,694). (Key: lever, 62;
pivot point of lever, 63;
shaft driven by gyrocompass,
69; shears, 67, 68; pin, 71;
spring, 73; dash pot, 74).

connected to the gyrocompass in such a way that the shaft's movements were like those of the ship's yawing. Pins on the shaft struck against the prongs of the shears to open them and hold them apart. Spring pressure closed the shears if they were not struck by the pins; the closing was retarded, however, by the resistance of a dashpot to prevent the action of the mechanism from being too abrupt. When the shears were open to a maximum degree as a result of a large and rapid yaw, the overthrow countering the high momentum of the ship was greatest. The overthrow of the rudder was controlled by a lever which was moved as a function of the opening of the shears (see figure 10.6). Sperry called the anti-yaw mechanism an "anticipator."

The anticipator was one of the prime subsystems of the patented gyropilot. The complete system was basically simple. Sperry used a gyrocompass repeater to establish a reference line upon which the course to be steered was based. Assuming that the compass performed perfectly, the ship turned around the fixed reference as it veered from course or yawed. The ship's angular displacement from the established reference was the error. As with the airplane stabilizer, Sperry used this error in combination with feedback from the rudder or control surfaces to generate an electric signal that controlled a servomotor. In the case of the ship gyropilot, the servomotor moved the helm and through the helm the steering engine and the rudder.

Not only was the anticipator a necessary complication in the system, as has been shown, but so were the provisions for automatically easing off the rudder. Sperry solved this problem in 1912 (see figure 10.7) by incorporating in the pilot double feedback (follow-up) operations. The first feedback occurred when the helm or steering engine put the rudder over in response to an error signal: the signal fed back to the gyropilot from the helm indicating that the error signal had evoked the desired response and that the signal should cut off. The duration of response—the amount of rudder movement—depended on the extent of course error. The ultimate objective, however, was not to move the rudder but to place the ship on the desired heading. As the ship began to swing back toward its course, another feedback signal eased off the rudder in proportion to the reduction in course error. The overthrow system was then activated, and the ship would finally settle on or near the proper heading.

To embody double feedback in the mechanism, a roller followed the motions of the helm and a drum followed the relative motion of the ship and gyrocompass. The surface of the controller drum had neutral, positive, and negative sections. According to the sector on which the roller rested, an electric signal was sent (or not sent) to the servomotor. If the roller rested on the positive section, the servomotor and the helm moved in one direction; if it rested on the negative section, the servo and helm moved in the other direction. For example, when the error signal was first generated, it moved the helm until feedback to the roller caused the roller to fall on the neutral sector and the signal stopped. Then, as the ship began to swing to its heading, the drums moved and another signal was generated, easing off the rudder.

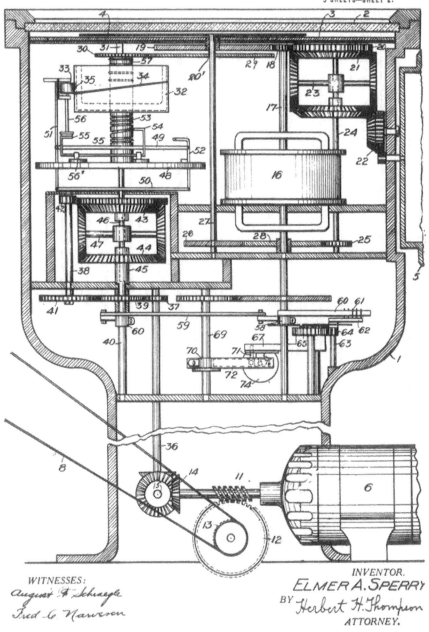

E. A. SPERRY,
NAVIGATIONAL APPARATUS.
APPLICATION FILED NOV. 13, 1914.

1,360,694.

Patented Nov. 30, 1920.
5 SHEETS—SHEET 2.

WITNESSES:
August H. Schraegle
Fred. G. Narveson

INVENTOR.
ELMER A. SPERRY
BY Herbert H. Thompson
ATTORNEY.

Figure 10.6 Sheet from Sperry marine gyropilot patent (Patent No. 1,360,694).
(Key: servomotor controlling helm, 6; gyrocompass repeater, 16; drum, 32; roller, 33; shears, 67, 68; dash pot, 74.)

282

Sperry used trains of gearing to combine the various inputs into the roller and the drum. The inputs included one from the gyrocompass repeater, another from a course-setting wheel, still another from the anticipator responding to the amplitude of the yaw, and the feedback input from the helm motor. These gear trains received independent signals as inputs, carried out logical operations, and produced functions of the signals as outputs.[24]

ii

Because Elmer Sperry did not use terms later common in discussing automatic controls, computers, information systems, guidance and control, automation, and cybernetics, his contribution to these fields is likely to be overlooked unless stressed. Professor Norbert Wiener, in his influential book *Cybernetics or Control, and Communication in the Animal and the Machine* (1948), noted at the time that a body of writing existed about communication, control, and statistical mechanics and that, for analysis to proceed, a common terminology and common approach to the related problems of the various fields were needed. To this end, he and his associates introduced the term "cybernetics" to cover the entire field of control and communication theory.[25] Others, however, have mistakenly assumed that Wiener established the subject and did the pioneer work in the fields covered by the term he originated. Such an assumption, however, is not fair to Elmer Sperry and the other pioneers in guidance and control.

The marine gyropilot and Sperry's other automatic guidance and control devices established him as a major figure in a field now central to modern scientific technology. He designed and made devices that Wiener and others only analyzed decades later. "Metal Mike" followed forty years of Sperry work in automatic control, a sequence extending back to Cortland and the automatically regulated electrical generator and arc light. For the marine pilot, Sperry and his engineers used components like those in his arc-light system, streetcar, electric automobile, gyrostabilizer, gyrocompass, airplane stabilizer, searchlight, and fire-control system. Some of his bright engineers thought his solutions intuitive, but he had labored in the field for four decades and the landscape was not unfamiliar.

Beginning in 1880, Sperry had built feedback, servomechanisms, and computation devices into his inventions; these devices were recognized decades later as the essence of cybernetics. The current regulator in his 1880 dynamo included feedback; the airplane stabilizer of 1914 incorporated

---

[24] The description of the gyropilot follows that in Sperry's patent filed November 13, 1914 (871,885), "Navigational Apparatus," issued November 30, 1920 (Patent No. 1,360,694). The early experimental, hand-built models undoubtedly differed from the patent description. The author has examined the gyropilot held by the Museum of History and Technology of the Smithsonian Institution. This was one of the first ten production models and judging from it, the gyropilot produced by the company for sale also differed from the patent design. The Museum of History and Technology model is Sperry Mark II, Model O, serial number 105 (museum accession number 103045 and catalog number 309,634).

[25] Wiener, *Cybernetics* (2nd edition; Cambridge: M.I.T. Press, 1965), p. 11.

the double feedback made necessary by the alignment error and the problem of monitoring the control surfaces. Servomechanisms were old acquaintances, for he had used them on his dynamo to rotate brushes around a commutator. Nor were analog computers unfamiliar: the gyrocompass of 1911 had a complex computer to make latitude, course, and speed corrections, and his battle-tracer for fire control performed logical operations on a number of inputs. Sperry used a drum controller to provide a programed sequence on his streetcar of 1893; the drum controller for "Metal Mike" was simpler. Sperry was an artist of automatic controls drawing upon a rich vocabulary and experience to compose his solution to the problems of the environment in which he chose to innovate.

There were other pioneers, such as Hannibal Ford and the Anschütz staff, working in the field and making contributions. They drew upon him and he upon them, for patents and technical literature kept the pioneers, the invisible college, informed. It would be interesting to compare Sperry's particular solutions and technological expressions with the others, but what can be suggested here is that although many similarities would be found, enough particularities would remain to permit one to speak of a Sperry "style" in automatic controls.

Sperry's place in the long history of automatic controls can be briefly summarized. The most notable of his controls involved feedback. Feedback has a history extending back to Ktesibios of Alexandria, who probably lived in the first half of the third century B.C., and to Heron of Alexandria, who lived three centuries later. The water clock of Ktesibios had a level control system, and the level control known to Heron resembles that in the water tank of the modern watercloset. Perhaps the oldest feedback device of purely European origin was the automatic temperature-control system invented by the Dutchman Cornelius Drebbel (1572–1632). Probably the best known of the feedback devices of the past was James Watt's steam-engine centrifugal governor. By Sperry's time there were thousands of feedback controls[26] because the practical use of electricity in the third decade of the nineteenth century made possible the use of the electromagnet to feed and regulate arc-light carbons and to regulate the current and voltage of generators.

Sperry began inventing feedback controls when they were being applied to electric light and power systems. Next, he invented controls for streetcars and electric automobiles, but he did not become substantially involved in the development of automatic controls for chemical processes, another major chapter in control history. He did, however, become the leader in America in the branch of the art later known as guidance and control, as he provided gyro-compasses and fire-control devices for the Navy. When men ventured into the air, Sperry helped them adapt with technology: his airplane stabilizers and instruments established his primacy in this remarkable chapter in the history of technology. Once his contribution is recognized, his aerial torpedo should place him among the pioneers of rocket and missile technology.

[26] On the history of feedback controls, see Otto Mayr, *Zur Fruhgeschichte Der Technischen Regelungen* (München: Oldenbourg, 1969)

The marine pilot and then the automatic airplane pilot added to Sperry's reputation in automatic controls, and the young men of his staff following his lead contributed much to these developments. His engineers and his company in the thirties and forties, after his death, continued to invent and develop anti-aircraft and naval gunfire-control systems that incorporated electronic components and complex computation. During World War II, the introduction and rapid development of the digital computer added a new dimension to automatic-control systems; after the war, of course, gyro guidance and control became essential for space flight. Then, with the control field expanding rapidly and taking on a mildly sensational character as "cybernation" and "automation," pioneers like Sperry were sometimes forgotten.

iii

Even though automatic controls had a rich and exciting future in the early twenties, and even though the lines he laid down had logical extentions, Elmer Sperry turned his attention to other areas. On numerous occasions, he had left a field after having established his position in it. In the twenties, he shifted his emphasis from automatic controls to a problem that had first captured his imagination as a young man in Chicago and that he never completely abandoned, even during the tumultuous war years.

More than a quarter of his patent applications during the decade after the war were on the internal-combustion engine; no other category absorbed even ten percent of his applications during this period. Sperry was determined to develop and bring to commercial fruition the compound-diesel engine, upon which he had first made a patent application in December 1892. (He proudly recalled that he had such an engine running before the Chicago World's Fair in 1893.)[27] A report made by a group of engineers in 1900 spurred him on, for they said that if the combustion engine could be successfully compounded, it would mean a most important gain in weight and size reduction—and would constitute a major technological breakthrough. Knowledge that Rudolf Diesel had failed and that German authorities then predicted compounding an "impossibility" reinforced his determination, especially in the postwar years when Sperry developed an antipathy for what he called "Teutonic arrogance."[28]

The successful use of diesel engines in submarines during the war and the demand for an engine using a less inflammable fuel for airplanes kept alive his interest. In 1919 he decided to concentrate his energies on the engine, making no fewer than three applications that year, four during the next two years, and eight in 1922. The concern among petroleum experts that the world was rapidly exhausting its oil supply (a fear relieved by the discovery of rich new deposits in Texas and Oklahoma a few years later) intensified his efforts. Sperry's long-term dedication to the development of the compound engine, which at times seems to have been misplaced, appears,

[27] Filed December 10, 1892 (454,752), "Gas Engine," issued June 8, 1915 (Patent No. 1,141,985).

[28] Elmer Sperry, "Compounding the Combustion Engine," paper read at the annual meeting of the American Society of Mechanical Engineers, December 6, 1921 (SP).

then, to have been compounded by psychological as well as economic motivations.

By 1921 there were a number of engineers and inventors throughout the world working on the compound engine. Sperry thought it wise then to publicize his efforts and achievements by reading a paper on "Compounding the Combustion Engine" at the annual meeting of the American Society of Mechanical Engineers. The paper, which caused a considerable stir, was abstracted and reprinted in leading technical and trade journals. In it, Sperry stated that his four-cycle engine was light compared with the normal diesel, its mechanical efficiency high, and its simplicity commendable. He believed that his results warranted the prediction that the engine would "at no distant date occupy a dignified place in the combustion-engine art." By this time, Sperry had built his fifth or sixth engine and was basing his predictions on its performance.

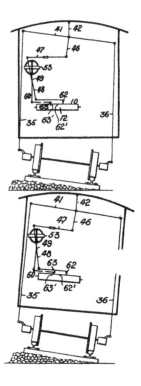

Sperry compared his compound internal-combustion engine to the compound steam engine, which used the expansive power of steam in only one cylinder, although in later engines the steam was led from a small high-pressure cylinder to a large low-pressure cylinder to utilize more fully the energy of the steam. The resulting thermodynamic efficiency, when the diesel was similarly compounded, would permit, Sperry anticipated, a substantial increase in material efficiency, or horsepower per unit weight. This was extremely desirable because the size and weight of the diesel had precluded its widespread use. He predicted that he would be able to reduce the weight from the 150 or more pounds per horsepower of the standard diesel to not more than three pounds per horsepower.

The qualified tone of Sperry's paper to the ASME indicated that while his accomplishments were substantial he still had a way to go along the development road. As was his custom, however, he put himself under pressure by committing himself to his ultimate objectives. He also used his enthusiasm and persuasiveness, based on the stated objectives, to raise capital for research and development. He had access to the funds of his gyroscope company, but he also resorted to using his patents and the commercial potential of his inventions to bring in capital from supporters interested in the establishment of a new company. In 1923 or 1924, he formed a development company, the Sperry Engine Company, to administer the funds and the project, but the company never reached commercial fruition.

There was a pattern to Sperry's efforts to innovate and exploit the engine commercially. Early in the decade, he stressed the engine's potential for marine installations; he then saw the engine as the prime component in a diesel-electric system for mainline railway propulsion. Finally, he concentrated on developing the engine for airplanes. Early in 1928, he received front-page publicity in the New York newspapers by announcing that his diesel engine had "been brought through a long period of secret experimentation to the point where it could be used in airplanes."[29] Sperry

[29] *New York Herald Tribune*, January 7, 1928. In an article, "Hydraulic Transmission for Diesel Locomotive," *Oil Engine Power*, (July 1925), pp. 411–12, he

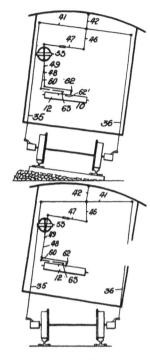

Figures 10.7
Diagrammatic representation
of the operation
of the track-elevation
recording device
under various conditions
(Patent No. 1,843,959).

stressed the safety that would result from using the heavy diesel fuel rather than gasoline, calling attention to recent accidents that, because of gasoline-fed fires, had taken the lives of passengers and pilots. Lawrence's death in an airplane accident, although not caused by fire, must have heightened Sperry's concern for safety; in trying to interest Edsel Ford in the use of the diesel in airplanes, Sperry wrote that he was willing to dedicate the rest of his life, with the active cooperation of "my two remaining sons," to ending the fire hazard by using the diesel fuel.[30]

Sperry continued to promote the engine even after his colleagues began to lose enthusiasm. A young engineer working with Sperry saw in 1926 that no one at the Sperry Gyroscope Company had any interest in the compound engine except Sperry.[31] Despite this lack of enthusiasm, Sperry promoted the project to numerous corporations and agencies. Although these promotional efforts failed and his test engines did not perform as well as anticipated, he persisted, and the essence of his persistence is encompassed in a brief note left in a diesel engine correspondence file: "Eliminate from our organization that pessimism which sees a difficulty in every opportunity, and engender that optimism which sees an opportunity in every difficulty." Some have estimated that Sperry spent more than $1 million in his effort to develop the engine.

The compound diesel was in fact Sperry's major failure. The judgment, however, should be qualified. Ralph L. Boyer, who worked closely with Sperry on the engine and later became vice president for engineering of a large manufacturer of diesels, believed that Sperry pioneered.[32] The major problem, Boyer recalled, was the lack of materials able to withstand the diesel's temperature and pressure extremes (the transfer valve between cylinders, for example, proved an unsurmountable problem). Yet, the fuel injection system used by Sperry was later adopted, and the supercharging done on the compound was years ahead of its time. The compound engine, Boyer decided, was technically but not commercially feasible, for the gain from compounding did not pay for the additional cost.

A notable success achieved in a relatively short time and with reasonable expenditure balanced Sperry's stimulating but frustrating diesel work. His success in devising a method for detecting hidden flaws in railroad rails and a means of determining the levelness and alignment of rails contributed substantially to railroad safety. American railroads continue to use Sperry rail cars to inspect more than 100,000 miles of track yearly and to detect more than 50,000 defective rails.[33]

Sperry developed a track-recorder car before the rail-detector car. In

not only wrote of his engine but also of the transmission he intended to use with it. Throughout the decade, Sperry received notice for his engine in professional and trade journals.

30 EAS to Edsel Ford, September 21, 1926 (SP).

31 R. L. Boyer to Robert B. Lea, October 2, 1961 (SP). Sperry took Boyer out of Ohio State to work closely with him on the engine when he learned that the young man had mechanical aptitude and had invented a compound engine.

32 R. Boyer to Robert Lea, October 2, 1961 (SP).

33 Information from Sperry Rail Service, a division of Automation Industries, Inc.

1912 Sir Henry W. Thornton, the general superintendent of the Long Island Railroad and later president of the Canadian National Railways, discussed with him the possibility of using gyroscopes to determine the alignment and levelness of railway track.[34] Wartime exigencies diverted Sperry's attention from a problem that his experience indicated could be solved easily with gyros. After the war, in the spring of 1923, the Santa Fe Railway asked Sperry, then the recognized gyro expert, whether the gyro could be used for this purpose. Like other railroads, the Santa Fe was competing for more traffic by increasing speed and efficiency, and poor rail and roadbed were limiting factors. Track walkers, or maintenance-of-way inspectors, often could not locate small ballast shifts, low joints, or slight changes in distance between rails.

In 1912 Sperry had anticipated using gyroscopes to establish a frame of reference and using a visual recording device (similar to his ship-roll-and-pitch recorder) to note deviations. In 1923, in response to the Santa Fe request, Sperry laid out, within weeks, the basic design of a track-recorder car; he was so sure of himself that he initiated construction before he had any orders. Subsequently, the design was refined and the car improved with the help of Leslie Carter and other Sperry development engineers. The track-recorder was a railway car of standard weight which carried two gyroscopes and a related system to "wash out" motions of the car body and to record deviations in track levelness. The recorder also noted other data, such as variation in the distance between the rails. Sperry liked to relate how the Santa Fe was able to run trains at twice the speed over stretches of track inspected and corrected with the aid of the Sperry car. The Santa Fe was certainly pleased, for when Sperry travelled across the country with engineers bound for the Japanese World Engineering Congress, the railroad surprised him by attaching a Sperry car to the special train so that he could observe its remarkable performance.

Sperry's work with the track recorder won the railways' respect, and the Santa Fe told him of another troublesome problem. A train wreck on the Lehigh Valley Railroad near Manchester, New York, on August 25, 1911, which took twenty-nine lives and injured many more persons, called attention to a grave menace. Subsequent examination of the broken rail causing the accident revealed piping, or a hidden flaw, due to faulty work at the rail mills. This type of flaw was thought then to be undetectable. "There is probably no defective member of which the railroad man stands in greater fear than the 'piped' rail," a contemporary journal observed.[35] The railroads had appealed to scientists and engineers to attack the problem, and although progress was made by the United States Bureau of

[34] This account of the rail-detector car and track-recorder car draws on a typescript in the Sperry Papers designated the Ruth-Ward manuscript. These two authors presumably revised the Leonard manuscript. Their section on the rail cars is authenticated by surviving source material in the Sperry Papers. "Ruth-Ward" seem to have had access to persons who were involved with Elmer Sperry in the development.

[35] "The Mercy of Steel and the Menace of Wood: A Study of Some Recent Railroad Accidents," *Scientific American*, CVIII (1913), p. 92.

Standards, no practical detector had been developed for the hidden flaws, later known as "internal transverse fissures." Without a detector, the railroads could only scrap all rails produced at the same mill and at the same time as a rail that had failed. An entire "heat" of rails might thereby be taken up and replaced in what can only be described as an effective but expensive solution.

Informed of the problem, Sperry studied the inadequacies of earlier detectors and then set out on a course of his own. First, he tried to invent a device based on the simple principle that the rail is a good conductor of electricity and that a conductor showed a potential, or voltage, drop along its length. He anticipated that the microscopic flaw in the rail would cause an irregularity in the potential drop. He then designed a device with contacts a short distance apart to move along the rails and measure the potential drop. If the potential drop were regular, the rail could be assumed sound. Using extremely heavy currents and an electronic amplifier to detect the small voltage variations, he and his engineers made a detector that performed excellently on new rails in the laboratory. When, however, the detector was tested in 1927 on the rails of the New York Central Railroad, the detector failed because the top of the rails had accumulated so much dirt that the detector could not make contact.

The American Railway Association, which had funded the development of his first detector car, then discontinued support, but Sperry did not abandon the project. His determination recalls an earlier incident that made a life-long impression on his young engineering assistant, Preston Bassett. Before he had even tested his detector, Sperry attended a meeting of railway engineers and metallurgists and informed them—to Bassett's amazement—that he had solved the problem. On the way back to Brooklyn, "Mr. Sperry was full of chuckles," remarking, " 'Well, now I've done it. Now we've got to make it work. I've burned our bridges.' " After the first device failed under test, Sperry explained to Bassett, " 'That's why I burn my bridges, so we don't give up too easily.' "[36]

After a year of working with his own funds and a small staff, Sperry succeeded in making a detector based on a different principle. Knowing that there was a uniform field of magnetic flux surrounding a rail, he decided to use detection coils which, when passed through the air above the rail, would locate wrinkles in the flux comparable to the variations in potential drop. When the new induction type of detector was mounted on a car, it worked; Sperry went on to devise a mechanism that automatically splashed white paint on the rail when the wrinkle was detected. In 1928 the new Sperry detector car entered regular service, sensing flaws and automatically marking them in a way that railroad men found uncanny but greatly reassuring.

The subsequent institutional, or organizational, history of the rail detector is reminiscent of earlier Chicago and Cleveland days. In 1926, when the detector was absorbing substantial development funds, Sperry formed the Sperry Development Company, bringing H. C. Drake and several other

[36] Unpublished biography by Bassett, pp. 101–2.

Sperry engineers into the new company to work on the rail detector and the diesel engine. Sperry assigned nongyroscope patents to the development company as assets to avoid draining money from his gyroscope company. Later, Sperry Products, Inc., was formed to take over the rail-detector assets, and a Sperry Rail Service was organized to build, lease, and operate the detector cars. By 1968, more than four-and-one-half million miles of rail had been tested and almost two-and-one-half million rail defects had been found.

iv

Sperry's inventive triumphs and his distinguished record of achievement stimulated journalists aware of the public's curiosity about inventors to interview him and to look for characteristics that might explain his inventive genius.[37] Many interviewers were impressed by the energy contained in his slight, wiry frame. He was compared to his own inventions, for they too held powerful forces under delicate control. Others were taken by the keen intelligence manifest in his features. His white hair, neatly trimmed mustache, and splendidly tailored suits, however, caused some to think of him as a plutocratic banker or famous diplomat rather than an inventor.

Sperry's conversation emphasized his energy and dogged industry. In response to the usual question about the secret of his inventive success, Sperry went beyond Edison and defined invention as 110 percent sweat, contrasting the months of persistent resolute work with the spark of genius initiating the effort. The spark without the persistence spelled the difference between the dilettante with bright ideas and the successful inventor. The determined effort to fulfill a vision could be sustained only, Sperry believed, if the inventor possessed immense enthusiasm. The inventor could also look forward to that time, "after long periods of research, patient experimentation, and repeated changes," of "great satisfactions," as when he first watched the gyrocompass start its swing toward the meridian. "On such occasions," he said, "there comes over me a welling up from within, a sort of elation, and life takes on a new and exalted aspect. That is living."[38]

Some saw a key to his inventiveness in his powers of concentration and visualization. He confessed that his power to think with mathematical abstractions was extremely limited, but in a remarkable way, he was able to

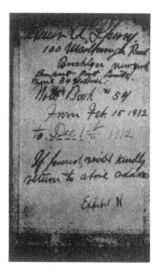

Figure 10.8
Opening page from one of the notebooks Sperry carried with him throughout his life.

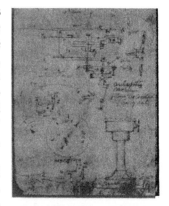

Figure 10.9
Sperry sketches.

[37] Among the articles and book chapters on Sperry are: Maurice Holland, "Brighter Than Sun: Elmer A. Sperry," in *Industrial Explorers* (New York, Harpers, 1928), pp. 56–75; Henry F. Pringle, "Profiles: Gadget-Maker," *The New Yorker* (April 19, 1930), pp. 24–26; C. H. Claudy, "The Romance of Invention: Elmer Sperry—The Man Who Harnessed the Motion of the Earth," *Scientific American*, CXXII (1920), pp. 420ff.; Frank Parker Stockbridge, "Sperry—Competitor of the Sun," *Popular Science Monthly* CXII (1928), pp. 24ff; the essays given by Gano Dunn, W. L. Saunders, and Bradley A. Fiske on the occasion of the award of the John Fritz Medal and later printed in *Mechanical Engineering*, XLIX (1927), pp. 101–16; and feature articles in the *Brooklyn Eagle*, February 16, 1930; *Chicago Daily News*, October 29, 1930.
[38] *A Narration of the Engineering and Scientific Achievements of Elmer Ambrose Sperry, John Fritz Medalist for 1927 . . . Together with Response by Dr. Sperry on the Occasion of the Presentation of the Medal, December 7, 1927.* A 54-page, privately published booklet (SP).

Figure 10.10
Sperry always carried
a circular slide rule
(of his own design)
and a pocket rule,
both shown here.
Also shown is a pair
of Sperry's eyeglasses.

visualize a machine and then operate it in his imagination. In the environment of his imagination, the machine underwent stresses and strains that helped him decide on its form and substance. He had an uncommon ability to conceive the forces affecting machines and also a vivid sense of inertia and friction in machines. Another unusual Sperry characteristic that attracted the attention of the most perceptive interviewers was the way he personified machines. He spoke of "putting the little fellow to work" and of harnessing "that brute"; he used the boy-meets-girl metaphor to explain ignition in his diesel engine; and he lamented the terrible burden of friction that one of the "little fellows" might bear.

In describing his style of invention, Sperry also emphasized the role of his ever-present notebooks. He told a *New Yorker* reporter in 1930 that he was on the last pages of number seventy-eight in the series begun in Cortland. The notebook sketches and notations made during sleepless nights, on trains, and during banquets were evidence of his preoccupation with the technological world. Sperry would also show interviewers the circular slide rule he had invented and his folding rule, devices that were his constant companions.

The portrait of an inventor that Sperry provided was influenced by his never failing sense of audience. He was aware of the popular image of the inventor and he did not offend that idealization too much in discussing his industriousness, his persistence, enthusiasm, risk-taking, and small eccentricities. He did not, however, enlighten his interviewers by discussing the complexities of technological affairs. He did not tell them, as is now quite obvious from his career, that he labored to keep abreast of the state of the art by hard reading in patents and technical papers. He did not bother to describe to them how he kept his sensitive antennae tuned to the needs of the technological systems to which he could respond with his kind of inventions. He also did not explain how a system of inventions could be introduced once he had established an entering wedge, as with his gyrocompass. Nor did he explain that he was an entrepreneur who presided over technological change from invention, through development, and into innovation: he was satisfied to have his interviewer confuse his roles as an inventor, engineer, and entrepreneur. Nor did he say that much of his perspiration came from finding the capital to finance his inventions and that he would often deny dividends in order to invest in research and development. Apparently, he did not think that the casual interviewer would appreciate the resourcefulness that came from his deep experience in the mechanical, electrical, and chemical fields which allowed him to try other approaches when one failed. These were Sperry's characteristics as an inventor that he left for others to note.

On the other hand, he did take pains to explain how he bridged the gap between the heroic inventor of the past and the research and development scientists and engineers of the twenties. Sperry realized that he had made a difficult transition from the isolated inventor working with an assistant or two—if funds were obtainable—to the director of a research and development company. Asked by one interviewer if he thought himself an "old-style inventor" or a research worker, Sperry replied that he was a research worker because he had the support of "a modern system" made up of a professional staff, a well-equipped laboratory, a patent attorney, and a small, specialized library.[39] "Research has done away with races to the Patent Office, with absentminded inventors who sometimes won glory and more often lost everything to smart promoters. It has eliminated, too," he lamented, "much of the exhilaration which came when uncertainty gave way to accomplishment."[40]

Sperry's ability to make the transition from inventor to researcher resulted in part from his appreciation of the role of science in invention and development. Sperry spoke of technology as applied science, but his original work involved, in fact, much more than the application of science. His inventions were often developed and in use before the principles underlying them were discovered and articulated by scientists. One characteristic remark of Sperry's was, "Whenever your theory and practice do not agree, your theory is faulty."[41] His emphasis was on the solution of problems not upon the method used in solving them. If the situation demanded trial-and-error, he readily responded; if science could be applied, he seized the opportunity. Sperry would not allow himself to be characterized, nor would he act, as an empiricist contemptuous of scientists, as Edison seems to have felt the need to do. It is possible that Sperry was more resourceful than Edison and better able to maintain his effectiveness as an inventor because he appreciated science.

Sperry especially admired the "practical" scientist or mathematician, whether ancient or modern. One of his heroes was Archimedes. Corresponding about Archimedes with Professor David Eugene Smith of Columbia University, a distinguished mathematician from Cortland County who had written a history of mathematics, Sperry pointed out with obvious approval, that "the Yankee of Antiquity" made a mathematical analysis only after achieving practical results.[42] Sperry was delighted to find that Smith shared his admiration for Archimedes and also that Smith, having a broader viewpoint than the professorial type, appreciated the practical mathematics. The admiration was mutual.[43]

Sperry could also appreciate pure science, as his enthusiastic and generous contribution to the work of Nobel physicist Albert A. Michelson indicates. Sperry occasionally saw Michelson during visits to his brother-in-law Herbert who lived near the University of Chicago, where Michelson was a professor of physics. During World War I, Sperry sought Michelson's advice on some technical problems, and after the war the physicist asked Sperry to construct some scientific instruments to be used in a new determination of the speed of light. Michelson wanted to use the Sperry high-intensity arc as a source of light and asked Sperry to build a high-speed, revolving,

[39] Holland, *Industrial Explorers*, p. 62.
[40] Pringle, "Gadget-Maker," p. 24.
[41] R. L. Boyer to R. B. Lea, October 2, 1961 (SP).
[42] EAS to David Eugene Smith, December 20, 1927, and January 4, 1928 (SP).
[43] David Eugene Smith to Elmer Sperry, Jr. (?), June 29, 1930 (SP).

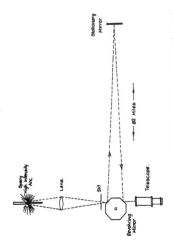

*Figure 10.11
Schematic
of Albert A. Michelson's
equipment for measuring
the speed of light.
The Sperry company
built the revolving mirror
as well as the
high-intensity searchlight.
From the* Sperryscope,
V, 1929.

multi-faceted mirror. From 1924 until 1926, the two corresponded about the problem of designing and constructing an eight- and a twelve-sided revolving mirror and about the use and maintenance of the arc.[44] Michelson provided the basic design for the revolving mirror, but Sperry suggested details and anticipated problems that Michelson did not foresee in the operation of a small precision device subjected to large dynamic forces. A major problem was grinding the surface of the mirrors to the high accuracy demanded and obtaining a running balance at about 40,000 rpm. Michelson valued Sperry's advice and his instruments, referring in an official communication to the "magnificent arc-light" lent to him by Sperry and "the gift of two revolving mirrors," which Michelson styled "beautiful specimens of workmanship."[45] The tone of Sperry's letters shows that he obviously enjoyed the problem and his participation in a significant episode in the history of pure science. The Sperry staff also found the work stimulating: Preston Bassett, whom Sperry dispatched to Mount Wilson to help set up the arc light, said that he "never enjoyed three days as much as these."[46] The engineers and machinists at the Sperry company, who achieved accuracies beyond that which "money can buy in this country," found the challenge exhilarating, and Sperry asked Michelson to send them a word as reward for the continuous work which brought at least one superintendent to the point of nervous exhaustion.

Characteristics that Sperry revealed to newspapermen and writers, the deeper traits and behavior trends established by the sources and the perspective of history, and relevant attitudes such as Sperry's toward science, all contribute toward an understanding of his inventiveness. In an essay on inventiveness, published in an anthology edited by the distinguished American historian Charles A. Beard, Sperry provided additional light on the character of his inventive genius.[47] Beard asked Sperry and other leading American scientists and engineers, including Ralph Flanders, Robert A. Millikan, Lee De Forest, Lillian Gilbreth, and Michael Pupin, to write responses to an earlier collection of essays entitled *Whither Mankind*.[48] In his chapter, "The Spirit of Invention in an Industrial Civilization," Sperry, after a half-century of the life of invention, wrote simply and candidly to define invention:

Think as I may, I cannot discover any time in which I have felt in the course of my work that I was performing any of the acts usually attributed to the inventor. So far as I can see, I have come up against situations that seemed

44 There are about fifty letters in the exchange between Sperry, Michelson, and their assistants, mostly on the revolving mirrors.
45 Michelson to Board of Trustees of the University of Chicago, quoted in a letter to EAS from the secretary of the board, February 19, 1926 (SP). Michelson also mentioned the Sperry instruments in his report entitled, "Measurement of the Velocity of Light Between Mount Wilson and Mount San Antonio," *Astrophysical Journal*, LXV (1927), pp. 35–56.
46 Preston Bassett to EAS, July 31, 1925 (SP).
47 Elmer Sperry, "Spirit of Invention in an Industrial Civilization," in *Toward Civilization*, ed. Charles A. Beard (New York: Longmans Green, 1930), pp. 47–68.
48 Charles Beard, ed., *Whither Mankind* (New York: Longmans Green, 1928).

to me to call for assistance. I was not usually at all sure that I could aid in improving the state of affairs in any way, but was fascinated by the challenge. So I would study the matter over; I would have my assistants bring before me everything that had been published about it, including the patent literature dealing with attempts to better the situation. When I had the facts before me I simply did the obvious thing. I tried to discern the weakest point and strengthen it; often this involved alterations with many ramifications which immediately revealed the scope of the entire project. Almost never have I hit upon the right solution at first. I have brought up in my imagination one remedy after another and must confess that I have many times rejected them all, not yet perceiving the one that looked simple, practical and hard-headed. Sometimes it is days and even months later that I am brought face to face with something that suggests the simple solution that I am looking for. Then I go back and say to myself, 'Now I am prepared to take the step. It is perfectly obvious that this is the way to do it and that the other ways all have their objections.' It usually transpires that the innovation-resisting public will find any amount of fault with the one that is finally chosen. But I have always been tolerably well fortified because I have the feeling that I have made a pretty thorough canvass of the methods which would in all probability occur to the other workers in this field.

In the same essay, Sperry also felt free to discuss the importance to the creative man of a trait more often associated with men in other walks of life. "Let us consider," he wrote:

courage in a somewhat broader aspect. Courage is one of the greatest world forces, if not the greatest. It is courage that has marked leadership in all times. The great world advances in all departments, including the arts, have from time immemorial depended on the courage of leaders. True it is that this word has almost always been associated with wars and warriors. Men dream dreams and have visions and we call them visionary. Once in a while some of these have the inquisitive faculty, but their projects still may die in an early stage. In this case the world has not been advanced. However, when one other factor is added, namely, indomitable courage, then the pioneer pushes his way through untold hardship, finds the mountain pass, and is the first to envision the whole new world lying beyond—the sunlit fertile valleys and vast unutilized resources. Thus it is that the inventor achieves and, also, thus it is that in unfolding the secret of any nation's advance Courage is given a much broader significance than is usually conveyed by narrative history.

Sperry's courage did not go unnoted by his professional colleagues. Willis R. Whitney for example, was impressed by how, under trial, Sperry showed external equanimity and even sanguineness, "no matter how deeply you are internally hurt. . . . You were always a marvel to me," the scientist continued, "you have often perhaps unwittingly helped me by that spirit."[49]

Sperry had known, he wrote, "untold hardships" in bringing about technological change. Rarely did he reveal other than a sanguine character, but in this essay of his mature years, he spoke out against tradition and dogma, "these downright deterrents to progress and blights on the energy and will that bring progress." Sperry was recalling the indifference he had

[49] Willis R. Whitney to EAS, March 12, 1930 (SP).

encountered, the unorthodox approaches he had made, and the tireless promotion to which he—a very proud and dignified man—had resorted so that his ideas might survive. "Conservatism," he wrote, "is simply an effort to gloss over . . . [dogma's] evils and does not excuse it from any of its iniquities." Paradoxically, Sperry damned dogma and tradition in the technological world, while supporting the social status quo. Yet, he had the wisdom to conclude his essay with the thought that, "if modern civilization is troubled in its soul about the so-called evils of the machine," it might be because it has not thought through its problems and presented them "effectively" to inventive minds. An effective presentation to Sperry was one that emphasized "reaching out" and building up, not lashing out and tearing down.

v

Sperry came to appreciate the Japanese as wonderfully inventive people who were, he believed, reaching out, and building up. The last years of his life were enriched by an interest and emotional attachment he developed for Japan and the Japanese. His correspondence about Japan and with Japanese friends reveals values and ideals that had shaped his professional life. He projected onto Japan many of the ideals that had long motivated his own activity, and he imagined in Japan a congenial environment.

When, during World War I, his interest first turned to Japan, his motivation was primarily to promote sales of company products there. The Japanese were in a period of naval expansion and wanted a navy as advanced as those of the occidental powers, which meant, among other things, gyrocompasses. Sperry hoped not only to sell them the compass but also to install gyrostabilizers on their warships. In this effort, Tom Morgan, the Sperry sales manager, and Elmer Sperry, Jr., visited Japan.

Other influences increased Sperry's interest. His daughter Helen visited Japan in 1917, and she, Morgan, and Elmer Sperry, Jr., all brought back enthusiastic reports. Sperry subsequently joined the Japan Society of New York, an organization which encouraged closer relations between the United States and Japan, and he became in 1922 a member of the society's fund-raising committees, along with Gerard Swope, president of General Electric. Sperry found that a number of other leading American industrialists and engineers shared his interest in rapidly industrializing Japan. Sperry was further drawn toward Japan by the interest the YMCA and Christian missionaries had in the country. Sperry had long been sympathetic to both of these activities and was increasingly involved in committee work on their behalf.

In 1922 Sperry visited Japan to attend the twenty-fifth anniversary celebration of the Japanese Society of Mechanical Engineers and the Japanese Society of Naval Architecture, the corresponding societies in America having made him, undoubtedly with his encouragement, their official representative. Named for the occasion an Honorary Vice President of the American Society of Mechanical Engineers and enjoying a world reputation among engineers and industrialists, Sperry was cordially welcomed by the

Figure 10.12 Luncheon party for Elmer and Zula Sperry at Baron Iwasaki's villa in 1922. Tom Morgan, later president of the Sperry Gyroscope Company, stands behind Zula Sperry; Admiral Hideo Takeda, Sperry's friend, is to Morgan's left. Also present were Hantaro Hagaoka (fourth from right), the physicist known at the Lord Kelvin of Japan; Masawo Kamo (third from right), professor and president of the Japanese Society of Mechanical Engineers; and Baroness Iwasaki (seated to Mrs. Sperry's right and in front of Baron Iwasaki). From the Sperryscope, III, 1923.

Japanese as a distinguished guest.

Zula accompanied Elmer on the two-month trip, and the reception they received in Japan was flattering. Besides being a guest of honor on ceremonial occasions, he was singled out for invitations to small dinners attended by leading personalities in the government, army, navy, industry, and the universities. Among those with whom Sperry became acquainted and whom he and Zula entertained at their farewell dinner at the Imperial Hotel were General Yamanashi, Minister of War; Vice Admiral K. Ide, Vice Minister of the Navy; Prince Tokugawa, President of the House of Peers; and Baron Hachiroemon Mitsui, head of the Mitsui family and of the giant Mitsui business enterprises.

Some of those whom Sperry met later became much closer to him through subsequent correspondence and visits. Mostly from the navy, the shipbuilding industry, and the science and engineering faculties of the universities, they represented the elite then presiding over the introduction of modern technology to Japan and were therefore influential in Japan during the twenties, a decade when parliamentary regimes primarily interested in developing the economic strength of the nation prevailed. This small group included Takuma Dan, the general director of the Mitsui firm, who had graduated in engineering from the Massachusetts Institute of Technology; Professor Hantaro Nagaoka, a professor of science and engineering at the Imperial Tokyo University and known as the Lord Kelvin of Japan; and Professor Baron Chuzaburo Shiba, professor of shipbuilding at the Imperial University, who had studied at London's Central Technical College (later the Imperial College of Science and Technology). Baron Vice Admiral Hideo Takeda, a retired naval engineer who headed the

Mitsubishi Shipbuilding Company and also the Mitsubishi Electrical Engineering Company, became a close friend of Sperry's. The two men were about the same age and found that their views on society and human nature were quite similar.

Sperry's reputation preceded him to Japan. A shipbuilding nation could not avoid appreciating his major inventions. His close relations with the United States Navy and his service on the Naval Consulting Board were also highly regarded in a country where naval-industrial relations were close. Having learned of his many patents in other fields and of his high standing as a professional engineer, his Japanese friends regarded him as second only to Thomas Edison as an inventor-engineer. As they came to know him better, they also came to appreciate the course of his career from simple rural beginnings to national prominence and to admire his moral values, which were not unlike their own.

Sperry left Japan in November with strong impressions of the country's rapid strides in technology. At a time when many Americans thought of Japan as imitative and second rate, Sperry's keen and experienced eye saw the unmistakable signs of technological excellence and maturity. Later, he carried this message to the world-wide engineering fraternity, which was also ill-informed about Japanese technology. Sperry had been especially impressed by two large dockyards, one of which he judged to be four times larger than any in America. He admired the systematic layout and operation of the yard, where they "take in Swedish pigiron at one end of the place (only it happens to be the middle) and put out a 33,000 ton battleship at the other end."[50] He observed that America had only two model ship basins for scientific experiments but that Japan had four. In Japan he also saw superior machine tools which were, he knew, the essence of precision manufacture and represented a heavy capital investment. He judged a Japanese-built, horizontal milling machine and a forging press to be larger than any in America. In a decade when the construction of high-voltage electrical-transmission networks, or grids, was a sign of advanced technology, Sperry found the Japanese construction of the finest kind, better than that he had seen in America. He also heard that the Japanese had three times more high-tension transmission lines per capita than the United States. The Japanese were no longer dependent upon the import of complex materials and machines such as generators, turbines, and armor plate.

After his return home, when he was giving his many talks on Japan to American engineering groups, he cited the physical evidence of technological excellence and he also praised the Japanese spirit that made these accomplishments possible. He thought there were no engineers in the world more studious, devoted, and enthusiastic. The explanation for their technological achievement was, he said, the absence of traditions and precedents, leaving "their minds . . . free to go straight to the mark."[51] He also attributed their technological achievement to the spirit of teamwork and

[50] Elmer A. Sperry, "Some Observations on a Trip to Japan," *Michigan Technic,* XXXVI (1923), p. 5.
[51] *Ibid.,* p. 6.

this spirit, in part, to the racial homogeneity of the nation. Sperry greatly admired another characteristic which he believed contributed to their technological progress: the orderliness, self-restraint, and dignity "on every hand."[52] Though no social scientist, Sperry probably perceived far better than most observers the spiritual foundations of technological development.

Sperry thought that the Japanese potential for technological greatness was manifested by their conquest of the natural environment and often referred to the cruel Japanese mountains that dominated the landscape and symbolized for him a hostile nature. These, he thought, had taken a toll for thousands of years, but in response to the challenge, the Japanese had developed techniques that made the environment livable. This capacity carried over into the modern era in which they used modern technology not simply to adapt to nature but to subdue her.[53] "I did not know," Sperry would say, "that 19/20 of the entire surface of Japan was covered with ragged, cruel mountains. . . . The vigor of the ceaseless effort to conquer these barriers has wrought marvels throughout the land and reacted strongly upon the people themselves."[54] These mountains may also have reminded Sperry of the rugged hills of Cortland which he knew as a child.

Baron Vice Admiral Hideo Takeda agreed with Sperry that the mountains had influenced Japanese character. Through his frequent letters to Sperry after the 1922 visit, Takeda deepened and extended the American's understanding of Japan. The two had first become acquainted during World War I, when Takeda, like so many other Japanese engineers, had visited the Sperry plant in Brooklyn. After Takeda's company, the Mitsubishi Shipbuilding Company (*Mitsubishi Zosen Kaisha*), became the licensee for Sperry products in Japan, business problems brought the men in closer contact. During the 1922 visit, Takeda accompanied Sperry on his government-sponsored visit to Korea. An extended correspondence followed and a warm relationship developed.

Takeda interested Sperry and seemed to reflect the strength of developing Japan. Son of a Samurai, he graduated from the Japanese Naval Engineering School in 1883 and then studied for a year in France. He rose to the rank of Vice Admiral of Engineers in 1913 and was decorated for wartime service in the wars with China and Russia. Before retiring in 1918 from active service to become the chairman of the Mitsubishi Shipbuilding Company, and later of the Mitsubishi Electrical Engineering Company, he was the director of the Naval Engineering School. Like most high-ranking naval officers, he spoke and wrote English (the Japanese Army officers used German).

Although mutual business interests were originally the basis for their relationship, Sperry discovered soon that they shared similar values. Their letters, extending over the years from 1922 to 1930, show the similarity of their views.[55] When Takeda praised frugality, simplicity, dignity, and

---

[52] *Ibid.*, p. 8.
[53] EAS to H. Takeda, February 14, 1923 (SP).
[54] Sperry, "Trip to Japan," p. 8.
[55] There are almost 50 long letters in the Takeda-Sperry correspondence.

serenity in mature men, Sperry warmly applauded. Takeda believed that older men, wise from experience, must guide the young, and Sperry wrote of the responsibility to improve and give "uplift." The Admiral lauded studiousness, attentiveness, politeness, courtesy, and unremitting hard work in young men, and Sperry heartily agreed. When Takeda counseled patience and resignation in the face of life's tribulations, Sperry gained strength in a time of personal sorrow. Elmer Sperry did not hesitate to tell his more articulate and philosophical friend Takeda that from him he drew inspiration and counsel. When Takeda asked himself whether he "had not betrayed Truth for the sake of the flesh and the dross of the world," Sperry acknowledged that "the material unfortunately is the realm in which I have my being."[56]

The Takeda-Sperry correspondence continued uninterrupted; Takeda was one of the last persons to whom Sperry wrote before his death. He never lost his admiration for Takeda and Japan but continued to find in both a manifestation and articulation of his own values and objectives. Japan revivified Sperry's youthful vision of America. In the twenties he was disillusioned by many aspects of the American social and political scene, often referring to the demoralizing influences of recent European immigration and holding up Manhattan Island as an example of all that was wrong with America—its loose morals, its alien radicalism, and its social discord and abandonment of the ethics of hard work and frugality.[57] By contrast, he found in the Japanese his own faith that technology would create an ideal world and the willingness, because of this faith, to work unquestioningly for material improvement through engineering and science.

On numerous occasions Sperry translated his attitude toward Japan into action. Besides his willingness to share his experience in technology with them, he contributed financially to the organizations promoting Japanese-American friendship. He often spoke and wrote about Japan, especially for audiences of engineers and industrialists. When the Congress of the United States abrogated the "gentleman's agreement" existing between the two countries, and totally excluded Japanese immigrants in 1924, Sperry wrote to many influential persons in the United States government, hoping to reverse the decision, as well as to his Japanese acquaintances, hoping to soften the blow to their pride. His major contribution toward better relations between Japan and the West was his promotion and organization of the World Engineering Congress held in Japan in the fall of 1929.

As the originator and American organizer of the World Engineering Congress, Sperry established himself as Japan's foremost friend among American engineers. His enthusiastic talks and lectures stimulated an interest that prepared the ground for his proposal for a congress there. After hearing Sperry speak on Japan, Calvin Rice, secretary of the American Society of Mechanical Engineers, predicted that every member of the audience would be eager to see Japan's remarkable technological progress. Sperry also

56 EAS to Takeda February 14, 1923 (SP).
57 Sperry, "Trip to Japan," p. 7; and *Suffolk County News*, May 10, 1929.

used his influence in several engineering organizations to promote a Japanese congress. With the cooperation of his friend Calvin Rice, he interested past presidents and the council of the ASME in the congress. As a member of the Engineering Foundation, an organization representing the major engineering societies, and as a member of the Division of Engineering and Industrial Research of the National Research Council, Sperry coordinated the interest of the engineering societies. He also used his many close contacts with leaders of the engineering world, such as Herbert Hoover and Thomas Edison. He had similar contacts in Japan.

A Sperry luncheon held early in 1925 at the Engineers Club in New York City for Dr. Masawo Kamo, the president of the Japanese Society of Mechanical Engineers, Calvin Rice, and the past presidents of the ASME, resulted in the formation of an informal ASME committee and a resolution that all of the great engineering societies of America sponsor a congress to be held in Japan within five years. Rice proceeded to enlist the cooperation of the other societies and Sperry, in March 1925, asked Kamo to inaugurate the effort there.[58]

Kamo, a professor of engineering at the Tokyo Imperial University, whom Sperry had met in 1922, was not only president of the Japanese mechanical engineers but also head of the Association for the Promotion of the Industrial Policies (*Kosei Kai*), an organization of several thousand engineers, industrialists, and educators who advised the government. Kamo elicited the aid of Takuma Dan, head of Mitsui and president of the Engineering Club of Japan, Baron Shiba, and others who knew Sperry.[59] These men and the Japanese Society of Engineers—an amalgamated engineering society (*Kogak Kai*)—persuaded the Japanese government to endorse officially and support financially the congress to be held in 1929.

In America, Sperry persuaded Herbert Hoover, then Secretary of Commerce, to endorse the congress and serve as honorary chairman of the American committee helping organize it. In response to the request for endorsement, Hoover replied succinctly and candidly that although he did not regard "these conferences as having any profound effect on advancing technology," they had other important values; in this case he thought it would be a well-deserved compliment to Japanese progress.[60] Sperry made the most of the support from the world-famous engineer and administrator. Letters were sent out over Hoover's signature inviting eighty-two prominent American engineers and industrialists to serve on the American committee. It was an indication of respect for him and the interest in Japan that eighty accepted, among them Edward Dean Adams, John J. Carty, Howard E. Coffin, Everette DeGolyer, Gano Dunn, W. F. Durand, Thomas A. Edison, John Hays Hammond, Samuel Insull, Dugald C. Jackson, Frank B. Jewett, Arthur E. Kennelly, Charles F. Kettering, Dexter S. Kimball, John W. Lieb, Arthur D. Little, A. A. Michelson, Ralph Modjeski, Robert A. Millikan, William Barclay Parsons, M. I. Pupin, E. Wilbur Rice, Jr., Charles M.

[58] EAS to M. Kamo, March 6, 1925; and cable, March 30, 1925 (SP).
[59] M. Kamo to EAS, April 22, 1925 (SP).
[60] Hoover to Charles F. Rand, May 20, 1925 (SP).

Schwab, Alfred P. Sloan, Jr., Lewis B. Stillwell, Ambrose Swasey, Gerard Swope, D. W. Taylor, Elihu Thomson, and Orville Wright. Most of these men were known personally to Sperry and were among the elite of American engineering in the 1920's. Elmer Sperry was named chairman of the committee; Hoover remained the honorary chairman.

The problems of organization encountered between 1925 and 1929 demanded Sperry's energy during a period when his health was not robust. Besides the many letters written to Americans and Japanese asking their cooperation, there were dinners to be held, funds to be raised, committees to be formed, and a program to be planned. Sperry freely gave his energy. For administrative expenses the Americans turned to private enterprise, a move indicative of the temper of the times. Almost $30,000 was contributed, mostly by large American corporations. Those contributing $1000 were the Sperry Gyroscope Company (the first contributor), Westinghouse Electric Manufacturing Company, Ford Motor Co., General Electric Co., Bethlehem Steel, Baldwin Locomotive Works, and others with interests in Japan. Samuel Insull, the Chicago power and light tycoon, gave $2000, and William S. Barstow, an electrical engineer and utility head, contributed $3000, the largest single contribution.[61]

Sperry wanted the congress to offset the adverse effects of the Japanese Exclusion Act, the small acts of prejudice that Japanese encountered in America, and, more generally, to promote peace and cooperation between America and Japan. He had a deep faith, as did many engineers then, in the ability of engineers to succeed in international relations where politicians failed. Sperry thought that engineers had a common sense, a rational approach, and a language that precluded misunderstanding. The engineer, Sperry predicted, "by his quiet work can be a great factor in bringing nations together." He wrote Takeda that

it is the engineer, through the fact that he speaks a common language in every tongue and not only has a common understanding but reaches common conclusions that must be looked to to draw the nations closer together; and when the engineers of the world have an appreciative understanding of each other, not only as to their common problems but their methods of solution, then in my judgment a mutual admiration and fondness for each other will spring into being, which no amount of political pressure or the devious ways of so-called diplomacy will ever be able to break through or change.[62]

The Japanese organizers of the congress also saw it as an opportunity to initiate and promote international cooperation and understanding among the engineers of the world, who were "so essential to the advancement of the welfare of mankind."[63] More than this, the Japanese interpreted the presence of the congress in Japan, the first major engineering congress to be held in the Far East, as a recognition of their technological achievement. "Japanese engineers," the official program commented, "have long hoped for and

[61] Maurice Holland to Frank Jewett, February 3, 1930 (SP).
[62] EAS to H. Takeda, April 9, 1929 (SP).
[63] *Program of the World Engineering Congress and World Power Congress* (Tokyo: Kosei-Kai Publishing Office, November 1, 1929), p. 29.

awaited the opportunity for such an international congress . . . and at last their hopes are to be realized."

The Japanese government, the engineering societies, and industry showed in many ways the importance they attached to the congress. Prince Yasuhito of Chichibu, eldest brother of Emperor Hirohito, acted as "patron" for the congress and attended, with his princess, several congress meetings and social occasions. Sperry was given to understand that never before had the royal family taken so deep and active an interest in any civil affair.[64] Sperry's Japanese acquaintances, Takuma Dan, Baron Shiba, and Masawo Kamo, were officers of the congress and helped to organize an unusually large program of 800 technical papers. Sperry found the organization of these sessions by the Japanese "nothing short of marvelous in perfection of every minutest detail." He found the 435 papers presented by the Japanese especially informative and vital and marveled at the ninety-one social events scheduled for the nine-day congress. Although modest, Sperry did not fail to note that he was singled out for honors and recognition by the Japanese before and during the congress. At the dinner given on October 2, 1929, in Washington by the Japanese ambassador for the American delegates to the congress, the ambassador said that Japan was particularly indebted to "Dr. Sperry" who, with his associates, "spared no effort in ensuring the success of the congress." As the head of the American delegation and an honorary vice president of the congress, Sperry gave a short address at the opening ceremony in Tokyo on October 29 and was told that his was the only speech that Prince Chichibu applauded. During a gala festivity at the premier's residence, he was escorted "away from the crowd" to a private reception for a few congress dignitaries and the premier, the prince, and princess. Sitting close to the royal couple and noting that all looked silent and glum, Sperry ventured to strike up a conversation with the princess and "the ice was broken."[65] As toastmaster at the dinner given by the American delegation for their Japanese hosts, he found himself seated at the speakers' table near the royal couple once again. Hearing that the prince was somewhat shy of public occasions and speaking, Sperry told him that he too found public speaking an awful challenge—which, in truth, was somewhat wide of the mark. Of all the festivities, Sperry decided that the dinner given in his honor on November 7, the last evening, the "greatest."[66] The *Mitsui Bussan Kaisha* (the export-import firm of the family), the *Mitsubishi Zosen Kaisha* (Takeda's shipbuilding company), and *Tokyo Keiki Seisakusho* jointly gave the dinner. The Japanese engineering and industrial leaders did not mute their praise of him, "ascribing all manner of virtues to me that we know I do not possess at all." The occasion caused him to write Zula:

I am extremely fond of these marvelous people. I have tried hard to analyze it. It may come from my worshipful attitude towards progress. The job of the engineer is to make progress, and the progress made ever since you and I

[64] EAS to H. W. Moody, February 15, 1913 (SP). This was the official letter-report that EAS sent to all the organizations he represented.
[65] EAS to Zula, November 2, 1929 (SP).
[66] EAS to Zula, November 10, 1929 (SP).

were here before has been so amazing as to almost paralyze one.[67]

vi

Sperry attributed continuing Japanese progress, in part, to mushrooming Japanese research and development facilities. His appreciation of the importance of industrial research in an industrial economy had been increased by his work for professional engineering organizations. In the twenties, he gave considerable time to engineering societies, which recognized him as a senior statesman of the profession. His association with them, it should be recalled, began during his Chicago days when he helped found the National Electric Light Association and the American Institute of Electrical Engineers. Because the organization of professional engineering in the United States was loose and ill-defined, Sperry's later contributions to the profession can be better understood if the structure of the profession is briefly examined.

No single society represented professional engineers in America. The best-established societies—the civil, mechanical, electrical, and mining and metallurgical engineers—had some claim to speak for the entire profession because of their encompassing scope. These, termed "founder societies," cooperated formally and informally to influence the character of the profession. Because the work done by most engineers was determined by the industrial corporation or the government agency for which they worked, the profession had comparatively little influence in this respect. The professional societies, however, did have more impact on the way in which the work was done. During the twenties, for example, the profession was particularly interested in promoting engineering research.

Sperry, recognized as a pioneer in engineering research and development, was chosen for positions of leadership in two organizations, the Engineering Foundation and the Division of Engineering and Industrial Research of the National Research Council; both were supported by the founder societies to promote research. The Engineering Foundation was founded in 1914 by a grant from Ambrose Swasey, the mechanical engineer and industrialist famous for his machine tools and scientific instruments; its purpose was "the furtherance of research in science and engineering. . . ."[68] Swasey succinctly described it as "a research institution under the direction of the Founder Societies." The Division of Engineering and Industrial Research was formed in 1919, when its parent body the National Research Council was reorganized on a permanent basis to promote scientific research in the national interests.[69]

[67] *Ibid.*
[68] The Engineering Foundation, *Annual Report for the Eighth Year* (New York City, 1923), p. 4; Swasey to W. L. Saunders, March 17, 1925 (copy of letter in files of Engineering Foundation, United Engineering Center, New York, N. Y.)
[69] *A History of the National Research Council* (Washington, D. C., 1933). The purpose of the Engineering and Industrial Research Division was to "encourage, initial, organize, and co-ordinate fundamental and engineering research in the field of industry." For several years, the similarity of purpose of the foundation and the

Figure 10.13 Meeting at the Franklin Institute in Philadelphia of Elihu Thomson, Frank J. Sprague, Charles Brush, Sperry, and E. W. Rice (left to right), to commemorate Charles Brush's introduction of his arc-light generator a half-century earlier.

Figure 10.14 A historic meeting of pioneers in 1929: Orville Wright, Amelia Earhart, Elmer A. Sperry, and Vilhjalmar Stefansson (the Arctic explorer).

Sperry was named a member-at-large of the board of the Engineering Foundation in 1921 and elected a vice chairman in 1925. The foundation, in 1925, specified as its preferred activity support of engineering research for the national engineering societies. It selected and approved research projects proposed by the engineering societies and decided upon the agency or individual to carry them out. The foundation provided financial support for the projects but did not assume responsibility for the administration of any of them. The Engineering Foundation, therefore, acted as a catalytic agent in the field of engineering research.

Sperry was a board member and then an officer of the foundation from February 1921, to February 1927. Other leading engineers who served on the board with Sperry were Edward Dean Adams, the developer of the Niagara Falls power complex; Gano Dunn, the electrical engineer and leader in scientific and engineering affairs; Frank B. Jewett of Bell Laboratories; Arthur D. Little, pioneer in the development of and articulate spokesman for industrial research; E. Wilbur Rice, long known to Sperry as technical director of General Electric; and W. L. Saunders, Sperry's colleague on the Naval Consulting Board.

While Sperry was helping to make policy, the Engineering Foundation stimulated and supported research projects of general interest to the engineering profession. Among the subjects investigated were reinforced concrete arches, arch dams, fatigue of metals, lubrication research, welding, wood finishing, thermal properties of steam, steel columns, and mining method.[70] Other organizations, private and professional, often contributed financially to the projects stimulated by the Engineering Foundation and maintained the projects after the foundation had done the "seeding."

In 1927 Sperry left the vice chairmanship of the Engineering Foundation to become chairman of the Division of Engineering and Industrial Research for the next three years. As chairman, he assumed the task of persuading American enterprise to support industrial research. Under Sperry, the division explained why research was important; later it stressed how to do research. To establish the importance of research, the division held general meetings to which leading industrialists were invited to hear members and guests discuss the advantages of research. During Sperry's tenure, the division also circulated a questionnaire to ascertain the state of industrial research in America. The administrator of the division, Maurice Holland, wrote a book, *Industrial Explorers* (New York, 1928), in which he presented "stories of the work, personalities, and peculiarities of twenty of the nation's leaders of industrial research." Among these were Sperry, Willis R. Whitney, and Frank Jewett. The division also promoted publicity in

division brought close cooperation between them. The Engineering Foundation endowed by the Swasey gift furthered research in science and engineering by providing administrative support for the inadequately funded division. Since both organizations had a large representation from the founder societies, coordination in the formulation of policy was possible. Engineering Foundation, *Engineering Foundation: A Half Century of Service, 1914–64* (New York, no date).

70 The Engineering Foundation annual reports, 1922–27. These were published annually by the foundation at the Engineering Societies Building, New York City.

the popular press, hoping thereby to persuade the stockholders of corporations to support managers committed to industrial research. These efforts undoubtedly contributed to the rapid development of America as the leading proponent of industrial research.

Besides the promotion of industrial research, the division, like the Engineering Foundation, spurred research projects by selecting problems of "broad fundamental character, or of importance to a group of industries or public utilities, or to some branch of the engineering profession."[71] During Sperry's tenure, highway research, welding, heat transmission, electrical insulation, and industrial lighting were among the projects carried out. The projects sometimes involved the establishment of a committee named by the division, the support of a paid director and office, conferences, numerous experiments, and the publication of results. Cooperation with trade organizations, engineering societies, and universities was not uncommon. The division also provided support for American participation in Sperry's World Engineering Congress.[72]

By the end of Sperry's three years as chairman, the work of establishing the need for research had been successful, and the division decided to turn to showing how best to do research. Success had been achieved in persuading smaller companies to follow the lead of larger ones in carrying on research, and trade organizations and other collective groups had been encouraged to sponsor projects of general interest. By the time of Sperry's death, industrial research in America had reached the stage at which the division was beginning to focus upon the need to involve universities.[73] This relationship would flourish within a few decades, but it was one in which Sperry and his company had not become involved.

Sperry was well suited to lead the division. In his speeches and informal talks, he reached a broad audience extending from presidents and managers of corporations, both large and small, to stockholders. His style of presentation did not offend the sophisticated and it captured the imagination of the inexperienced. He was equally at ease speaking on research at a professional meeting or to a cross-section of laymen at a formal after-dinner speech. More important, his experience and personal characteristics were such that he had the respect of research scientists as well as the less theoretical and more empirical development engineers. He provided an excellent bridge at a time when America was moving from unorganized invention to organized science-based research.

Sperry's peers recognized his contribution to the profession and his excellence as an inventor, engineer, and entrepreneur by electing him in 1928 president of the venerable and influential American Society of Mechanical Engineers. In this office, he followed such prominent engineers as

---

[71] *Division of Engineering and Industrial Research of the National Research Council* (New York, 1929); a pamphlet.

[72] Annual Report, Division of Engineering and Industrial Research, National Research Council," May 1929; mimeograph.

[73] "Executive Committee Meeting of the Division of Engineering and Industrial Research, July 17, 1930," National Research Council Archives, Washington, D.C.

Figure 10.15 In 1926, the leading engineering societies of America awarded Sperry the John Fritz medal.

Robert Henry Thurston, Coleman Sellers, John Fritz, Ambrose Swasey, Frederick Winslow Taylor, George Westinghouse, and Charles M. Schwab. Although the office was an honor, it was not without its responsibilities. Sperry attended meetings of the governing council every two months or so until his voyage to Japan. He received reports and routine proposals from Calvin W. Rice, the administrative secretary and his long-time acquaintance. He approved the budget, heard technical and other reports of the many committees of the society, considered publications for the society, helped appoint representatives to conferences, and discussed the work of the professional divisions of the ASME. The decisions that he and his fellow council members made required experience in dealing with the diverse and loosely structured activities of the society. In addition to these duties, Sperry had the heavy responsibilities of final arrangements for American participation in the World Engineering Congress in Japan. The society gave him its support in this task.[74]

The profession also honored him by awarding him the Fritz Medal for 1927. The John Fritz medal, awarded annually by the founder societies "for scientific or industrial achievement," was the highest award in the engineering profession. Named for the pioneer in the American steel industry,

[74] This summary of Sperry's duties has been obtained from a reading of the minutes of the ASME "Council Minutes" while he was president. These are on file in the United Engineering Center in New York City.

the prestigious medal had been awarded to Lord Kelvin, Westinghouse, Bell, Edison, Alfred Nobel, Elihu Thomson, George Goethals, Orville Wright, Marconi, and thirteen other distinguished engineers and inventors before Sperry. The award cited Sperry for his "development of the gyro-compass and the application of the gyroscope to the stabilization of ships and aeroplanes."

When Sperry first heard of the award, he was so pleased that he shared his first reactions with a passing office boy upon whom the information but not the enthusiasm was lost. The award ceremonies included a large banquet for the distinguished members of the engineering world and speeches on Sperry delivered by Gano Dunn, W. L. Saunders, and Admiral Bradley A. Fiske. Dunn, a past-president of the American Institute of Electrical Engineers and chairman of the National Research Council, spoke of "Sperry's Early Life and Pioneer Work"; Saunders, who besides being chairman of the Naval Consulting Board was a past-president of the American Institute of Mining and Metallurgical Engineers, discussed the "Activities of Elmer Ambrose Sperry in the Work of the Naval Consulting Board . . . and in Mining"; and Sperry's old friend Admiral Fiske presented the "Contributions of . . . Sperry to the Naval Arts."[75]

Other honors from the profession came to Sperry. In 1914 and 1916, he had won the Collier Trophy awarded by the National Aeronautic Association. In 1927 the Council of the American Society of Mechanical Engineers awarded him the Holley Medal, bestowed "for some great and unique act of genius of an engineering nature that has accomplished a great and timely public benefit." Sperry was cited in the award for achievements and inventions advancing the naval arts, including the gyrocompass that "freed navigation from the dangers of the fluctuating magnet compass." Then in 1929 the American Iron and Steel Institute voted him the American Iron and Steel Institute Medal for his paper, "Non-Destructive Detection of Flaws." Also in 1929 the Franklin Institute awarded him the Elliott Cresson Medal, an annual award for original research, invention or development, or "unusual skill or perfection in workmanship." Sperry received the award for navigational and recording instruments contributing to safety, comfort, and economy.

Science did not ignore Sperry as it did Edison for many years. The National Academy of Sciences elected Sperry a member in 1925. Several years later, when the academy finally elected Edison, Sperry remarked to a colleague as the meeting adjourned, "All during the discussion about Edison I kept saying to myself, 'Sperry, how the hell did you get in.' "[76] The academy members at the time numbered about 100, most of whom were scientists. There were only 17 engineers, but among these were Herbert Hoover, Frank Jewett, John J. Carty, Gano Dunn, and Sperry's old friend, Admiral David W. Taylor. Frank Jewett wrote that, as an "intimate personal friend," he was deeply gratified with the academy's choice.[77] Sperry

[75] *Narration of the Engineering and Scientific Achievements of Elmer Ambrose Sperry.*
[76] Pringle, "Gadget-Maker," p. 24.
[77] Jewett to EAS, April 30, 1925 (SP).

was "deeply sensible of the great honor . . . conferred on me . . . and exceedingly proud to be numbered with such a distinguished group, many of whom are friends of mine."[78]

vii

Although Sperry was primarily concerned with professional affairs, his interests broadened with age. In his business diary, along with notations of appointments related to his work, one finds an interesting variety of nonprofessional notes. There is, however, a pattern to the responsibilities that he assumed outside of the professional sphere. He never abandoned his religious commitments. The YMCA had played a major role in his development; he now assumed a substantial role in the continued development of the YMCA. Sunday school had meant much to him as a boy; as an adult, he contributed to the Sunday school movement. As a young man writing a paper for a Cortland Normal professor, he had extolled Yankee genius and industrial America; an older Sperry had unbounded enthusiasm for Japanese ingenuity and genius and showed this in his support of societies promoting friendship and cultural exchange between America and Japan. Nor did he forget Cortland, its people and its needs.

The calendar for his last few months of life is indicative of his philanthrpoic commitments and also of the variety and fullness of his life. No longer physically strong, he was able, nevertheless, to attend over a two-month period, YMCA directors' board and luncheon meetings, a dinner in his honor given by the Trustees of the Brooklyn YMCA, a luncheon conference of the World Sunday School Association and of the council of the Federation of Churches, and a directors' meeting of the Japan Society, as well as to entertain Japanese visitors. Throughout the last decade of his life such meetings appeared regularly on his calendar.

Sperry never gave up his role as lecturer and after-dinner speaker. He frequently lectured on the gyro and its uses to such universities as Harvard, Princeton, Columbia, Chicago, Michigan, and the Sorbonne. His spring lecture to the graduating class at the U.S. Naval Academy became an annual affair. Besides these talks, he frequently appeared before clubs and religious organizations to share his knowledge of Japan. He once wrote to a friend that, "having spent five nights within a week on a sleeper, I do not feel as though I should ever get rested and my natural impulse is to say 'Never Again.' . . . [Lectures] certainly do use up a great deal of vital force."[79] He did not, however, refuse his friend's request for another.

Sperry was generous with his time and money. He left about a third of his fortune to the YMCA. Perhaps he sentimentalized and simplified, but he said that the root of his bequests was his gratitude to the YMCA for having introduced him to the engineering and industrial world by making possible his trip to the Philadelphia Exposition of 1876. He might as easily have gone on to speak of the way in which the Y library, with its technical journals, and the Y speakers, with their technical interest, cap-

[78] EAS to David White, Home Secretary, National Academy of Sciences, May 7, 1925 (EAS file, National Archives, Washington, D.C.).
[79] EAS to B. B. Lamme of Westinghouse, December 21, 1916 (SP).

tured his imagination and inspired him. Sperry also appreciated the YMCA because it served young men in their character-forming years. He spoke of the organization as his "college," and his contribution to its endowment must have made many colleges envious. In 1929 he gave $125,000 to the Brooklyn and Queens YMCA, added several hundred thousand in 1930, and provided for a total trust of $1,000,000 through his will. Edward Sperry was one of the three trustees of the fund, the income of which was for the first ten years to be used for the Brooklyn and Queens YMCA, for which Sperry had been a director and trustee for more than twenty years. After the ten-year period, the income was to be used for any YMCA project the trustees might designate, although his stated preference was still the Brooklyn and Queens organization. In 1936 the Flatbush Branch dedicated the Sperry Memorial Building to which half of the trust fund income had been applied for the first ten years.

Sperry remembered not only the YMCA from his Cortland period but also the town of Cortland and its inhabitants. Especially close to him was his Aunt Helen Willett, his father's sister, who had mothered him when he was a boy. He often went out of his way during his busiest times to call upon "dear Aunt Helen" and regularly sent her money to help meet expenses. His interest in Cortland and its citizens resulted in his election, in 1923, as president of the very active Cortland County Society of New York. When the town launched a drive to build a free library, his was the largest single contribution in the $100,000 raised, and he delighted in cajoling his former patent lawyer into making a sizable contribution. The lawyer had first approached the inventor as a lover of books who would certainly want to help build the library; Sperry quickly pointed out that a lawyer certainly depended more upon books than he and prodded his friend into making a larger gift. Sperry laid the cornerstone when the new Cortland Free Library building was dedicated in June 1927, and he took pleasure in recalling how "Cortland county tolerated me under foot' and how he made his first inventions there. He described his gift as a way to "pay back . . . the people of Cortland."[80]

As a member of the executive committee of the World Sunday School Association, the inventor helped raise funds for a Sunday School building in Japan and furthered missionary work in both Japan and China. Believing that "whatever I have attained that is worth while . . . is not due to myself but to my early training both in the Sunday School and in my Christian home," he strove to spread the Christian faith.[81] To do this, Sperry generously supported C. H. Robertson, a missionary to China, who was admired by Sperry and other leaders of the world of engineering and science because he had been an engineering professor and a dean at Purdue University until he decided, around 1902, to go to China as a science lecturer for the international committee of the YMCA. At Purdue he had been a superb athlete and student leader. Discerning in China a strong interest

[80] Cortland *Standard*, June 24, 1924, p. 5.
[81] Address by Sperry at the Sunday School Association dinner at the Imperial Hotel, Tokyo, September 29, 1922 (SP).

in western science and technology, Robertson believed that he could bring a message of science and demonstrate a "life of Christian power." Robertson needed expensive scientific equipment for his lectures, and during a visit to New York in 1910, after a period in China, he learned of Sperry's fascinating work with the gyro. Immediately he called Sperry, who invited him to "come right down." The conversation extended long into the evening, and Robertson later wrote, "so began one of the most delightful acquaintances of my far-traveled life."[82]

Sperry subsequently supported Robertson by providing equipment—worth $5000 in 1920 alone—and sponsoring fund-raising dinners for him. At one luncheon organized by Sperry in 1926, Edward Dean Adams, Ambrose Swasey, Frank Jewett, and the presidents of the American Telephone and Telegraph Company and the Westinghouse Company were present to hear Robertson. Robertson had much to tell, for he had spent years in China during unsettled times and in Russia during the civil war that followed the Russian revolution. As a friend and supporter of Robertson's, Sperry received letters from him in Russia and China reporting hardships as well as remarkable successes with his dramatically demonstrated lectures on the radio, the gyroscope, and other scientific and technological matters.

The YMCA, the World Sunday School Association, and Cortland by no means cover the range of Sperry's philanthropic activities. With a small group of prominent engineers, he helped to organize the Museum of Peaceful Arts in New York City, a museum modeled after the famous *Deutsches Museum* in Munich. His list of charities numbered more than fifty at the time of his death, among them the Brooklyn Academy of Music, Stevens Institute of Technology, Cornell University, New York University (College of Engineering), and other universities. He also served as president of the Prospect Park South Association (Brooklyn) and was on the board of the Polytechnic Institute of Brooklyn. For his many professional achievements and his public services, Northwestern University, Lehigh University, and Stevens Institute awarded him honorary degrees. Colgate University intended to award a degree to him in June 1930, but illness prevented this.

Sperry consistently affirmed his Christian commitment in public service and in philanthropy; his views on politics, social questions, and morals also revealed consistency and integrity of character. He strove for order, regularity, harmony, and efficiency among men, as he had in the world of machines. He sought in human affairs the stability he built into his control and guidance devices and looked to rules of conduct for the steady reference his compass provided for ships. It might seem paradoxical that a creative man ardently desired order, but he made his machines and processes to provide order. Furthermore, he knew that technological progress was unlikely in times of social and political unrest. Perhaps this explains the abhorrence for radicals and agitators expressed in his later life.

Sperry associated his own values of order, regularity, harmony, and efficiency with the engineering profession in general. He also thought of

[82] C. H. Robertson to Edward Sperry, June 25, 1930 (SP).

himself and other engineers as the makers of a new world of material affluence and international peace. The engineer had, Sperry believed, begun to enjoy the prestige and dignity owed him by society because of his contributions to social welfare. "The engineer," he wrote, "has now come to be recognized not only as an extremely useful citizen, but a primary producer of fundamental importance. . . . The engineer is now established as one of the greatest contributors to increased comfort and luxury both in and out of the home." Because engineering and engineers were "in a way, Esperanto," Sperry believed that the nations should look to engineering as a great force drawing men of all nationalities into a closer bond of friendship. The engineering spirit, universally manifest, was to him "the cultivation of the spirit of cooperation, and the inspiration which gives vision and incentive for new effort and greater achievement."[83] The World Engineering Congress in Japan was to him evidence of this spirit; he did not see as inconsistent the contributions that engineers made to the development of weapons. He thought the engineer preeminently "an Apostle of Peace. Wanton destruction is to him anathema."[84]

Like many successful engineers Sperry admired the engineering method and thought it applicable to solving political and social problems. He believed that the engineer's outstanding trait was his ability to deal with and build on "facts" alone. Engineers learned quite early that "nothing is more disastrous than a failure correctly to perceive, interpret, and utilize exact facts. . . . This habit of mind," Sperry was certain, "makes him not only the best type of citizen in his own country, but the best and most reliable medium for lasting contacts with his colleagues and co-workers abroad in all nations." Sperry assumed that a rational approach based on facts, had a universal validity and authority. The method and approach of the engineer, while illuminated by science, was not, Sperry thought, constrained by erroneous theory. He contrasted the empiricism and open-mindedness of the engineer with the mouldy theorizing and dogmatism of the academic mind and method. He found that politicians, diplomats, and economists could learn much from the hard-nosed, realistic, and practical approach of the engineer.

For a man of Sperry's views, the times were propitious. Herbert Hoover, "the great engineer," was rising rapidly in the twenties to the peak of political power. After Hoover became president in 1929, Sperry believed that the eyes of the world had turned to the United States, where for the second time its fate would be in the hands of "an Engineer-President." (To Sperry and other engineers, George Washington, the land surveyor, was the first.) Hoover's professional characteristics and moral principles recommended him to Sperry. A Stanford University graduate in engineering, Hoover was a manager and consulting engineer for railways and mines in the United States, Mexico, Australia, Russia, China, and India from 1896 until 1914. He had written the *Principles of Mining* (1909) and other

[83] Elmer Sperry, "Recognition of the Engineer and the American Engineering Societies," a paper read around 1929 (SP).
[84] *Ibid.*, p. 6.

Figure 10.16 Sperry in 1922 with Herbert Hoover, then Secretary of Commerce, and Charles Schwab, head of the Bethlehem Steel Co. From Sperry Memorial Book Committee, Dr. Sperry as We Knew Him *(Yokohama: Nichi-Bei Press, 1931).*

shorter essays on engineering and economics. Further, he won the John Fritz medal two years after Sperry. During World War I, he became world-famous for his organization of the American Relief Committee, Committee for the Relief of Belgium, and other agencies aiding war victims in Europe.

Sperry esteemed Hoover and his views so highly that he circulated excerpts from Hoover's speeches among friends and acquaintances. It was not long after a speech of Hoover's at a Roosevelt birthday dinner in 1919 that Sperry became an active campaigner for Hoover for president.[85] In the 1919 speech, Hoover struck a responsive chord in Sperry by asking for a return to the foundations of the republic and for a revitalization of traditional institutions. Hoover rejected phrase-makers who would solve problems by "isms," including Americanism, Bolshevism, trade unionism, internationalism, capitalism, and "a dozen others." The phrase-makers and the loose thinkers, in his opinion, imported their ideas and social diseases from Europe. He sought an analysis of problems, not "hunches,"a thorough re-examination of American principles, and the formulation of new policies through "straight thinking." Equality and opportunity were two of his key words.

In the fall, Sperry was circulating a letter in support of Hoover. Sperry

[85] The speech was given at the Waldorf Astoria on October 27 before the Rocky Mountain Club and reported in the *New York Sun* the following day.

also wrote that "we should return to *first* principles," and that the next president should be an engineer—a real man "trained to think about real things from a practical standpoint and by direct methods." He added that "in the engineering profession,"

we are all for him, and we feel that the time has come when we must break away from academic and idealistic theories and settle down to practical realities, with a firm determination that the forces that array themselves against personal initiative, efficiency, and productivity must be overwhelmed, and that justice must rule regardless of politics and the bold front put up by those in the wrong.[86]

Sperry gave Hoover substantial support. As a member of the executive committee of the Brooklyn Hoover Republicans, he worked for Hoover's nomination. He also responded to Hoover's request that he serve on the New York City Committee of Hoover's Relief Council, which was raising funds for the starving children in Europe. Hoover promised Sperry that a committee meeting would be short and to the point—"the facts on which your judgment is sought will be presented by me in not more than fifteen minutes."[87] The two men drew closer together because of their common interest in the Far East. In January 1928, Sperry became an honorary vice president of the Hoover-for-President Engineers' National Committee, along with Elihu Thomson and Ambrose Swasey. Harry Guggenheim, Michael Pupin, John Hays Hammond, and C. A. Stone of Stone and Webster were also officers.[88]

On occasion, Sperry gave Hoover his views on political and social questions. Sperry resisted trade unions in the belief that they interfered with the harmonious workings of industry, but in 1928, just before Hoover won the nomination, Sperry suggested that Hoover should take a positive attitude toward labor. "I have been an employer of labor since I was in my teens," Sperry wrote,

and through the last fifty years I have been a rather close observer of the change and evolution of its attitude towards capital. With a few exceptions, this attitude has grown steadily towards one of fine cooperation, which has now become pronounced. I believe that our attitude towards labor, instead of being patronizing, should emphasize and dwell upon the fine progressive spirit that has brought about these results. This, of course, has come about through efforts of both employer and employee, but as I review the situation, I firmly believe that labor is entitled to the greater credit of the two. Its path has been fraught with many more misgivings. It has been longer and harder and steeper, but it has come through. So my idea is that we should recede from any patronizing attitude and give them the fullest credit for these fine, noble acts.[89]

Sperry's attitude was undoubtedly influenced by the affection his employees at the Sperry Gyroscope Company had for him; letters and ceremonies in

[86] EAS to John Hays Hammond, October 24, 1919 (SP).
[87] Herbert Hoover to EAS, December 9, 1920 (SP).
[88] *New York Times,* January 27, 1928.
[89] EAS to Herbert Hoover, September 7, 1928 (SP).

his honor testify to this. Hoover thanked him for the letter, read it with interest, promised to give it attention, and expressed, as he had before, his appreciation for Sperry's long-standing friendship and support.[90]

Sperry also telegraphed and wrote Hoover in an effort to reinforce his stand on prohibition. Prohibition was a passionate cause of Sperry's. He never drank and delighted in calling the local ice-cream parlor his bar. He thought it amusing to invite unsuspecting engineer friends to this bar; fortunately, their reactions have been left unrecorded. Sperry's views on prohibition, often publicly expressed, are little credit to his reason but testify to his ardor. In a widely printed interview given in 1929 when he was president of the ASME, he said that in industries whose heads did not observe prohibition and who sought to "tear down" the Volstead Act "there has been no progress." Where the leaders were public and private prohibitionists, "there has been magnificent progress—in the product and efficiency of the industry." Sperry did not give the sources of the facts with which he was "building," but he attributed European industrial sloth to drink.[91]

Sperry also had some opinions about Manhattan Island, and one reporter exploited them. The interview bore the sensational headline, "Manhattan Was Always A Menace To Rest Of U.S. Declares E. A. Sperry."[92] Sperry surveyed for the reporter Manhattan Island's history, one influenced by Tories during the Revolutionary War, Copperheads during the Civil War, Mugwumps in the postwar period, and speakeasy proprietors in the twenties. In Sperry's mind, Manhattan was associated with what he called an "Old World" element that had consistently resisted the tenets of Americanism. He had not, however, abandoned all hope, for Manhattan's underworld and "Old World" would be "bucking up" against the new president, Herbert Hoover, "who is determined to enforce the law and who does not know such a thing as failure." The reporter than heard a discourse on the virtues of engineers who deal with "pure fact and truth" as constrasted with lawyers, especially Manhattan lawyers, who were attempting to nullify the prohibition laws.[93]

In the twenties, Sperry frequently criticized the "Old World." The butt of his criticism was often Germany, which he viewed as populated by engineers who tended to patronize Yankees. In Europe, however, Sperry found what he believed to be one glimmer of revitalization: while on tour in Italy

[90] Herbert Hoover to EAS, September 12, 1928; see also Hoover to EAS, June 16, 1928 (SP).
[91] *The Chicago Daily News,* April 15, 1929.
[92] *Suffolk County News,* May 10, 1929.
[93] It is also an indication of his character that Sperry subsequently regretted the interview and the way it was published. He wrote to the president of the newspaper chain responsible and complained that a "dignified" interview had been lowered in tone and made sensational. EAS to Frank Gannett of the Gannett Newspapers, May 11, 1929 (SP). Sperry thought that in his effort to support a worthy cause, prohibition, he had been exploited. He admitted, however, that the material had been submitted to him and that a hasty departure on a trip had not given him time to go over it carefully. One cannot help wondering about the reaction Sperry may have had from Manhattan friends after delivering such broadsides.

in 1926, he met Mussolini. Sperry reported that during their interview Mussolini "danced about a lot" but expressed extreme interest when Sperry pulled a small gyro from his pocket and showed the dictator how ships were stabilized. After this experience, Sperry, who had "gained a most favorable impression of this marvelous man," decided that he was not in the slightest sense a dictator "but simply a wise leader full of patriotism and untiring in 'up-building' his country."[94] Later Sperry wrote to Mussolini, "You are the outstanding advocate and example of the dignity, and even sanctity of *work* which is the underlying national ideal in America." As a result, Sperry added, a strong bond of sympathy existed between American engineers and Mussolini, and "our spontaneous reaction" is to aid so "noble a cause as yours."[95] Mussolini wired Sperry his thanks.

Sperry's admiration for Mussolini was matched by his animosity toward agitators who would disturb the political and social order. In letters to friends, he expressed his opposition to the Soviet experiment, his distaste for labor agitators in America, and his contempt for "parlor Bolsheviki." Sperry's strongest reaction was consistently against the destructiveness of the agitator, "a man that tears down instead of building up." As a lover of order, Sperry reacted instinctively against those who would create the chaotic conditions in which he could not work. On several occasions, this animosity toward the "alien" agitators degenerated into prejudice. He wrote, for example, "If only the leader has enough hay and stubble on his face of the black variety, low enough brow, and is wide of feature and thick of lip, [his followers] . . . go into raptures."[96] Because he found so much that was alien in the east, Sperry sentimentalized the great "Chatauqua-loving" west, home of the average American. Although he presided over revolutionary change in the world of technology and often admired the nonconformist engineer and inventor, he could not transfer these attitudes to the political and social world.

viii

Sperry was reluctant to abandon the demanding schedule imposed upon him by his work—and good works—but age demanded occasional relaxation. After the war, an operation injured a leg nerve; a cane and a weakened golf drive gave lasting evidence of this.[97] Noting his health and the demands he placed upon himself, concerned friends often urged him to slow down. He insisted, however, on an exhausting pace, until the tempo was obviously too much and Zula persuaded him to take more frequent vacations.

After the war, in the winter season, Elmer and Zula often vacationed in Asheville, or Pinehurst, North Carolina, in Florida, and several times in the Carribean. They chose hotels where people of similar interests and style congregated; thus Sperry often found himself on the golf links or at dinner with leading engineers, engineering and science professors, and industrial-

[94] EAS to Admiral Earle, September 10, 1926 (SP).
[95] EAS to Mussolini, January 25, 1927 (SP).
[96] EAS to F. Creagh-Osborne, March 20, 1920 (SP).
[97] EAS to W. D. Tisdale, October 17, 1924 (SP).

Figure 10.17 The Sperry house in Bellport, Long Island.

ists. These were not the prominent figures that the public associated with big business or industry but men interested more in the technological than in the financial and managerial aspects of the industrial world.

In the summers, Zula established a headquarters at the Goldthwaite Inn at Bellport, Long Island, and the children and Elmer moved in and out as work and other activities allowed. Fifty-eight miles from Manhattan and even closer to Brooklyn, Bellport, on the south shore of Long Island, had become a resort by the end of the nineteenth century. It had been a harbor and shipbuilding community, but a great storm about 1840 closed the inlet, and the village languished until the railroad and the vacationers began to arrive. Some bought sizable homes on the tree-lined streets and others came to stay at the hotels, inns, and boarding houses. Bellport soon became the Sperry's second home.

When the Sperrys first stayed at the Goldthwaite Inn, Edward was entering Cornell University, Elmer, Jr., was about to enter Phillips Exeter, Lawrence was to be sent to Evans School, and Helen had graduated from Packer Collegiate School. They discovered other young people who also enjoyed sailing on the bay and surf bathing on Fire Island, three miles across the bay. Zula found the American plan at the inn left her time for her family; and Elmer commuted from Brooklyn. Bellport was not South-hampton, but the Sperrys preferred the more modest houses, the flower gardens, the shaded streets, the dignified informality, and the upper middle-class professional people who visited or resided there.[98]

They liked it so well that in 1919 they bought two lots on the highest bluff overlooking the bay, and Elmer Sperry found time to help the local contractor design the summer home, which was finally ready in 1924. The

[98] Charles H. Towne, "Loafing Down Long Island," *Century*, CII (1921), pp. 136–37; C. H. Towne, "Along Long Island," *Arts and Decoration*, XXIII (1925), pp. 21–23; and Henry Hazelton, *The Boroughs of Brooklyn and Queens, Counties of Nassau and Suffolk* (New York: Lewis Historical Publishing Co., 1925), II, pp. 819–20.

large three-floor clapboard home had a handsome copper roof chosen by Sperry because he greatly admired the patina copper acquired by weathering. Determined to incorporate his ideas into the house, he also designed a retractable *porte-cochere* over the circular drive. Zula and Elmer had an apartment on the first floor overlooking the water; children and then grandchildren filled the rest of the house throughout the summer. To the rear of the house was Sperry's study and office and a telephone that kept him in very close contact with his engineers at Sperry Gyroscope.

Sperry's friends saw another facet of his personality when they visited him at Bellport.[99] Sperry often played golf on the Bellport course, taking the game as "a complete diversion, never thinking of business while trying to beat the other fellow."[100] But his inventiveness did seduce him into contriving an ingeniously supported golf bag, and his engineering sense caused him to calculate the force and trajectory of his drives. Usually, however, he controlled his instincts and delighted in the out-of-doors and the companionship. Persons remembering him at Bellport also report that he had time for children, taking pleasure in gathering together a group and piloting them to the local movie. After the show, he took his young charges to the village drugstore for a round of chocolate ice-cream sodas, his favorite drink. He also liked to interest his visiting grandchildren in science, and one of them remembered a small collection of metals on his desk used to explain metals and their uses. The grandchildren also liked the beamy old catboat in which the entire Sperry family sailed each Sunday morning for surf swimming on the beach at Fire Island.

One of Sperry's favorite Bellport neighbors was Anne Lloyd, an amateur poetess. When a poetry competition was announced as a part of the jubilee celebration of Edison's invention of the incandescent lamp, Sperry told her of the inventor's persistence, courage, and touch of genius, and her poem won the Edison prize. Anne Lloyd also wrote a poem about Zula Sperry, which her husband felt captured the essence of her character. Another Sperry favorite who visited Bellport was Maurice Marechel, a French cellist, who married a close friend of Helen Sperry's and often visited America. Sperry took great interest in the musician, for he learned from Marechel about the problems of composing music, not unlike those problems Sperry had known in inventing. He helped Marechel arrange to record difficult passages for self-criticism, and when Marechel gave concerts in Chicago and New York, Sperry insisted that all of his friends buy tickets.

Marechel formed a strong impression of Elmer and Zula Sperry and later wrote:

I do not believe I ever met in any other part of the world a so beautiful and well balanced union of two people, Mother Sperry, the soul of the house, the solidity, the calm, the thoughtfulness for everybody, the sensation of security and rest, [sensed] as soon as you met her quiet smile at the door.

[99] This description of Bellport life is drawn from Preston Bassett's unpublished biography and from my interview with Perry Bigelow in 1966. Bigelow, who joined the Sperry Company as a young man, was a Bellport resident.
[100] EAS to W. D. Tisdale, October 17, 1924 (SP).

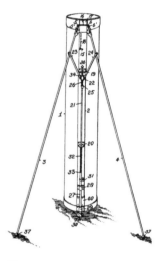

Figure 10.18
Sperry's golf-bag support,
Supporting Means for Golf
Bag (Patent No. 1,686,774).

Figure 10.19 Elmer and Zula Sperry embark for Europe aboard the Majestic.

Father Sperry, with adorable nice little hat, his eyes full of light, the fire of his mind which was stimulating and warming, a real fire of optimism and genius. They gave so much to us. They really represented so much: America for me, America with the best she has.[101]

In the twenties Elmer and Zula twice toured Europe and the Near East. Sperry lectured to interested naval officers and established business contacts but also spent time sightseeing and relaxing. Neither Europe nor the Near East, however, invoked in him the enthusiasm he felt for Japan. Zula, writing of these trips with more enthusiasm than he, noted in her diary their stay in the spring of 1926 at the Savoy in London, dinner with "the baron" at the Breidenbacher Hof in Dusseldorf, dining with distinguished company at the Europa in Milan before "father" lectured to the naval college, and a dinner with Signor Guglielmo Marconi at the Grand Hotel in Rome. Her daughter-in-law Winnie, Lawrence's widow then living in Paris, took her to Ciro's and to the Russian Ballet. Later they shopped for hats at Reboux, had tea with Ruth Draper, and then showed the grandchildren Versailles.

One experience captured Elmer Sperry's imagination during a Mediterranean cruise about a year later. On visiting the Museum of Antiquities in Athens, he discovered an "unmistakable" application of the "pure-gyroscopic

101 Letter from Maurice Marechel to Helen Sperry Lea, quoted in the Preston Bassett manuscript.

principle" in a bas-relief on a stela of Periclean Greece.[102] The relief showed a boy running alongside a hoop; his gesture toward the hoop was such that Sperry considered it the first complete demonstration of pure gyroscopic reaction he had encountered. Later, in Paris, Sperry also immensely appreciated seeing the original Foucault gyro, which the curator allowed him to operate: "Of all the French courtesies that have ever been extended to me, it seems . . . that this was the most lovely thought and expression."[103] An invitation to lecture on his gyroscopic applications at the Sorbonne also "overwhelmed" him. He was somewhat apprehensive about not treating the subject mathematically, but the "Dean backed us up entirely in treating the subject as we did."[104]

Another experience that Sperry fully appreciated was an encounter with Helen Keller, the author and teacher who symbolized, to many people, the triumph of human will over adverse circumstances. Sperry first wrote to her expressing his admiration for her "wonderful" accomplishments and sensitive spirit, and at the same time regretting that she abhorred, as he judged from her writings, "the mechanistic world which results from truly great scientific and engineering achievements." Confessing that he was "positively in love" with the world of science and mechanics, he asked whether he might share with her his interest in the gyroscope whose action was "a remarkable manifestation of natural law."[105] Responding enthusiastically to his invitation to visit the Brooklyn factory, she assured him, "You are no stranger to me," and said that she thought it wonderful that the "creator" of the gyroscope would give her insight into his life's work. She added that it was not the mechanistic world from which she shrank but the disregard of the things of the spirit by some of the men who created and lived in this world. Knowing that Sperry was a "noble exception," she anticipated "seeing" his face with her "ten eyes."[106]

For him, her visit to the Sperry company was "the greatest, the most inspirational," of its history; for her, the visit was a "wonder-filled" afternoon. Communicating through her companion, Sperry had her feel the motions of the compass as he described its actions and principles; he conveyed to her the power of the gyrostabilizer steadying a ship at sea; and he had her stand in the warm beam of the powerful searchlight. She explained to him that she imagined light as "like thought—a bright, amazing thing perceived by the eyes of the mind. I am conscious of an infinite variety of images, relations and degrees of brightness and darkness, void and fullness, space, height, depth and conceptual harmonies which I transmit into sound and color." He had "shown" her with the gyroscope, she recalled, the motions of the earth, and "I felt like Galileo who, when he knelt before his judges, cried, 'It moves.'" When Sperry "waved his thought-wand," she imagined an airport beacon directing the flight of an airplane through the

[102] EAS to the editor of the *National Geographic Magazine,* June 6, 1927 (SP).
[103] EAS to Maurice Marechel, April 19, 1927 (SP).
[104] *Ibid.*
[105] EAS to Helen Keller, August 27, 1929 (SP).
[106] Helen Keller to EAS, October 3, 1929, and January 29, 1930 (SP).

dark. She felt that Sperry would be remembered for a compass and a star, for "if there were no other immortality, you would live forever in that achievement."[107] In parting, Sperry said that he hoped he had given her "another garden of beauty" in which she, in her imagination, might stroll.[108] Seldom before had Sperry so candidly revealed his sentiments as on this occasion.

Sperry also revealed his emotions in his verses. He left a few poems in his files that prove a familiarity with meter and a richness in vocabulary. Though the verses were idle-hour compositions sometimes done on the backs of business letters and professional society bulletins, there are several drafts of each indicating his persistence. Some of them deal with nature, referring occasionally to its forbidding aspects, but often conveying appreciation for its less obvious wonders. In the "Mountain Pine," he differs with those who love the beech, the ash, and the oak "and reject the pine as of gloomy and forbidding mien," for the pine is lonely, stately, constant, faithful, dignified, and possessed of delicacy and taste. In a celebration of the "Bobolink," Sperry exercises a vocabulary of musical terms and concepts probably acquired from Zula. In "The Kite," the restraint, or string, makes the kite's flight possible. In "The Lure of the Unsolved," the unsolved becomes the mistress and he the lover, "fond love, strong and strongest love—the UNSOLVED." Dignity, restraint, and a love of the unknown and the unsolved were familiar Sperry characteristics.

ix

To the end of his life, Sperry pursued his rich and varied career: there was no long illness, no sharp curtailment of activities. Only a few months before his death, his schedule included appointments with John D. Rockefeller, Jr., and John D. Rockefeller III; President Hoover and the Japanese Ambassador; sessions with industrialists and engineers about his inventions; a meeting of the American Engineering Council; and a reception at which he was to receive the Gary Medal. He was also preparing a chapter for the Charles Beard anthology and carrying on his work for his development company. He had been troubled by heart symptoms, his postwar operation still caused him to use a cane, and his doctor had warned him of overexertion, but one of his engineers observed that though his body was strained, his mind retained its force and brilliance.

But there had been changes and irreversible losses. He resigned as president of the Sperry Gyroscope Company on July 28, 1926, to become chairman of the board. On December 20, 1928, Clement M. Keys and a group of associates called on Sperry. Keys, very much in the news at the time, had announced two weeks earlier the formation of North American Aviation, a $25,000,000 investment and holding corporation.[109] Having

107 Helen Keller to EAS, February 27, 1930 (SP).
108 EAS to Helen Keller, February 24, 1930 (SP).
109 Elsbeth E. Freudenthal, *The Aviation Business: From Kitty Hawk to Wall Street* (New York, 1940), quoted in *The History of the American Aircraft Industry:*

taken over the business and financial management of Glenn Curtiss' company some years earlier, Keys was establishing a giant conglomerate of aviation companies and wanted to acquire the Sperry company because of its renewed interest in aircraft instruments.[110] Encouraged by his son Elmer, Jr., Sperry sold the Sperry Gyroscope Company to Keys. Terms of the transaction were not made public, but newspapers reported that the Sperry company was known to be worth many millions of dollars.[111] Sperry told the press that his company "has been working more and more into the aeronautical field"; Keys said he needed Sperry capabilities to promote aviation by reducing flying risks through instrumentation, lighting, and other technology. Keys was also interested in the Sperry bombsight and anti-aircraft system.

The understanding between Sperry and Keys provided for Sperry's engineers. Keys said that he would not have bought the Sperry company if he had felt it necessary to reorganize it.[112] As for his own future, Elmer Sperry thought that the new arrangement simplified matters by leaving him with the development and research company where he wanted to concentrate his energies, "the commercial side of my business having never appealed to me in the slightest."[113]

The renewed emphasis on aviation must have reminded Elmer Sperry acutely and painfully of working with Lawrence years earlier to advance the field. Lawrence had piloted his own company successfully through the slow postwar years, and in his advanced position and with his experience, he seemed destined to seize the opportunities that more propitious years would bring. One can also imagine Lawrence, in the late twenties, achieving a triumph like Lindbergh's trans-Atlantic flight. But Lawrence died flying across the English Channel on December 13, 1924.

He was flying the small Sperry Messenger airplane out over the channel to France when a witness heard the plane's engine running badly. About noon, another saw the plane fall into the water where coast guard boats soon located the floating wreckage. Lawrence's body was not found until almost a month later.

Elmer Sperry did not show outwardly his inward distress. He was concerned about Zula and Lawrence's young wife, Winnie, both of whom, he found, "have been perfectly marvelous and have stood up under it in such a way that everybody has wondered." Zula could not, however, mask her anguish entirely, for her hair turned white.[114] Perhaps the eulogy and the

An Anthology, ed. G. R. Simonson (Cambridge: M.I.T. Press, 1968), pp. 78–79. By 1933 Keys' North American Aviation had acquired control of, or held substantial stock in, Curtiss-Wright Corporation, Douglas Aircraft, Ford Instrument Company, and Berliner-Joyce Aircraft. Besides these manufacturing companies, North American controlled or owned large stockholdings in major air lines.

[110] *New York Times,* December 24, 1928.

[111] *New York Times,* December 24, 1928, and interview with Elmer Sperry, Jr., February 23 and 24, 1966.

[112] C. M. Keys to EAS, January 23, 1929 (SP).

[113] EAS to W. B. Mayo, January 16, 1929 (SP).

[114] EAS to Jean Goodman, March 5, 1924 (SP).

Figure 10.20 The last meeting of Elmer Sperry and Lawrence and his wife. Lawrence and his airplane, shown aboard the Mauretania, *crashed in the channel in December.*

tributes made the pain more bearable; they at least proved once again that Lawrence's perilous efforts for aviation had not been without result. Two formations of Mitchell Field airplanes, following the funeral cortege to Brookville Cemetery, dropped flowers *en route* and over the grave. The honorary pall bearers and mourners included men who a half-century later are remembered as legendary pioneers. General William ("Billy") Mitchell characterized Lawrence as one of the most brilliant minds and greatest developers in the world of aviation, but for the father there was the sad notation made each December in his diary, "the day Lawrence was lost."

Zula passed away on March 11, 1930. She and Elmer had sailed for a vacation in the warmer climate of the Carribean, but during the voyage on the *SS Mauretania,* she contracted pneumonia. Taken to an Anglo-American hospital in Havana, she seemed on the way to recovery when a blood clot stopped her heart. Elmer was with her and saw that she died without pain. Always the precise observer, he reported her passing carefully to her sons and daughter and to her relatives, but his state was revealed more fully by his turning to the loving task of writing to all her friends and relatives to collect reminiscences in hopes of preparing a little book about her. It was no surprise to her husband that the words "serene," "quiet," "sweetness of disposition," "love of music," "poise," "balance," "comprehending," "composed," "children," and "family" often appeared in these letters. He also found consolation in a poem[115] written about Zula by his friend, Anne Lloyd, that began

> Who that ever saw it,
>   Can forget the grace
> Of the spirit shining
>   In her face?

[115] "Who that ever touched it,/Failed to understand/All the loving language/Of her hand?/Who that ever heard it,/Neglected to rejoice/In the gentle music/Of her voice?/Who that ever felt it,/Missed the smallest part/In the benediction/Of her heart?"—Anne Lloyd.

Elmer Sperry did not recover as he had earlier from physical and spiritual shocks. He survived her death by less than a hundred days, and during most of these, he was inactive. His responsibilities as a past president of the American Society of Mechanical Engineers brought him to attempt a train trip to Washington to help celebrate its fiftieth anniversary but *en route* on April 7, 1930, he suffered a slight stroke. Enforced rest at home followed until a serious gall-stone operation was deemed necessary. After the operation, he failed rapidly, and died in St. John's Hospital, Brooklyn, on June 16, 1930, at 3:45 P.M.

His mind never failed. So that he might concentrate, he asked Lea, his son-in-law, to read to him in the hospital. They began with "A" in the encyclopedia, but they did not reach "Z." As he lay in the hot Brooklyn hospital room, Sperry even invented a device to solve a pressing problem. He had ice placed before his electric fan so that cool breezes would play upon him.

The world knew that a major figure of the era had passed. Newspapers across the country wrote about him, and the obituaries in the scientific and technical journals were numerous and laudatory. The President of the United States, presidents of the professional societies, cabinet members, admirals, generals, and the leaders of the world of technology attended the funeral or sent expressions of sorrow. In Japan, the highest political and industrial leaders, including Admiral Takeda, held a special memorial service.[116]

ELMER A. SPERRY.

"The lot of every one of us," wrote HORACE, "is tossing about in "the urn, destined sooner or later "to come forth and place us in "Charon's boat for everlasting "exile." This time the lot has come forth for one whose going the whole planet will regret. A passenger on the high seas, remembering the fate of Palinurus, wrote of this great inventor whose mechanical helmsman, called familiarly the "Metal Mike," now guides great ships to their desired havens:

Now is old Palinurus' occupation gone;
It has been taken o'er by one named SPERRY
Who has installed a "Metal Mike" instead—
He'll soon be putting one on Charon's ferry.

Unhappily for us who still remain this side the River of Woe, Mr. SPERRY has gone to his "aeturnum exsilium," but he has left among men an everlasting fame, and imagination allows one to think of his inventive spirit making suggestions to the ferryman about improving service in the crossing for the benefit of those who have to take it later. For Mr. SPERRY was ever thinking of how he might make the dwellers on earth a little more at ease whether on sea or land or in the air. For those who travel by sea he provided not only the pilot who whatever betides holds his rudder true, but also a "stabilizer" to prevent the rolling of the ship and a device for signaling to prevent collisions. For those who travel by air he has helped to maintain the equilibrium of their planes and to lessen the peril of fire and to penetrate the fog. Toward the end of his long series of inventions he perfected an instrument for detecting the slightest imperfection in a steel rail.

What his born genius for invention would have done for a slower-going civilization one cannot imagine, so closely has it been associated with the swifter agents of this mobile age. He would have been put among the Titans in the ancient Greek age. But with all his seemingly miraculous achievements, he who harnessed the motion of the earth to do his bidding was a gentle human being, generous in his sympathies and beloved in his own person. America, claiming him for her own and proud of his contributions, which were recognized by the highest honor bestowed by his own profession in his own country, remembers with special satisfaction that he was acclaimed by his fellow-engineers of the world when he presided at the International Congress of Electrical Engineers in Japan last year.

Figure 10.21
New York Times,
June 17, 1930.

[116] The memorial service is included in the commemorative volume, *Dr. Sperry As We Knew Him* (Yokohama, 1930?). This volume in Japanese and English, of over 400 pages is composed of reminiscences of, and biographical sketches by, his Japanese friends. It is well illustrated. Elmer Sperry is buried beside Zula in the Green-Wood Cemetery in Brooklyn. Edward Sperry died November 6, 1945; Elmer Sperry, Jr., died December 21, 1968.

APPENDIX A   Elmer Sperry Patent Applications Maturing as Patents

	DATE FILED	APPLICATION NUMBER	TITLE	DATE ISSUED	PATENT No.
1880	Dec. 22, 1880	not recorded	Dynamo-Electric Machine	June 27, 1882	260,132
1881	April 15, 1881	30,916	Electric Arc Lamp	Sept. 9, 1884	304,966
	April 28, 1881	not recorded	Dynamo-Electric Machine	Dec. 12, 1882	268,956
	Sept. 7, 1881	not recorded	Vehicle Wheel	Sept. 5, 1882	263,982
	Oct. 24, 1881	not recorded	Armature for Dynamo-Electric Machine	Aug. 1, 1882	261,965
1882	June 28, 1882	not recorded	Valve for Steam-Engine (with A. J. Belden)	Oct. 17, 1882	266,217
	Sept. 30, 1882	not recorded	Machine for Cutting Screw Threads	Feb. 26, 1884	294,092
1883	Feb. 17, 1883	not recorded	Combined Annunciator and Alarm System	Oct. 16, 1883	286,970
	Mar. 14, 1883	88,130	Regulator for Dynamo-Electric Machines	Dec. 7, 1886	353,986
	Mar. 20, 1883	not recorded	Electric Regulator	April 29, 1884	297,866
	Mar. 30, 1883	not recorded	Regulating Device for Electric Lamps	July 1, 1884	301,175
	June 9, 1883	not recorded	Commutator and Brush for Electric Motors and Generators	April 29, 1884	297,867
	June 9, 1883	199,988	Annular Armature for Dynamo-Electric Machines	Dec. 7, 1886	353,989
	Oct. 13, 1883	108,912	Electric Motor	Dec. 7, 1886	353,987
	Nov. 1, 1883	201, ?	Electric Regulator	Dec. 7, 1886	353,990

	Date Filed	Application Number	Title	Date Issued	Patent No.
1884	Mar. 21, 1884	201,599	Vitrified Asbestos [sic]	June 15, 1886	343,651
	Mar. 21, 1884	125,005	Dynamo Electric-Machine	Dec. 28, 1886	354,946
	Oct. 28, 1884	146,698	Regulator for Dynamo-Electric Machines	Dec. 7, 1886	353,988
	Oct. 31, 1884	146,947	Dynamo-Electric Machine	Dec. 28, 1886	354,945
1885	No Applications				
1886	April 3, 1886	197,595	Electric Generator or Motor	June 7, 1892	476,426
	May 8, 1886	201,586	Lightning Arrester	Aug. 30, 1887	369,036
	May 21, 1886	202,898	Apparatus for Distributing Electric Power	April 26, 1887	361,843
	May 22, 1886	not recorded	Safety Device for Incandescent Lamps	April 26, 1887	361,844
	June 3, 1886	204,031	Electric Car Brake	Mar. 29, 1887	360,060
	Nov. 23, 1886	219,654	Electrical Switch	Jan. 22, 1889	396,439
1887	Dec. 12, 1887	257,716	Gas Engine	Aug. 5, 1890	433,551
1888	Mar. 24, 1888	268,447	Electric Battery	Feb. 19, 1889	397,945
	April 5, 1888	269,656	Automatic Grounding Device	Mar. 26, 1889	400,264
	Oct. 8, 1888	287,481	Electric Mining-Machine	May 23, 1893	497,832
	Oct. 22, 1888	288,812	Regulator for Dynamo-Electric Machines	Feb. 26, 1888	398,668
	Oct. 22, 1888	288,813	Electric-Arc Lamp	June 18, 1889	405,440
	Dec. 7, 1888	292,891	Electric Motor and Regulator Thereof	Jan. 28, 1890	420,117
1889	Feb. 11, 1889	299,477	Dynamo	June 18, 1889	405,441
	April 1, 1889	305,581	Lightning Arrester	Jan. 7, 1890	418,824
	April 1, 1889	305,584	Electric Drilling Machine	June 11, 1889	405,187
	April 1, 1889	305,585	Electric Mining Machine	June 11, 1889	405,188
	April 1, 1889	305,583	Galvanic Battery	Dec. 17, 1889	417,290
	April 1, 1889	305,582	System of Electrical Distribution for Mines	Feb. 10, 1891	446,030
	April 12, 1889	306,991	System of Trans. by Electricity in Mines	Aug. 12, 1890	434,363

Date Filed	Application Number	Title	Date Issued	Patent No.
Sept. 9, 1889	323,477	Gas Pressure Regulator	Dec. 31, 1889	418,245
Sept. 9, 1889	323,480	Pick for Mining Machines	May 27, 1890	428,787
Sept. 9, 1889	323,479	Rotating Part of Dynamos and Motors	Aug. 12, 1890	434,096

1890

Date Filed	Application Number	Title	Date Issued	Patent No.
Jan. 10, 1890	336,591	Dynamo Electric Machine	Mar. 1, 1892	469,725
Mar 17, 1890	344,219	Power-Gearing for Vehicles	Aug. 12, 1890	434,097
April 18, 1890	348,459	Electric Arc Lamp	July 19, 1892	479,029
June 7, 1890	354,604	Electric Arc Lamp	Aug. 9, 1892	480,525
July 2, 1890	357,519	Electric Switch	Feb. 10, 1891	446,031
July 11, 1890	358,401	Mining Machinery	June 23, 1891	454,500
July 14, 1890	358,594	Power Gearing for Vehicles	Aug. 12, 1890	434,098

1891

Date Filed	Application Number	Title	Date Issued	Patent No.
Jan. 15, 1891	377,921	Mineral Drilling Machine	Sept. 15, 1891	459,596
Feb. 10, 1891	382,152	Commutator	June 9, 1891	453,822
Feb. 16, 1891	381,551	Electric Mine Car	July 5, 1892	478,138
Mar. 19, 1891	385,581	Fuse-Block for Electric Circuits	June 7, 1892	476,570
April 8, 1891	388,117	Electric Locomotive	July 5, 1892	478,139
May 4, 1891	391,469	Mining Machine	July 5, 1892	478,141
June 11, 1891	395,856	Electric Locomotive	Mar. 8, 1892	470,516
June 24, 1891	397,299	Electric Trolley	Jan. 12, 1892	466,807
July 9, 1891	398,921	Trolley Wire Support	Mar. 22, 1892	471,151
July 25, 1891	400,666	Track for Vehicles	Jan. 12, 1892	466,809
Sept. 17, 1891	463,260	Electro-Conducting Bearing for Trolley or Other Wheels	Oct. 3, 1893	505,994
Oct. 23, 1891	409,583	Electric Locomotive	July 25, 1893	502,020
Nov. 16, 1891	411,983	Dynamo Electric Machine or Motor	July 5, 1892	478,142
Dec. 17, 1891	415,397	Armature for Electric Machines	July 19, 1892	479,030

1892

Date Filed	Application Number	Title	Date Issued	Patent No.
Jan. 8, 1892	417,419	Revolving Armature for Electric Machines	July 11, 1893	501,194
April 1, 1892	427,382	Trolley Wire Switch	Aug. 15, 1893	503,443
April 1, 1892	427,384	Electric Railway Trolley	July 25, 1893	501,968
April 18, 1892	429,571	Trolley Wire Splice	July 5, 1892	478,140
April 25, 1892	493,049	Automatic Regulator for Dynamo Electric Machines	Jan. 16, 1894	513,062
April 29, 1892	431,132	Electric Locomotive	July 11, 1893	501,195
April 29, 1892	431,133	Clutch for Shafts	Feb. 26, 1895	534,676
Dec. 10, 1892	454,752	Gas Engine	June 8, 1915	1,141,985

	DATE FILED	APPLICATION NUMBER	TITLE	DATE ISSUED	PATENT NO.
1893	Jan. 28, 1893	460,026	Coupling Device	Aug. 18, 1896	565,935
	Mar. 6, 1893	464,651	Power Transmitting Device	July 27, 1897	587,019
	June 5, 1893	476,612	Motor Truck	Mar. 5, 1895	535,303
	Aug. 21, 1893	483,611	Electrical Controller	Nov. 28, 1893	509,776
	Aug. 23, 1893	483,820	Electrical Controller	Feb. 27, 1894	515,374
	Nov. 11, 1893	490,698	Electrical Controller	Mar. 12, 1895	535,511
	Nov. 11, 1893	490,697	Mechanical Movement	July 7, 1896	563,424
	Nov. 11, 1893	490,696	Power Gearing for Trucks	Sept. 8, 1896	567,418
1894	Jan. 30, 1894	498,510	Power Transmitting Mechanism	June 23, 1896	562,498
	Jan. 30, 1894	498,511	Electric Brake	Feb. 26, 1895	534,974
	Feb. 5, 1894	499,230	Apparatus for Arresting Motion of Electrically Propelled Mechanism	Feb. 26, 1895	534,975
	April 13, 1894	507,379	Mounting for Electric Motors	Mar. 5, 1895	535,304
	June 6, 1894	513,635	Car Wheel	Feb. 26, 1895	534,976
	June 6, 1894	513,634	Power Gearing for Electric Cars	May 19, 1896	560,375
	June 8, 1894	534,977	Electric Brake	Feb. 26, 1895	534,977
	June 9, 1894	514,035	Power Gearing for Electric Car	July 7, 1896	563,425
	June 29, 1894	516,053	Power Gearing for Electric or Other Motors	Aug. 18, 1896	565,936
	July 3, 1894	516,460	Coupling for Electric or other Power Transmissions	June 23, 1896	562,499
	July 3, 1894	516,461	Coupling Device	June 23, 1896	562,500
	July 21, 1894	518,169	Controller for Electric or Other Motors	Sept. 4, 1894	525,394
	July 24, 1894	518,417	System and Apparatus for Control of Electric Machines	Sept. 4, 1894	525,395
	Oct. 15, 1894	525,916	Steam Engine	Dec. 17, 1895	551,480
	Oct. 16, 1894	526,036	Arc Rupturing Device	April 9, 1895	537,130
1895	Feb. 16, 1895	538,665	Electric Brake	Aug. 18, 1896	565,937
	Feb. 16, 1895	538,663	System of Control for Electric Motor	Feb. 25, 1896	555,291
	Feb. 16, 1895	538,664	System of Control for Electric Motors	Aug. 25, 1896	566,426
	Feb. 16, 1895	538,662	Power Gearing for Electric Cars	June 2, 1896	561,354
	Feb. 19, 1895	539,019	Electric Brake	Feb. 16, 1897	577,119
	Feb. 20, 1895	539,146	Electric Controller	June 23, 1896	562,501

	Date Filed	Application Number	Title	Date Issued	Patent No.
	Feb. 21, 1895	538,297	System of Control for Electric Motors	Oct. 13, 1896	569,305
	Feb. 21, 1895	539,298	Valve	Feb. 27, 1912	1,018,591
	April 12, 1895	545,434	Power Transmitting Device	Mar. 31, 1896	557,162
	July 5, 1895	554,942	Power Transmitting Device	Sept. 10, 1895	545,920
1896	Feb. 18, 1896	579,794	Motor Vehicle	Dec. 20, 1898	616,153
	Mar. 20, 1896	584,061	Controller for Electric Cars	May 26, 1896	560,658
	Mar. 20, 1896	584,062	Controller for Electric Cars	June 16, 1896	562,100
	April 13, 1896	587,333	Power Transmitting Gearing for Electric Railway Trucks	Aug. 18, 1896	565,938
	April 23, 1896	588,721	Controller for Electric Railway Cars	Aug. 18, 1896	565,939
	June 20, 1896	596,253	Electric Brake	Nov. 17, 1896	571,409
	June 26, 1896	597,042	Electric Car Brake	Dec. 29, 1896	574,120
	July 31, 1896	601,148	Electric Controller	Nov. 17, 1896	571,410
	Sept. 29, 1896	606,489	Electric Controller	Feb. 16, 1897	577,081
1897	May 17, 1897	636,924	Actuating Device for Railway Appliances	Oct. 30, 1900	660,805
1898	Aug. 9, 1898	689,008	Electric Vehicle	Jan. 9, 1900	640,968
	Aug. 19, 1898	689,057	Storage Battery	April 3, 1900	646,922
	Aug. 19, 1898	689,059	Gearing for Motor Vehicles	Mar. 27, 1900	646,081
	Aug. 25, 1898	689,462	Motor Gearing	Mar. 20, 1900	645,902
	Sept. 6, 1898	690,359	System and Apparatus for Controlling	Oct. 31, 1899	635,815
	Sept. 14, 1898	24,093	System of Control for Electrically Propelled Trains	June 30, 1903	732,130
	Sept. 17, 1898	691,150	System and Apparatus for Controlling Vehicles	Oct. 31, 1899	635,816
1899	April 22, 1899	714,054	Battery Jar or Receptacle	Mar. 27, 1900	646,325
	Sept. 11, 1899	730,196	Apparatus for Governing and Controlling marine or Other Engines	Oct. 23, 1900	660,318
	Sept. 11, 1899	730,118	Motor Carriage	Feb. 13, 1900	643,257

Date Filed	Application Number	Title	Date Issued	Patent No.
Sept. 13, 1899	730,320	Electric Battery and Mounting Same	July 1, 1902	703,673
Sept. 16, 1899	730,692	System of Electric Circuits and Brakes for Vehicles	Jan. 16, 1900	641,412
Sept. 30, 1899	732,147	Element for Storage Batteries	May 22, 1900	649,998
Sept. 30, 1899	732,148	Electric Storage Battery	May 15, 1900	649,491
Oct. 7, 1899	732,859	Cellulose Envelope for Elements of Storage Batteries	April 3, 1900	646,923
Oct. 7, 1899	732,860	Envelope for Storage Batteries	May 8, 1900	649,003
Oct. 17, 1899	733,873	Actuating Device for Railway Appliances	Oct. 30, 1900	660,825
Oct. 30, 1899	735,167	Motor Vehicle	Feb. 13, 1900	643,258
Oct. 30, 1899	735,168	Motor Vehicle Brake	Mar. 20, 1900	645,903
Nov. 6, 1899	735,968	Storage Battery Tray or Case	Dec. 10, 1901	688,749
Nov. 6, 1899	735,971	Connection for Batteries	July 1, 1902	703,674

1900   No Applications

1901	May 25, 1901	61,849	Envelope for Battery Electrodes	Mar. 25, 1902	696,209
	May 25, 1901	81,848	Armored Element for Electric Batteries	Mar. 24, 1903	723,326
	Aug. 12, 1901	71,710	Storage Battery	Mar. 24, 1903	723,327
	Aug. 12, 1901	71,712	Manufacture of Envelopes for Storage Batteries	Mar. 24, 1903	723,329
	Aug. 12, 1901	71,711	Storage Battery	Mar. 24, 1903	723,328
	Nov. 25, 1901	83,491	Separator for Storage Batteries	May 26, 1903	729,100

1902	Feb. 20, 1902	94,925	Process of Manufacturing Elements for Storage Batteries	Nov. 4, 1902	713,020
	Oct. 20, 1902	127,933	Electric Railway	Dec. 16, 1902	716,125
	Nov. 12, 1902	130,998	Electric Railway System	Oct. 18, 1904	772,679
	Nov. 26, 1902	132,847	Electric Locomotive	April 9, 1907	849,432
	Dec. 8, 1902	134,254	Rack Rail for Locomotives	Oct. 18, 1904	772,680
	Dec. 5, 1902	134,035	System of Electric Generation, Distribution, and Control	Jan. 26, 1904	750,497

	DATE FILED	APPLICATION NUMBER	TITLE	DATE ISSUED	PATENT NO.
1903	Jan. 17, 1903	139,399	System of Electrical Generation, Distribution, and Control	Jan. 26, 1904	750,471
	Mar. 5, 1903	146,280	Power Transmitting Mechanism	Jan. 26, 1904	750,498
	Mar. 5, 1903	146,281	Power Transmitting Mechanism	Jan. 26, 1904	750,499
	Mar. 9, 1903	147,028	System of Power Transmission	April 20, 1915	1,136,058
	July 1, 1903	163,838	System of Electrical Generation, Distribution, and Control	Jan. 26, 1904	750,500
	Nov. 24, 1903	182,498	Frictional Driving Mechanism	Aug. 16, 1904	767,604
	Dec. 5, 1903	183,967	Electrolytic Cell	May 26, 1914	1,097,826
	Dec. 21, 1903	185,995	Logarithmic Calculator	Oct. 25, 1904	773,235
1904	Jan. 27, 1904	190,878	Storage Battery Electrode	Feb. 7, 1905	781,795
	Jan. 28, 1904	191,053	Storage Battery Electrode and Process of Making Same	Nov. 1, 1904	773,685
	Feb. 11, 1904	193,158	Storage Battery	Nov. 1, 1904	773,686
	Mar. 31, 1904	200,889	Rack Rail	Oct. 25, 1904	772,971
1905	Jan. 18, 1905	241,680	Combustion Engine	May 15, 1917	1,226,132
	Feb. 25, 1905	247,236	Centrifugal Machine	Dec. 31, 1912	1,048,905
	June 6, 1905	263,957	Regulating Apparatus	Oct. 23, 1906	833,760
	Nov. 23, 1905	288,780	Method of Making Stannic Chloride	Mar. 17, 1908	882,354
	Nov. 25, 1905	289,581	Method for Determining and Reducing Tin Compounds and Other Products	Dec. 17, 1907	874,040
	Dec. 7, 1905	290,768	Apparatus for Making Stannic Chloride	Jan. 21, 1908	877,243
1906	Jan. 27, 1906	298,254	Method of Purifying Stannic Chloride	Jan. 21, 1908	877,244
	Mar. 5, 1906	304,417	Method of Effecting Reactions Between Solids and Gases	Jan. 21, 1908	877,246
	Mar. 5, 1906	304,416	Apparatus for Effecting Reactions [Stannic Chloride]	Jan. 21, 1909	877,245

Date Filed	Application Number	Title	Date Issued	Patent No.
Mar. 5, 1906	304,329	Method of Making Stannic Chloride	Jan. 21, 1908	877,247
April 13, 1906	311,559	Process of Producing Stannic Chloride	Nov. 26, 1907	872,205
April 13, 1906	311,560	Process of Detinning	Dec. 10, 1907	873,699
Aug. 30, 1906	332,701	Lead Pigment and Similar Compounds	Oct. 1, 1907	867,436
Sept. 15, 1906	334,754	Electrolyte Refining of Tin	Dec. 24, 1907	874,707
Oct. 8, 1906	337,883	Reclaiming Tin Scrap	Feb. 18, 1908	879,596
Dec. 29, 1906	350,093	Apparatus for Effecting Reactions	April 21, 1908	885,391
Dec. 31, 1906	350,154	Electric Lighting	Dec. 17, 1907	973,804

1907

Date Filed	Application Number	Title	Date Issued	Patent No.
Jan. 9, 1907	351,403	Method of Detinning	Mar. 31,1908	883,500
Jan. 22, 1907	353,561	Process for Detinning and Recovering	Nov.26, 1907	872,092
Feb. 20, 1907	358,490	Method of Making Stannic Chloride	Jan. 21, 1908	877,248
April 27, 1907	370,681	Process of Preparing Pure Tin Compounds	May 12, 1908	887,538
Aug. 24, 1907	389,953	Method for Detinning and Producing Tin Components and Other Products	Dec. 31, 1907	875,632
Oct. 4, 1907	395,866	Process of Dehydrating Moist Chlorine	Dec. 1, 1908	905,602
Oct. 17, 1907	397,834	Process of Preparing Merchantable Iron and Tin Compounds from Tin Plate Scrap	Sept. 1, 1908	897,796
Dec. 2, 1907	404,767	Steadying Device for Vehicles	Dec. 29, 1908	907,907

1908

Date Filed	Application Number	Title	Date Issued	Patent No.
Jan. 4, 1908	409,359	Electric Heater	Nov. 10, 1914	1,116,855
Feb. 7, 1908	414,674	Electric Locomotive	Sept. 1, 1908	897,312
Mar. 13, 1908	420,946	Process of Preparing Merchantable Iron from Tin Plate Scrap	Oct. 13, 1908	901,266
April 11, 1908	426,613	Process of Preparing Merchantable Iron from Tin Plate Scrap	Dec. 8, 1908	906,321
May 21, 1908	434,048	Ship Gyroscope	Aug. 17, 1915	1,150,311
Aug. 1, 1908	446,437	Process of Producing Clear Stannic Chloride Solutions	Aug. 23, 1910	967,990
Nov. 6, 1908	461,289	Vehicle Control	May 4, 1915	1,137,804

	DATE FILED	APPLICATION NUMBER	TITLE	DATE ISSUED	PATENT NO.
1909	Jan. 26, 1909	474,255	Cathode	June 14, 1910	961,549
	Sept. 25, 1909	519,533	Ship Gyrocompass Set	Oct. 2, 1917	1,242,065
1910	No Applications				
1911	June 21, 1911	634,595	Gyroscopic Navigation Apparatus	Feb. 5, 1918	1,255,480
	June 21, 1911	634,594	Gyroscopic Compass	Sept. 17, 1918	1,279,471
1912	July 11, 1912	708,809	Gyroscopic Apparatus	June 13, 1916	1,186,856
	Aug. 8, 1912	714,024	Apparatus for Determining Periodic Motion	Sept. 2, 1913	1,071,815
1913	Nov. 14, 1913	801,020	Gearing	Feb. 26, 1918	1,257,417
1914	April 30, 1914	835,522	Automatic Gun-Pointing	Aug. 28, 1917	1,238,503
	May 12, 1914	838,060	Speed and Direction Indicator for Aircraft	Aug. 6, 1918	1,274,622
	May 21, 1914	839,952	Periscope Azimuth Indicator	Oct. 31, 1916	1,203,151
	July 14, 1914	850,874	Gyroscopic Stabilizer	Aug. 14, 1917	1,236,993
	July 17, 1914	851,477	Aeroplane Stabilizer	Feb. 8, 1921	1,368,226
	Aug. 31, 1914	859,329	Multiple Turret Target Indicator	Mar. 4, 1919	1,296,439
	Oct. 10, 1914	866,011	System of Gunfire Control	Oct. 19, 1920	1,356,505
	Nov. 11, 1914	871,436	Electric Arc Light	April 29, 1919	1,302,488
	Nov. 13, 1914	871,885	Navigational Apparatus	Nov. 30, 1920	1,360,694
	Dec. 18, 1914	877,953	Plotting Indicator	Feb. 13, 1917	1,215,425
1915	Jan. 6, 1915	716	Ship Stabilizing and Rolling Apparatus	July 10, 1917	1,232,619
	Jan. 7, 1915	962	Internal Combustion Engine	Aug. 15, 1916	1,194,889
	Feb. 3, 1915	5,819	Navigational Instrument	April 15, 1919	1,300,890
	Feb. 9, 1915	7,027	Electric Drive for Gyroscopes	April 15, 1919	1,301,014
	Mar. 12, 1915	13,986	Dead Beat Inclinometer	Jan. 25, 1921	1,366,430
	April 2, 1915	18,861	Driving and Governing Means for Torpedoes	May 17, 1921	1,378,291
	June 28, 1915	36,615	Method of Operating Flaming Arc Lights for Projectors	May 22, 1917	1,227,210
	Aug. 23, 1915	46,819	Repeater System for Gyro Compasses	Mar. 4, 1919	1,296,440

Date Filed	Application Number	Title	Date Issued	Patent No.
Aug. 26, 1915	47,550	Gyroscopic Apparatus for Torpedoes	Feb. 20, 1923	1,446,276
Sept. 24, 1915	52,333	Electric Battery	Nov. 18, 1919	1,321,947
Oct. 25, 1915	57,815	Torpedo Gyroscope	Aug. 5, 1919	1,312,084
Dec. 2, 1915	64,616	Stabilizing Gyroscopes	Nov. 9, 1920	1,358,258
Dec. 22, 1915	68,116	Method of Operating Electrodes for Searchlights	Nov. 2, 1920	1,357,827

1916

Date Filed	Application Number	Title	Date Issued	Patent No.
Jan. 25, 1916	74,075	Controlling Mechanism for Ships' Gyroscopes	June 1, 1920	1,342,397
Mar. 15, 1916	84,292	Gyroscopic Apparatus for Torpedoes	July 4, 1922	1,421,854
April 13, 1916	88,490	Method of Searchlight Ventilation	Oct. 14, 1919	1,318,701
May 5, 1916	95,724	Light Extinguisher for Arc Lamps	Dec. 14, 1920	1,362,574
May 25, 1916	99,715	Military Searchlight	April 11, 1922	1,412,757
July 15, 1916	109,564	Electrode for Searchlights and Method of Making Same	June 1, 1920	1,342,398
July 22, 1916	110,753	Arc Striking and Extinguishing Apparatus	May 31, 1921	1,379,881
Sept. 30, 1916	123,155	Means for Governing the Rolling of Ships	April 17, 1923	1,452,482
Nov. 23, 1916	132,995	Electrical and Gyroscopic Apparatus for Torpedoes	Aug. 26, 1919	1,314,157
Nov. 24, 1916	133,230	Gyroscopic Roll and Pitch Recorder	Dec. 6, 1921	1,399,032
Dec. 2, 1916	134,562	Wind Driven Generator for Aircraft	Dec. 21, 1920	1,362,753
Dec. 13, 1916	136,917	Rotor for Gyroscopic Stabilizers	Oct. 7, 1919	1,318,302

1917

Date Filed	Application Number	Title	Date Issued	Patent No.
Jan. 27, 1917	144,867	Electric Indicator for Vibrations of the Air	May 8, 1923	1,459,085
Jan. 30, 1917	143,443	Observation Apparatus for Submarines	Oct. 18, 1921	1,393,844
Feb. 7, 1917	147,071	Feeding Mechanism for Searchlights	Dec. 14, 1920	1,362,575
Mar. 31, 1917	158,780	Mulitple Gyro Ship Stabilizer	Oct. 27, 1925	1,558,514
April 3, 1917	159,396	Torch Searchlight	Jan. 17, 1922	1,403,876
April 9, 1917	160,877	Director Firing System	April 22, 1930	1,755,340
April 20, 1917	163,342	Rotor for Gyroscopes	Aug. 15, 1922	1,426,336
May 17, 1917	169,160	Correction Device for Repeater Compasses	Jan. 10, 1922	1,403,062

Date Filed	Application Number	Title	Date Issued	Patent No.
May 31, 1917	171,846	Submarine Net and Method of Laying Same	Mar. 9, 1920	1,333,224
June 7, 1917	173,276	Means for Detecting Submarine Boats	Aug. 29, 1922	1,427,560
June 9, 1917	173,688	Means for Tracing and Locating Submarine Boats	Mar. 9, 1920	1,333,238
July 9, 1917	179,586	Signalling Apparatus for Detecting Submarines	Aug. 15, 1922	1,426,337
July 9, 1917	179,587	Visual Signalling Means	Aug. 15, 1922	1,426,338
July 18, 1917	181,204	Multiple Expansion Internal Combustion Engine	Dec. 23, 1919	1,325,810
Aug. 11, 1917	185,719	Alarm Device for Gyroscope	Feb. 24, 1925	1,527,932
Aug. 11, 1917	185,718	Stabilizing Gyroscope	Aug. 5, 1919	1,312,085
Sept. 14, 1917	191,329	Cooling and Lubrication System for Bearings	Dec. 21, 1920	1,362,754
Sept. 21, 1917	192,571	Submarine Mine	Sept. 2, 1924	1,506,784
Oct. 11, 1917	195,901	Gyroscope	Oct. 18, 1921	1,393,845
Nov. 8, 1917	200,860	Submarine Contact Mine	April 28, 1925	1,535,633
Nov. 19, 1917	202,706	Sighting Device for Ordnance	April 11, 1922	1,412,758
Dec. 8, 1917	206,134	Reflecting Surface for Optical Instruments	Nov. 5, 1918	1,283,943
Dec. 18, 1917	207,786	Wireless Controlled Aerial Torpedo	Feb. 17, 1931	1,792,937
Dec. 22, 1917	208,390	Automatic Sighting Mechanism for Aircraft Guns	Jan. 15, 1924	1,481,248

1918

Date Filed	Application Number	Title	Date Issued	Patent No.
Jan. 12, 1918	211,472	Aeroplane Gun	Sept. 23, 1924	1,509,267
Mar. 13, 1918	222,208	Electrode Holder for Searchlights	Oct. 22, 1918	1,282,133
Mar. 18, 1918	222,614	Fuel Injecting and Igniting Means for Oil Engines	Oct. 17, 1922	1,432,214
April 19, 1918	229,486	Universally Mounted Searchlight	Dec. 19, 1922	1,439,028
April 19, 1918	229,485	Searchlight for the Guidance or Detection of Aircraft	June 26, 1923	1,459,902
May 20, 1918	235,449	Signalling System for Warships	Feb. 12, 1924	1,483,489
July 15, 1918	211,989	Gyroscopic Inclinometer for Airplanes	Dec. 1, 1925	1,563,934

Date Filed	Application Number	Title	Date Issued	Patent No.
July 26, 1918	246,856	Dirigible Gravity Points	July 19, 1921	1,384,868
Aug. 8, 1918	248,834	Launching Mechanism for Aeroplanes	May 9, 1922	1,415,847
Sept. 16, 1918	254,342	Lubricating and Cooling System for Gyroscopes	Aug. 9, 1921	1,387,018
Sept. 18, 1918	254,534	Position Indicator for Aircraft	Jan. 13, 1925	1,522,924
Oct. 11, 1918	257,776	Indicator for Aircraft	Oct. 4, 1932	1,880,994
Nov. 19, 1918	263,217	Automatic Pilot for Aeroplanes	June 6, 1922	1,418,335
Nov. 26, 1918	264,160	Wire Wound Gyro Rotor	Aug. 15, 1922	1,426,339
Dec. 12, 1918	266,369	Speed Measuring Device	Dec. 21, 1926	1,611,253
Dec. 26, 1918	268,319	Osmotic Diaphram	Feb. 8, 1921	1,368,227
1919 Feb. 7, 1919	275,634	Method and Apparatus for Separating Foreign Substances from Dead Matter	Dec. 27, 1921	1,401,743
Feb. 12, 1919	276,594	Shoal-Water Indicator and Ship's Log	Feb. 10, 1925	1,525,963
April 14, 1919	290,059	Mercury-Cooled Transfer Valve	Dec. 16, 1924	1,519,272
July 21, 1919	312,193	Valve Mechanism for Internal Combustion Engines	May 7, 1929	1,711,703
Sept. 18, 1919	324,731	Gyroscopic Navigational Apparatus	Feb. 20, 1923	1,445,805
Sept. 19, 1919	324,747	Apparatus for Testing Gyroscopic Compasses	Nov. 3, 1925	1,560,435
Sept. 26, 1919	326,659	Oil-Burning Engine	Dec. 4, 1928	1,693,966
Nov. 20, 1919	339,300	Method of Treating White Lead	April 24, 1923	1,452,620
Dec. 5, 1919	342,659	Synchronous Transmission System	Sept. 18, 1923	1,468,330
Dec. 17, 1919	345,660	Engine and the Transmission of Power Therefrom	Jan. 1, 1929	1,697,292
1920 Jan. 3, 1920	349,345	Fire Control System	Mar. 18, 1924	1,487,282
Jan. 24, 1920	353,739	Turn Indicator	Feb. 21, 1922	1,407,491
Feb. 21,1920	360,319	Means for Spinning Up Gyroscopes on Aircraft	Feb. 19, 1924	1,483,992
Mar. 12, 1920	365,144	Wireless Repeater System	Sept. 5, 1922	1,428,507

Date Filed	Application Number	Title	Date Issued	Patent No.
Mar. 12, 1920	365,145	Gyrocompass Relay Transmitter	April 26, 1927	1,626,123
April 10, 1920	372,809	Self-Syncronous Transmission System	Mar. 22, 1932	1,850,640
April 13, 1920	373,496	Cooling Motion-Picture Projectors and Films	Jan. 1, 1924	1,479,630
June 14, 1920	388,729	Accelerating Means for Automatic Vehicle	Sept. 21, 1926	1,600,651
July 2, 1920	393,557	Method of Gunfire Control for Battleships	April 17, 1923	1,452,484
Aug. 10, 1920	402,657	Internal Combustion Engine Cylinder	Aug. 28, 1928	1,682,357
Aug. 12, 1920	403,049	Means for Indicating Loss of Synchronism in Transmission Systems	May 17, 1927	1,629,236
Aug. 23, 1920	405,407	Gyroscopic Compass	June 24, 1924	1,499,321
Sept. 9, 1920	409,190	Cooling Means for Gyromotors	Oct. 14, 1924	1,511,240
Sept. 10, 1920	409,308	Supporting Means for Golf Bags	Oct. 9, 1928	1,686,774
Sept. 18, 1920	411,319	Synchronizing Mechanism	Jan. 24, 1928	1,656,962
Oct. 12, 1920	416,505	Repeater Compass	Feb. 8, 1927	1,617,310
1921 Jan. 8, 1921	435,837	Means for Preventing Racing of Ships' Engines	July 31, 1928	1,678,714
March 3, 1921	449,412	Valve Mechanism for Combustion Engines	May 3, 1924	1,627,210
May 4, 1921	466,774	Aviation Beacon	Aug. 11, 1925	1,548,958
June 4, 1921	474,909	Drag Rudder for Gravity Bomb	May 12, 1925	1,537,713
June 7, 1921	475,640	Gravity Bomb	Sept. 2, 1924	1,506,785
July 20, 1921	486,181	Piston for Internal-Combustion Engines	Aug. 16, 1927	1,639,062
July 21, 1921	486,182	Cooling Means for Internal Combustion Engines	Sept. 18, 1928	1,684,803
July 21, 1921	486,526	Gyroscopic Line of Sight Stabilizer	Oct. 23, 1928	1,688,559
Aug. 30, 1921	496,808	Searchlight Cooling and Ventilating Means	Sept. 6, 1927	1,641,301
Oct. 14, 1921	507,592	Method and Means for Balancing Masses	Sept. 21, 1926	1,600,568
Oct. 14, 1921	507,591	Gravity Bomb	April 26, 1927	1,626,363
Dec. 1, 1921	519,053	Wakeless Torpedo	Oct. 23, 1928	1,688,761

Date Filed	Application Number	Title	Date Issued	Patent No.
1922 Jan. 14, 1922	529,314	Piston for Internal Combustion Engine	Oct. 23, 1928	1,688,403
Jan. 20, 1922	530,687	Combustion Engine Locomotive	May 5, 1931	1,803,876
Jan. 23, 1922	531,297	Logarithmic Calculator	May 29, 1928	1,671,616
Jan. 24, 1922	531,329	Oil Engine	Jan. 13, 1931	1,788,412
Mar. 28, 1922	547,384	Combustion Engine for Driving Ships	Aug. 28, 1928	1,682,358
Mar. 29, 1922	547,599	Recorder for Ships	Mar. 22, 1932	1,850,978
Apr. 26, 1922	556,784	Means for Cooling the Pistons of Heat Engines	Nov. 15, 1924	1,648,968
Apr. 26, 1922	556,783	Beacon System for Night Flying	April 16, 1929	1,709,377
May 5, 1922	558,643	Transmission System for Automatic Vehicles	Oct. 23, 1928	1,688,691
May 25, 1922	563,735	Fuel Injecting and Igniting Means for Oil Engines	Mar. 18, 1930	1,751,254
Aug. 5, 1922	579,863	Improvement in Combustion Chamber for Combustion Engines	July 19, 1932	1,867,682
Aug. 21, 1922	583,084	Wireless Repeater System	April 6, 1926	1,579,669
Nov. 14, 1922	600,829	Crankless Engine	May 21, 1929	1,714,145
1923 Mar. 17, 1923	625,831	Gyroscopic Pendulum	Dec. 6, 1927	1,651,845
Oct. 22, 1923	669,940	Combustion Engine	July 19, 1932	1,867,683
Nov. 6, 1923	673,210	Ship's Signaling or Broadcasting Device	Aug. 9, 1927	1,638,417
June 1, 1923	642,745	Searchlight for the Guidance and Detection of Aircraft	May 15, 1928	1,669,882
1924 Jan. 9, 1924	685,271	Sounding Device	June 10, 1930	1,763,377
Feb. 26, 1924	690,930	Track Recorder System	Feb. 9, 1932	1,843,959
June 18, 1924	720,753	Locking Device for Gyroscopes	Mar. 5, 1929	1,704,489
Oct. 7, 1924	742,093	Two-Cycle Supercharging Combustion Engine	July 17, 1928	1,677,305
Nov. 18, 1924	750,695	Method and Means for Imparting Intelligence	Nov. 3, 1931	1,830,041
1925 Mar. 27, 1925	18,673	Automatic Steering Mechanism for Dirigible Aircraft	July 12, 1932	1,867,334
July 18, 1925	44,505	Stabilizing Device [vehicles such as autos]	Oct. 21, 1930	1,778,734

	Date Filed	Application Number	Title	Date Issued	Patent No.
	July 24, 1925	45,958	Control Gyro	Jan. 13, 1931	1,788,807
	July 25, 1925	45,996	Automatic Accelerator	May 26, 1931	1,806,652
	Sept. 21, 1925	57,674	Method of Refining Crude Fuel Oil	Jan. 15, 1929	1,699,379
	Dec. 17, 1925	75,931	Borehole Inclination Recorder	July 7, 1931	1,812,994
1926	April 3, 1926	99,678	Automatic Pilot	Aug. 11, 1931	1,818,103
	Nov. 27, 1926	151,175	Automatic Steering Device	Dec. 29, 1931	1,838,965
1927	April 13, 1927	183,275	Means for Preventing Pitching of Ships	April 14, 1931	1,800,365
	Aug. 27, 1927	215,980	Fissure Detector for Steel Rails	May 5, 1931	1,804,380
	Nov. 25, 1927	235,519	Automatic Launching Device for Airplanes	Nov. 5, 1929	1,734,353
1928	April 12, 1928	269,347	Self-Synchronous Transmission System	Mar. 22, 1932	1,850,780
	May 9, 1928	276,212	Engine	April 26, 1932	1,855,929
	June 4, 1928	282,820	Means for Preventing Vibration of Crank Shafts	July 19, 1932	1,867,684
	Aug. 10, 1929	298,771	Fissure Detector for Metals	Aug. 25, 1931	1,820,505
	Oct. 12, 1928	312,019	Roadway Inspecting Means	Dec. 22, 1931	1,837,633
	Nov. 21, 1928	320,886	Fissure Detector for Magnetic Material	July 19, 1932	1,867,685
	Nov. 22, 1928	321,128	Means for Controlling Dirigible Aircraft	Aug. 11, 1931	1,818,104
	Nov. 28, 1928	322,320	Variable Pitch Propellor	May 2, 1933	1,907,014
1929	Aug. 17, 1929	386,538	Flaw Detector for Irregular Objects	Aug. 30, 1932	1,874,067
	Nov. 20, 1929	408,459	Marine Aircraft	Aug. 16, 1932	1,871,476
1930	June 26, 1930	463,893	Variable Pitch Propellor	Aug. 30, 1932	1,874,719

APPENDIX B   Engineering and Scientific Papers of Elmer Sperry*

1886   "The Best Form of Dynamo," *Proceedings of the National Electric Light Association*, I, 200-206

1892   "Electricity in Bituminous Mining," *American Institute of Electrical Engineers, Transactions*, IX, 375-400

1894   "The Electric Brake in Practice," *American Institute of Electrical Engineers, Transactions*, XI, 682-709

1899   "Electric Automobiles," *American Institute of Electrical Engineers, Transactions*, XVI, 509-25

1903   "Axle-Lighting," *American Institute of Electrical Engineers, Transactions*, XXI, 155-62

1903   "The Use of Pyroxyline in Electric Storage Batteries," *American Electrochemical Society, Transactions*, III, 169-73

1906   "Electrochemical Pocesses as Station Load Equalizers," *American Electrochemical Society, Transactions*, IX, 147-50

1908   "Utilization of Power Stations for Electrochemical and Electrothermal Processes During Periods of Low Load," *American Electrochemical Society, Transactions*, XIV, 259-61
"Preparation and Uses of Anhydrous Stannic Chloride in Silk Dyeing," *Journal of the Society of Chemical Industry*, XXVII, 312-14

1909   "The Manufacture of Anhydrous Chlorin from Moist Dilute Gases and Its Industrial Application in Chlorine Detinning," *Journal of Industrial and Engineering Chemistry*, I, 511-18

1910   "The Gyroscope for Marine Purposes," *Society of Naval Architects and Marine Engineers, Transactions*, XVIII, 143-54

---

* See also, "List of the More Important Scientific Papers of Elmer Ambrose Sperry," in J. C. Hunsaker, *Biographical Memoir of Elmer Ambrose Sperry, 1860-1930* (Washington, D.C.; National Academy of Sciences, 1955), pp. 28-30.

1912 "Active Type of Stabilizing Gyro," *Society of Naval Architects and Marine Engineers, Transactions,* XX, 201-15

1913 "Engineering Applications of the Gyroscope," *Franklin Institute Journal,* CLXXV, 447-82
"Some Graphic Studies of the Active Gyro Stabilizer," *Society of Naval Architects and Marine Engineers, Transactions,* XXI, 181-87
"Safe Flight and Automatic Stabilization of Aeroplanes," *New York Electrical Society, Transactions,* VI, No. 6, New Series, pp. 80-93

1915 "Recent Progress with the Active Type of Gyro Stabilizer for Ships," *Society of Naval Architects and Marine Engineers, Transactions,* XXIII, 43-48

1916 "The Commercial Gyroscopic Compass," *Society of Naval Architects and Marine Engineers, Transactions,* XXIV, 207-14
"The Electrically-Driven Gyroscope and Its Uses," *New York Electrical Society, Transactions,* VII, 13-26

1917 "Aerial Navigation Over Water," *Society of Automotive Engineers, Transactions,* XII, 153-65
"Gyro Ship Stabilizers in Service," *Society of Naval Architects and Marine Engineers, Transactions,* XXV, 293-99

1919 "Non-Rolling Passenger Liners—Observations on a Large Stabilized Ship in Service, Including the Plant and Economies Effected by Stabilization," *Society of Naval Architects and Marine Engineers, Transactions,* XXVII, 99-108

1921 "Compounding the Combustion Engine," *American Society of Mechanical Engineers, Transactions,* XLIII, 677-716

1922 "Automatic Steering," *Society of Naval Architects and Marine Engineers, Transactions,* XXX, 53-57

1923 "The Gyro Ship-Stabilizer," *Journal of the Society of Naval Architects,* XXXII, 225-48

1927 "The Light Supercharged Diesel Engine for Use in Air Service," *Mechanical Engineering,* XLIX, 723-26

1928 "The Non-Destructive Detection of Hidden Flaws," *American Iron and Steel Institute, Yearbook, 1928,* pp. 312-41

1929 "Non-Destructive Testing," *American Society for Steel Treating, Transactions,* XVI, 771-90
"Non-Destructive Testing of Welds," *Journal of the American Welding Society,* VIII, 48-61

INDEX